Water Distribution Operator Training Handbook

Water Distribution Operator Training Handbook

Fourth Edition

William C. Lauer

Water Distribution Operator Training Handbook

Disclaimer
This book is provided for informational purposes only, with the understanding that the publisher, editors, and authors are not thereby engaged in rendering engineering or other professional services. The authors, editors, and publisher make no claim as to the accuracy of the book's contents, or their applicability to any particular circumstance. The editors, authors, and publisher accept no liability to any person for the information or advice provided in this book or for loss or damages incurred by any person as a result of reliance on its contents. The reader is urged to consult with an appropriate licensed professional before taking any action or making any interpretation that is within the realm of a licensed professional practice.

ISBN: 978-1-58321-954-6
eISBN: 978-1-61300-241-4

Project Manager/Senior Technical Editor: Melissa Valentine
Cover Design: Melanie Yamamoto/Cheryl Armstrong
Production: Glacier Publishing Services, Inc.

Library of Congress Cataloging-in-Publication Data
Water distribution operator training handbook / [edited] by William C. Lauer. --
Fourth edition.
pages cm
Includes bibliographical references and index.
ISBN 978-1-58321-954-6 (alk. paper)
1. Water--Distribution--Handbooks, manuals, etc. 2. Water-supply engineering--Fieldwork--Handbooks, manuals, etc. 3. Water-supply engineers--Training of--Handbooks, manuals, etc. I. Lauer, Bill.
TD481.W373 2013
628.1'44--dc23

2013021236

Printed on
Recycled Paper

Contents

Preface

A safe, reliable drinking water supply is one of the most important resources for a community. Sufficient quantities of water must always be available at adequate pressure, and the water must be safe for human consumption at all times. A water distribution operator is legally and morally responsible to see that these goals are accomplished.

The distribution operator is typically the first line of contact between the general public and the water utility. Therefore, they must conduct public contacts in a manner that will maintain a good image of the utility and instill customer confidence in the service being provided.

The approach for this fourth edition was to ensure that it included the operator certification knowledge requirements included in the Association of Boards of Certification (ABC) Need-To-Know criteria. In addition, operator requirements for several state certification boards (e.g., California, Pennsylvania, and Texas) were also added where there were differences. Any topics regarding these certification requirements that were not already included in the previous edition of the book were added. All of the numerous references and illustrations were reviewed and updated.

In addition, several new chapters were added to cover topics that have emerged since publication of the last edition. The regulatory review chapter was revised extensively. New chapters were added describing the management approach to distribution system operation and the operational practices operators can use to improve system performance. The disinfection of pipelines and storage facilities is now included as its own chapter. The math calculations that distribution system operators need to know are also included.

The result of these additions and revisions make this book an indispensable companion reference for water distribution system operations personnel. Every water system operator should have a copy to use when preparing for certification exams as well as for everyday use. In order to maintain the publication at a reasonable size, some subjects receive only superficial coverage. However, sources of additional, specific information are available elsewhere. The publications referenced at the end of each chapter are available from the American Water Works Association (AWWA) and other sources at nominal cost.

In addition to this book, operators should have an up-to-date copy of their state's public water supply regulations. Although basic requirements for water system operation are dictated by regulations under the Safe Drinking Water Act, in many cases states have modified the federal regulations and some may actually be more stringent. Each state also has many additional requirements that can only be found by reviewing the state regulations.

Finally, many thanks are extended to David Plank, Gay Porter DeNileon, and Melissa Valentine at AWWA for their help and support during the publication process. And a special thanks to Marcellus Jones, Jr. for his review of this book.

William C. Lauer

Publication History

The first edition of this handbook was published by AWWA in 1976, using, in part, material prepared by members of the Pacific Northwest Section of AWWA and contributions from many additional AWWA members.

The second edition was completely revised under the auspices of the AWWA Operations and Maintenance Committee, with Kanwal Oberoi, chairman. The technical editor was Harry Von Huben.

William C. Lauer of AWWA provided a technical review of the third edition.

The fourth edition was an extensive revision, published in 2013, by its author William C. Lauer.

Chapter 1

Distribution System Regulations

Drinking water regulations have undergone major and dramatic changes during the past two decades, and trends indicate that they will continue to become more stringent and complicated. It is important that all water system operators understand the basic reasons for having regulations, how they are administered, and why compliance with them is essential. The reader should recognize that regulatory requirements are constantly changing. It is the operator's responsibility to keep current on all regulatory requirements.

FEDERAL REGULATIONS

Although the regulations required by the Safe Drinking Water Act (SDWA) are of prime interest in the operation and administration of water distribution systems, operators must also adhere to regulations required by several other federal environmental and safety acts.

Safe Drinking Water Act Requirements

Requirements under the SDWA are quite extensive, and complete details can be found in publications (and websites) listed in the bibliography at the end of this chapter. The SDWA includes a number of (current and proposed) rules including the following:

- Surface Water Treatment Rule (SWTR)
- Total Coliform Rule (TCR)
- Interim Enhanced Surface Water Treatment Rule (IESWTR)
- Long-Term 1 Enhanced Surface Water Treatment Rule (LT1ESWTR)
- Long-Term 2 Enhanced Surface Water Treatment Rule (LT2ESWTR)
- Ground Water Rule (GWR)
- Total Trihalomethane (TTHM) Rule
- Stage 1 Disinfectants/Disinfection By-products Rule (Stage 1 DBPR)
- Stage 2 Disinfectants/Disinfection By-products Rule (Stage 2 DBPR)
- Lead and Copper Rule (LCR)
- Public Notification (PN) Rule

- Filter Backwash Recycle Rule (FBRR)
- Unregulated Contaminant Monitoring Rule (UCMR)

The following discussion will primarily center on requirements that affect the operation of water distribution systems.

Prior to 1975, review of public water supplies was done by each state, usually by the state health department. The SDWA was passed by Congress in 1975 for a combination of reasons. One of the primary purposes was to create uniform national standards for drinking water quality to ensure that every public water supply in the country would meet minimum health standards. Another was that scientists and public health officials had recently discovered many previously unrecognized disease organisms and chemicals that could contaminate drinking water and might pose a health threat to the public. It was considered beyond the capability of the individual states to deal with these problems.

The SDWA delegates responsibility for administering the provisions of the act to the US Environmental Protection Agency (USEPA). The agency is headquartered in Washington, D.C., and has 10 regional offices in major cities of the United States. Some principal duties of the agency are to

- Set maximum allowable concentrations for contaminants that might present a health threat in drinking water—these are called maximum contaminants levels (MCLs);
- Delegate primary enforcement responsibility for local administration of the requirements to state agencies;
- Provide grant funds to the states to assist them in operating the greatly expanded program mandated by the federal requirements;
- Monitor state activities to ensure that all water systems are being required to meet the federal requirements; and
- Provide continued research on drinking water contaminants and improvement of treatment methods.

State Primacy

The intent of the SDWA is for each state to accept primary enforcement responsibility (primacy) for the operation of the state's drinking water program. Under the provisions of the delegation, the state must establish requirements for public water systems that are at least as stringent as those set by USEPA. The primacy agency in each state was designated by the state governor. In some states the primacy agency is the state health department, and in others it is the state environmental protection agency, department of natural resources, or pollution control agency. USEPA has primacy in any state (e.g., Wyoming) that has not accepted this role.

Classes of Public Water Systems

The basic definition of a *public water system* in the SDWA is, in essence, a system that supplies piped water for human consumption and has at least

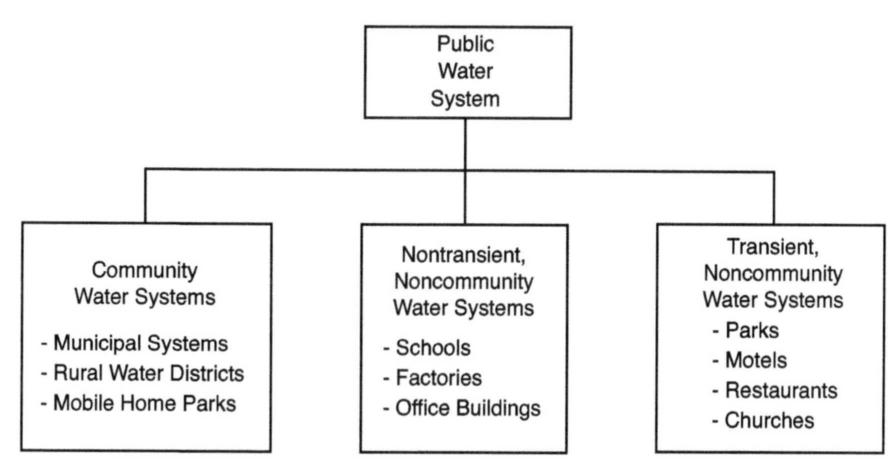

Source: Drinking Water Handbook for Public Officials (1993).

Figure 1-1 Classification of public water systems

15 service connections or serves 25 or more persons for 60 or more days of the year. Examples of water systems that would not fall under the federal definition are private homes, groups of fewer than 15 homes using the same well, and summer camps that operate for fewer than 60 days per year. These systems are, however, generally under some degree of supervision by a local, area, or state health department.

USEPA has further divided public water systems into three classifications (Figure 1-1):

1. *Community public water systems* serve 15 or more homes. Besides municipal water utilities, this classification also covers mobile home parks and small homeowner associations that have their own water supply and serve more than 15 homes.
2. *Nontransient, noncommunity public water systems* are establishments that have their own private water systems, serving an average of at least 25 persons who do not live at the location, but the same people use the water for more than 6 months per year. Examples are schools and factories.
3. *Transient, noncommunity public water systems* are establishments such as parks and motels that have their own water systems and serve an average of at least 25 persons per day, but these persons use the water only occasionally and for short periods of time.

The monitoring requirements for community and nontransient, noncommunity systems include all contaminants that are considered a public health threat. Transient, noncommunity systems are only required to monitor for nitrate, nitrite, and microbiological contamination.

Regulation of Contaminants

The National Primary Drinking Water Regulations (NPDWRs) specify MCLs or a treatment technique requirement for contaminants that may be found in drinking water and could have an adverse health effect on humans.

Specific concentration limits for the chemicals are listed, and all community and nontransient, noncommunity systems must test for their presence. If a water system is found to have concentrations of chemicals present that are above the MCL, the system must either change its water source or treat the water to reduce the chemical concentration. Primary regulations are mandatory and must be complied with by all water systems to which they apply.

The National Secondary Drinking Water Regulations apply to drinking water contaminants that may adversely affect the aesthetic qualities of water, such as taste, odor, or color. These qualities have no known adverse health effects, but they seriously affect public acceptance of the water. Secondary regulations are not mandatory but are strongly urged by USEPA. Some state regulatory agencies have made some of the secondary limits mandatory in their states.

Public Notification

The SDWA mandates that the public be kept informed of noncompliance with federal requirements by requiring that noncomplying systems provide public notification. If public water systems violate any of the operating, monitoring, or reporting requirements, or if the water quality exceeds an MCL, the system must inform the public of the problems. Even though the problem may have already been corrected, an explanation must be provided in the news media describing the public health significance of the violation.

The language and methods of providing public notification are mandated by USEPA to ensure that the public is fully informed. If a system is required to provide public notification, the state primacy agency will provide full instructions.

Water distribution operators should understand that, although public notification is intended to keep the public informed, if it is caused by a simple mistake such as forgetting to send in the monthly samples, it can cause some embarrassment for the system's staff. To avoid this situation, careful attention must be given to state requirements. If there is any problem in meeting any of the requirements, it should be discussed with the state agency representative.

If an operator is required to provide public notification, it should be made as positive as possible. Although the basic wording is mandatory, other wording can be added to keep it from sounding completely negative to the public. Such wording can be discussed with the primacy agency representative.

Monitoring and Reporting

To ensure that the drinking water supplied by all public water systems meets federal and state requirements, system operators are required to regularly collect samples and have the water tested. The regulations specify minimum sampling frequencies, sampling locations, testing procedures, methods of keeping records, and frequency of reporting to the state. The regulations also mandate special reporting procedures to be followed if a contaminant exceeds an MCL.

All systems must provide periodic monitoring for microbiological contaminants and some chemical contaminants. The frequency of sampling and the chemicals that must be tested for depend on the size of the water system, the source of water, and the history of analyses.

State policies vary on providing laboratory services. Some states have the laboratory facilities available to perform all required analyses or, in some cases, a certain number of the required analyses for a system. Most states charge for all or some of the laboratory services. Sample analyses that are required and cannot be performed by a state laboratory must be taken or sent to a state-certified private laboratory.

If the analysis of a sample exceeds an MCL, resampling is required, and the state should be contacted immediately for special instructions. There is always a possibility that such a sample was caused by a sampling or laboratory error, but it must be handled as though it was actually caused by contamination of the water supply.

The results of all water analyses must be periodically sent to the state. Failure to have the required analyses performed or to report the results to the state will usually result in the system having to provide public notification. States typically have special forms for submitting the data and specify a number of days following the end of the monitoring period by which the form must be submitted. The minimum information that must be provided in the form is listed in Table 1-1. State regulators may also require other information for their own records and documentation.

Table 1-1 Laboratory report summary requirements

Type of Information	Summary Requirement
Sampling information	Date, place, and time of sampling
	Name of sample collector
	Identification of sample
	• Routine or check sample
	• Raw or treated water
Analysis information	Date of analysis
	Laboratory conducting analysis
	Name of person responsible for analysis
	Analytical method used
	Analysis results

Table 1-2 Record-keeping requirements

Type of Records	Time Period
Bacteriological and turbidity analyses	5 years
Chemical analyses	10 years
Actions taken to correct violations	3 years
Sanitary survey reports	10 years
Exemptions	5 years following expiration

There are also specific requirements for the length of time a water system must retain records. Table 1-2 lists the record-keeping requirements mandated by USEPA.

Water Quality Monitoring

Although most water quality monitoring is related to ensuring proper quality of the source water or treatment processes, many of the samples are collected from the distribution system. Thus, sample collection often becomes a duty of distribution system personnel. The reason for collecting samples from the distribution system is that there are some opportunities for water quality to change after it enters the distribution system, and under the requirements of the SDWA, it is the duty of the water purveyor to deliver water of proper quality to the consumer's tap.

Methods of Collecting Samples

Two basic methods of collecting samples are grab sampling and composite sampling. A *grab sample* is a single volume of water collected at one time from a single place. To sample water in the distribution system, a faucet is used to fill a bottle. This sample represents the quality of the water only at the time the sample was collected. If the quality of the water is relatively uniform, the sample will be quite representative. If the quality varies, the sample may not be representative.

A *composite sample* consists of a series of grab samples collected from the same point at different times and mixed together. The composite is then analyzed to obtain the average value. If the composite sample is made up of equal-volume samples collected at regular intervals, it is called a *time composite sample.* Another method is to collect samples at regular time intervals, but the size of each grab sample is proportional to the flow at the time of sampling. This is called a *flow-proportional composite sample.*

Although composite sampling appears to be a good idea because it provides an average of water quality, it cannot be used for most analyses of drinking water quality because a majority of parameters are not stable over a period of time.

Sample Storage and Shipment

Care must always be taken to use the exact sample containers specified or provided by the laboratory that will be doing the analyses. Most sample containers are now plastic to avoid the possibility of glass breaking during shipment. Some samples for organic chemical analysis must be collected in special glass containers because some of the chemical might permeate the walls of a plastic container.

Sample holding time before analysis is quite critical for some parameters. If a laboratory receives a sample that has passed the specified holding time, it is supposed to declare the sample invalid and request resampling. Some samples can be refrigerated or treated once they arrive at the laboratory to extend the holding time, allowing the laboratory a few more days before the analyses must be completed.

Many laboratories do not work on weekends, so this should be taken into consideration when sending samples. Bacteriological analyses must, for example, be performed immediately by the laboratory. The best time to collect and send these samples is on a Monday or Tuesday so they will reach the laboratory by mid-week. Samples should be sent to the laboratory by the fastest means available, such as first-class mail or special carrier.

Sample Point Selection

Samples are collected from various points in the distribution system to determine the quality of water delivered to consumers. In some cases, distribution system samples may be significantly different from samples collected as the water enters the system. For example, corrosion in pipelines, bacterial growth, or algae growth in the pipes can cause increases in color, odor, turbidity, and chemical content (e.g., lead and copper). More seriously, a cross-connection between the distribution system and a source of contamination can result in chemical or biological contamination of the water.

Most samples collected from the distribution system will be used to test for coliform bacteria and chlorine residual. The two primary considerations in determining the number and location of sampling points are that they should be

1. Representative of each different source of water entering the system (i.e., if there are several wells that pump directly into the system, samples should be obtained that are representative of the water from each one); and
2. Representative of the various conditions within the system (such as dead ends, loops, storage facilities, and each pressure zone).

The required number of samples that must be collected and the frequency of sampling depend on the number of customers served, the water source, and other factors. Specific sampling instructions must be obtained from the state primacy agency.

Sample Faucets

After representative sample points have been located on the distribution system, specific locations having suitable faucets for sampling must be identified. If suitably located, public buildings and the homes of utility employees are convenient places to collect samples. Otherwise, arrangements must be made to collect samples from businesses or private homes.

The following types of sampling faucets should not be used:

- Any faucet located close to the bottom of a sink, because containers may touch the faucet
- Any leaking faucet with water running out from around the handle and down the outside
- Any faucet with threads, such as a sill cock, because water generally does not flow smoothly from them and may drip contamination from the threads
- Any faucet connected to a home water-treatment unit, such as a water softener or carbon filter
- Drinking fountain

It is also best to try to find a faucet without an aerator. If faucets with aerators must be used, follow the state recommendations on whether or not the aerator should be removed for sampling.

Each sample point must be described in detail on the sample report form—not just the house address, but which faucet in which room. If resampling is necessary, the same faucet used for the first sample must be used.

When it is necessary to establish a sampling point at a location on the water system where no public building or home gives access for regular sampling, a permanent sampling station can be installed (Figure 1-2).

Courtesy of Gil Industries, Inc.

Figure 1-2 Example of a permanent sampling station

Sample Collection

For collection of bacteriological and most other samples, the procedure is to open the faucet so that it will produce a steady, moderate flow. Opening the faucet to full flow for flushing is not usually desirable because the flow may not be smooth and water will splash up onto the outside of the spout. If a steady flow cannot be obtained, the faucet should not be used.

The water should be allowed to run long enough to flush any stagnant water from the house plumbing, which usually takes 2 to 5 minutes. The line is usually clear when the water temperature drops and stabilizes. The sample is then collected without changing the flow setting. The sample container lid should be held (not set down on the counter) with the threads down during sample collection and replaced immediately. The sample container should then be labeled.

The exception to the above-mentioned procedure is sampling for lead and copper analysis. These are to be first-draw samples and require special procedures.

Bottles to be used for collection of bacteriological samples should not be rinsed before they are filled. These bottles are usually prepared with a small quantity of thiosulfate at the bottom to immediately stop the action of the residual chlorine in the water.

Special-Purpose Samples

It is occasionally necessary to collect special samples, particularly in response to customer complaints, such as taste and odor issues. To check on this type of complaint, one sample should be collected immediately as the tap is opened to be representative of water that has been in the plumbing system, then a second sample should be collected after the line has been flushed. It is sometimes helpful to collect both hot- and cold-water samples in this manner. These samples can be used to identify whether the problem is in the customer's plumbing system or coming from the water distribution system. Many customer complaints of taste, odor, or color are found to be from their own water heaters, water softeners, or home water-treatment devices.

Laboratory Certification

It is imperative that the monitoring of all water systems be consistent; therefore, all laboratory analyses must be performed by experienced technicians under carefully controlled conditions. For this reason, compliance sample analyses are acceptable to the state only if they have been performed by a certified laboratory. The only exceptions are measurements for turbidity, chlorine residual, temperature, and pH, which may be performed by a person acceptable to the state, using approved equipment and methods.

Most states operate certified laboratories that can accept some or all of the samples from water systems. The states also certify private laboratories that may be used for performing water analyses. Most large water utilities have their own certified laboratories because of the great number of samples that must be processed.

Consumer Confidence Reports

One of the very significant provisions of the 1996 SDWA Amendments is the consumer confidence report (CCR) requirement. The purpose of the CCR is to provide all water customers with basic facts regarding their drinking water so that individuals can make decisions about water consumption based on their personal health. This directive has been likened to the requirement that packaged food companies disclose what is in their food products.

The reports must be prepared yearly by every community water system. Water systems serving more than 10,000 people must mail the report to customers. Smaller systems must notify customers as directed by the state primacy agency.

A water system that only distributes purchased water (satellite system) must prepare the report for their consumers. Information on the source water and chemical analyses must be provided to the satellite system by the system selling the water (parent system).

Some states are preparing much of the information for their water systems, but the system operator must still add local information. Water system operators should keep in mind that CCRs provide an opportunity to educate consumers about the sources and quality of their drinking water. Educated consumers are more likely to help protect drinking water sources and be more understanding of the need to upgrade the water system to make their drinking water safe.

USEPA Regulation Information

Current information on USEPA regulations can be obtained by contacting the Safe Drinking Water Hotline at 800-426-4791. Also see the Office of Ground Water and Drinking Water Web page at http://water.epa.gov/drink.

STATE REGULATIONS

Under the provisions of primacy delegation, each state must have requirements applying to public water systems that are at least as stringent as those set by USEPA. States occasionally establish requirements that are more stringent. Federal requirements are only for factors that USEPA considers directly related to public health. So, in addition to the federal requirements, each state also establishes other requirements to ensure proper water system operation.

Operator Certification

One requirement of the 1996 SDWA Amendments is that USEPA must establish minimum standards for state operator certification programs.

Most states have had some form of certification for water system operators, but, unfortunately, each state has its own idea of how operators should be classified, so there has been little national consistency.

Among the more important requirements are that each water system must at all times be under the direct supervision of a certified operator, operators must have a high school or equivalent education and pass an examination to receive certification, and the state must establish training requirements for certification renewal. Most states have a separate certification class for distribution system operators.

Cross-Connection Control

The states also generally promote cross-connection control programs for all water systems. Many states have their own cross-connection control manuals and assist water systems in setting up local programs. Cross-connection control is covered in detail in chapter 13.

Construction Approval

The SDWA requires states to review plans for water system construction and improvements. In general, plans and specifications for the proposed work must be prepared by a professional engineer and submitted for approval before work begins. State engineers review the plans for suitability of materials, conformance with state regulations, and other factors.

Some states allow small distribution system additions without approval or allow approval after construction. State regulations should be reviewed to ensure compliance with requirements.

Technical Assistance

One of the staff functions of the state drinking water program is to provide technical assistance to water system operators. Field staff with training and experience are usually available to provide advice and assistance. If possible, they will provide advice over the phone, but if the problem is of sufficient magnitude, they will arrange personal visits. Staff may also, on some occasions, suggest other sources of information or assistance.

Enforcement

Because of the direct relationship between drinking water quality and public health, it is rare for anyone to purposely disregard state and federal regulations. Most violations of regulations are caused by not understanding requirements or forgetting something that must be done.

The SDWA requires states to use enforcement actions when federal requirements are violated. Then if the state does not take appropriate action, USEPA is prepared to step in and do it. Minor infractions are handled by public notification, but intentional disregard for requirements can result in substantial monetary fines.

REQUIREMENTS OF SPECIAL INTEREST TO DISTRIBUTION SYSTEM OPERATORS

Distribution system regulations address three main areas of concern: microbiological safety, disinfection by-products (DBPs), and lead. The microbiological safety of the water reaching customers' taps is of primary concern, and this was the initial focus of the distribution system regulatory requirements.

DBPs, such as total trihalomethanes (TTHM), are created by chemical reactions between disinfectants (like chlorine) and other substances in the water. High levels in water may increase the risk of cancer for some individuals over a lifetime. Therefore, MCLs and monitoring requirements are included in the appropriate rules. These requirements are changing as more is learned about the levels of concern.

Lead is hazardous if consumed in high amounts, particularly for children. Water with certain characteristics may dissolve lead from solder or plumbing fixtures (or lead service lines) and may pose a risk to consumers. Therefore, special tap sampling requirements are mandated to determine the need to stabilize the water or perhaps replace lead water services. The applicable regulatory rules are discussed in more detail in the following sections.

Total Coliform Rule

The objective of the TCR is to promote routine surveillance of distribution system water quality to search for contamination from fecal matter and/or disease-causing bacteria. All points in a distribution system cannot be monitored, and complete absence of fecal matter and disease-causing bacteria cannot be ensured. The TCR is a regulatory approach for the implementation of monitoring programs sufficient to verify that public health is being protected as much as possible, as well as allowing utilities to identify any potential contamination problems in their distribution system. The rule requires monthly sampling at each distribution sampling point.

If a routine monthly sample is total coliform (TC) positive, the utility must determine fecal coliform (FC) or *Escherichia coli* (EC) in the same sample and must also perform verification monitoring by collecting a second sample and reanalyzing TC and FC/EC within 24 hours. The system is not in compliance if either of the following occurs: (1) if analysis and reanalysis of a given sampling location is TC positive (TC[+]) both times and FC[/EC+] at least one of these times or (2) if more than 5 percent of all monthly samples for a 12-month period are TC[+].

The TCR and the Revised Total Coliform Rule (RTCR) that was finalized in 2013 impact all systems. The RTCR requires public water systems that are vulnerable to microbial contamination to identify and fix problems. The RTCR also establishes criteria for systems to qualify for and stay on reduced monitoring, thereby providing incentives for improved water system operation.

The RTCR also changed monitoring frequencies for some systems. It links monitoring frequency to water quality and system performance and provides criteria that well-operated small systems must meet to qualify and stay on reduced monitoring. It also requires increased monitoring for high-risk small systems with unacceptable compliance history and establishes some new monitoring requirements for seasonal systems such as state and national parks.

The RTCR further establishes a maximum contaminant level goal (MCLG) and an MCL for *E. coli* and eliminated the MCLG and MCL for total coliform, replacing it with a treatment technique for coliform that requires assessment and corrective action. The rule establishes an MCLG and an MCL of zero for *E. coli*, a more specific indicator of fecal contamination and potentially harmful pathogens than total coliform. USEPA has removed the MCLG and MCL of zero for total coliform. Many of the organisms detected by total coliform methods are not of fecal origin and do not have any direct public health implication.

Under the treatment technique for coliform, total coliform serves as an indicator of a potential pathway of contamination into the distribution system. A public water system that exceeds a specified frequency of total coliform occurrence must conduct an assessment to determine if any sanitary defects exist and, if found, correct them. In addition a public water system that incurs an *E. coli* MCL violation must conduct an assessment and correct any sanitary defects found.

The rule eliminated monthly PN requirements based only on the presence of total coliform. Total coliform in the distribution system may indicate a potential pathway for contamination but in and of itself does not indicate a health threat. Instead, the rule requires PN when an *E. coli* MCL violation occurs, indicating a potential health threat, or when a public water system fails to conduct the required assessment and corrective action.

Disinfectants/Disinfection By-product Rules

There are several rules that, together, address the issues created by the formation of various potentially harmful compounds by the addition of some disinfectants. Chlorine, for example, can form trihalomethanes (THMs) if certain organic substances are present. The concentration of some by-products can increase in the distribution system. Therefore, the rules require testing samples collected at sites throughout the system. Some important aspects of these rules for distribution system operators are given in the following sections.

Stage 1 Disinfectants and Disinfection By-products Rule

The Stage 1 DBPR applies to community water systems and nontransient, noncommunity systems, including those serving fewer than 10,000 people, that add a disinfectant to the drinking water during any part of the treatment process.

The rule includes the following key provisions:

- Maximum residual disinfectant levels (MRDLs) for three disinfectants—chlorine (4.0 mg/L), chloramines (4.0 mg/L), and chlorine dioxide (0.8 mg/L);
- MCLs for TTHM—0.080 mg/L; haloacetic acids (HAA5)—0.060 mg/L; and two inorganic DBPs—chlorite (1.0 mg/L) and bromate (0.010 mg/L); and
- A treatment technique for removal of DBP precursor material (*enhanced coagulation*).

Stage 2 Disinfectants and Disinfection By-products Rule

The rule tightened requirements for DBPs, but compliance is not achieved by modifying the numerical value of the MCLs or by requiring monitoring of new constituents. Instead, the rule makes compliance more difficult than under the Stage 1 DBPR by (1) changing the way the compliance value is calculated and (2) changing the compliance monitoring locations to sites representative of the greatest potential for THM and HAA formation. These changes were incorporated to attempt to account for peak spatial occurrence in the system. This change in focus reflects concerns of utilities and regulators caused by the potential for reproductive and developmental health effects associated with repeated exposure over a 12-month period at peak locations within the system.

The compliance value in the Stage 2 DBPR is called the locational running annual average (LRAA), and it is calculated by separately averaging the four quarterly samples at each monitoring location. Compliance is based on the maximum LRAA value (see Table 1-1). Furthermore, the Stage 2 DBPR included several interim steps that led to the replacement of many existing Stage 1 DBPR monitoring locations with new locations representative of the greatest potential for consumer exposure to high levels of TTHM and HAA5.

The Stage 2 DBPR requires that facilities maintain compliance with the Stage 1 DBPR using the existing monitoring locations during the first three years after the final version of the Stage 2 DBPR was published. In the time period between the third and sixth year after the Stage 2 DBPR was published, compliance continues to be based on maintaining 80/60 (TTHM and HAA5) or lower for the running annual average; it also includes a requirement for maximum LRAA at existing Stage 1 monitoring locations. These time periods during the Stage 2 DBPR are called "Stage 2A" and "Stage 2B."

The long-term goal of the Stage 2 DBPR is to identify locations within the distribution system with the greatest potential for either TTHM or HAA5 formation and then base compliance on maintenance of LRAA at or below 80/60 for each of these locations. Many of these locations were identified during the initial distribution system evaluation (IDSE). Consequently, the IDSE and the Stage 2A were actually just transition phases between the Stage 1 DBPR and the eventual long-term requirements of Stage 2B.

The IDSE included monitoring, modeling, and/or other evaluations of drinking water distribution systems to identify locations representative of the greatest potential for consumer exposure to high levels of TTHM and HAA5. The goal of the IDSE was to evaluate a number of potential monitoring locations to justify selection of monitoring locations for long-term compliance (i.e., Stage 2B) with the Stage 2 DBPR.

One item to note regarding the Stage 2 DBPR as it applies to TTHM and HAA5 is that the goal is to find the locations in the distribution system where average annual levels of these DBPs are highest. TTHM formation increases as contact time with free or combined chlorine increases, although formation in the presence of combined chlorine is limited. Therefore, establishing points in the distribution system with highest potential for TTHM formation is related to points with maximum water age. Utilities that have not performed a tracer study in the distribution system to determine water age should consider doing so.

By contrast, peak locations for HAA5 are more complicated because microorganisms in biofilm attached to distribution system pipe surfaces can biodegrade HAA5. Consequently, increasing formation of HAA5 over time is offset by biodegradation, eventually reaching a point where HAA5 levels decrease over time, even to the point where they drop to zero.

In chloramination systems, HAA5 formation is limited. In fact, ammonium chloride is added as a quenching agent in HAA5 compliance samples in order to halt HAA5 formation prior to analysis (see *Standard Methods for the Examination of Water and Wastewater*, latest edition). Therefore, little additional HAA5 formation occurs after chloramination to offset HAA5 biodegradation occurring in the distribution system.

Surface Water Treatment Rule

This rule is primarily directed at the treatment of water from surface water sources. It was originally intended to protect the public from exposure to *Giardia lamblia*. The rule was expanded by the Interim Enhanced Surface Water Treatment Rule (IESWTR) to include *Cryptosporidium*. The Long-Term 1 and 2 Enhanced Surface Water Treatment Rules strengthen the requirements for microbial protection of all sizes of water systems. Portions of these rules affect distribution systems, so it is important to describe the rules and to highlight these requirements.

Interim Enhanced Surface Water Treatment Rule

The IESWTR applies to systems using surface water, or groundwater under the direct influence of surface water, that serve 10,000 or more persons. The rule also includes provisions for states to conduct sanitary surveys for surface water systems regardless of system size. The rule builds on the treatment technique requirements of the SWTR with the following key additions and modifications of importance in distribution systems:

- Disinfection profiles must be prepared by systems with TTHM or HAA5 annual distribution system levels of 0.064 mg/L or 0.048 mg/L, respectively, or higher. The disinfection profiles will consist of daily *G. lamblia* log inactivation over a period of one to three years. These will be used to establish benchmarks for microbial protection to ensure that there are no significant reductions as systems modify disinfection practices to meet the Stage 1 DBPR.
- Systems using groundwater under the direct influence of surface water are now subject to the new rules dealing with *Cryptosporidium*.
- Inclusion of *Cryptosporidium* in the watershed control requirements for unfiltered public water systems.
- Requirements for covers on new finished water reservoirs.
- Sanitary surveys, conducted by states, for all surface water systems regardless of size.
- The rule includes disinfection benchmark provisions to ensure continued levels of microbial protection while facilities take the necessary steps to comply with new DBP standards.

Sanitary Surveys

Sanitary surveys are a requirement of the IESWTR. A sanitary survey is "an onsite review of the water source, facilities, equipment, operation, and maintenance of the public water system for the purpose of evaluating the adequacy of such source, facilities, equipment, operation, and maintenance for producing and distributing safe drinking water" (USEPA 1999). These surveys are usually performed by the state primacy agency and are required of all surface water systems and groundwater systems under the direct influence of surface water.

Sanitary surveys are typically divided into eight main sections, although some state primacy groups may have more.

1. Water sources
2. Water treatment process
3. Water supply pumps and pumping facilities
4. Storage facilities
5. Distribution systems
6. Monitoring, reporting, and data verification
7. Water system management and operations
8. Operator compliance with state requirements

Sanitary surveys are required on a periodic basis usually every three years. Surveys may be comprehensive or focused according to the regulatory agency requirements.

Long-Term 1 Enhanced Surface Water Treatment Rule

The LT1ESWTR strengthened microbial controls for small systems (i.e., those systems serving fewer than 10,000 people). The rule also prevents significant increase in microbial risk where small systems take steps to implement the Stage 1 DBPR. The rule also addresses disinfection profiling and benchmarking.

Long-Term 2 Enhanced Surface Water Treatment Rule

The update to the Surface Water Treatment Rule is called the LT2ESWTR, and it supplements SWTR requirements contained in the IESWTR for large surface water systems (>10,000 persons) and the LT1ESWTR for small systems (<10,000 persons).

One of the key elements of the LT2ESWTR was the use of *Cryptosporidium* monitoring results to classify surface water sources into one of four USEPA-defined risk levels called "bins." Facilities in the lowest bin (bin 1) are required to maintain compliance with the current IESWTR. Facilities in higher bins (bins 2 to 4) are required to either (1) provide additional *Cryptosporidium* protection from new facilities or programs not currently in use at a facility or (2) demonstrate greater *Cryptosporidium* protection capabilities of existing facilities and programs using a group of USEPA-approved treatment technologies, watershed programs, and demonstration studies, collectively referred to as the "Microbial Toolbox."

Implementation of the LT2ESWTR was phased over many years according to system size. Four size categories were established (schedule 1–4, with 4 being the smallest <10,000 population) for implementing the rule. The rule for schedule 4 systems allows filtered supplies to perform initial monitoring for fecal coliform to determine if *Cryptosporidium* monitoring is required.

One of the most potentially useful and cost-effective tools for utilities that was used to comply with the LT2ESWTR and demonstrate the true *Cryptosporidium* removal capability of an existing system is the demonstration of performance (DOP) credit. It was especially advantageous for facilities in bin 2. The DOP study can be conducted on an entire treatment process or a specific segment of the process. It can include monitoring of ambient aerobic spores in full-scale treatment processes or in pilot-scale spiking studies using *Cryptosporidium*, aerobic spores, or some other suitable microbial surrogate. The *Long-Term 2 Enhanced Surface Water Treatment Rule: Toolbox Guidance Manual* (USEPA 2003) describes cases where the DOP credit is likely the most cost-effective solution if the facility is assigned to bin 2, and the DOP credit can also be useful as a low-cost safety factor if the facility is assigned to bins 3 or 4.

Lead and Copper Rule

The LCR (promulgated in 1991 and revised in 2007) seeks to minimize lead and copper at users' taps. The rule establishes action levels for lead (0.015 mg/L) and copper (1.30 mg/L) for the 90th percentile of the samples measured at customer taps. Monitoring for a variety of water quality parameters is required. In addition to monitoring, all large systems are required to conduct corrosion studies to determine optimal lead and copper corrosion control strategies.

If the action triggers are exceeded, the system is required to evaluate several approaches: public education, source water treatment, corrosion control practices, and possibly lead pipe replacement. Corrosion control can include pH/alkalinity adjustment, corrosion inhibitor addition, and calcium adjustment.

This rule can affect disinfection strategies because some of the control measures for lead and copper involve water chemistry adjustments (specifically pH control). These adjustments can affect the formation of DBPs and disinfection effectiveness. Therefore, corrosion control measures employed to comply with the LCR must also be considered in the selection of an overall disinfection strategy.

The objective of the LCR is to control corrosiveness of the finished water in drinking water distribution systems to limit the amount of lead (Pb) and copper (Cu) that may be leached from certain metal pipes and fittings in the distribution system. Of particular concern are pipes and fittings connecting the household tap to the distribution system service line at individual homes or businesses, especially because water can remain stagnant in these service lines for long periods of time, increasing the potential to leach Pb, Cu, and other metals. Although the utility is not responsible for maintaining and/or replacing these household connections, they are responsible for controlling pH and corrosiveness of the water delivered to consumers.

Details of the LCR include the following:

- The LCR became effective Dec. 7, 1992.
- The action level for Pb is 0.015 mg/L and for Cu is 1.3 mg/L.
- A utility is in compliance at each sampling event (frequency discussed in the following paragraphs) when <10 percent of the distribution system samples are above the action level for Pb and Cu (i.e., 90th percentile value for the sampling event must be below the action level).
- Utilities found not to be in compliance must modify water treatment until they are in compliance. The term *action level* is used rather than MCL because noncompliance (i.e., exceeding an action level) triggers a need for modifications in treatment.
- The utility must sample each entry point into the distribution system during each sampling event.

After identifying sampling locations and determining initial tap water Pb and Cu levels at each of these locations, utilities must also monitor other water quality parameters (WQPs) at these same locations as needed to monitor and evaluate corrosion control characteristics of treated water. The only exemptions from analysis of these WQPs are systems serving less than 50,000 people for which Pb and Cu levels in initial samples are below action levels.

Pb, Cu, and WQPs are initially collected at 6-month intervals, and then this frequency can be reduced if action levels are not exceeded and optimal water treatment is maintained. Systems that are in noncompliance and are performing additional corrosion-control activities must continue to monitor at 6-month intervals, plus they must collect WQPs from distribution system entry points every 2 weeks.

Each utility must complete a survey and evaluate materials that comprise their distribution system, in addition to using other available information, to target homes that are at high risk for Pb/Cu contamination.

Revisions to the LCR were enacted in 2007. These clarifications to the existing rule were made in seven areas:

1. Minimum number of samples required
2. Definitions for compliance and monitoring periods
3. Reduced monitoring criteria
4. Consumer notice of lead tap water monitoring results (Within 30 days of learning the results, all systems must provide individual lead tap results to people who receive water from sites that were sampled, *regardless of whether the results exceed the lead action level.*)
5. Advanced notification and approval of long-term treatment changes
6. Public education requirements (Community water systems must deliver materials to bill-paying customers and post lead information on water bills, work in concert with local health agencies to reach at-risk populations [children, pregnant woman], deliver to other organizations serving "at-risk" populations, provide press releases, and include new outreach activities.)
7. Reevaluation of lead service lines (sample from any lead service lines not completely replaced to determine impact on lead levels.)

The local regulatory agency can be consulted for those revisions that are applicable to a particular system.

Ground Water Rule

USEPA promulgated the final Ground Water Rule (GWR) in October 2006 to reduce the risk of exposure to fecal contamination that may be present in public water systems that use groundwater sources.

The GWR establishes a risk-targeted strategy to identify groundwater systems that are at high risk for fecal contamination. The rule also specifies when corrective action (which may include disinfection) is required to protect consumers who receive water from groundwater systems from bacteria and viruses.

A sanitary survey is required, by the state primacy agency, at regular intervals depending on the condition of the water system as determined in the initial survey. Systems found to be at high risk for fecal contamination are required to provide 4-log inactivation of viruses. Increased monitoring for fecal contamination indicators may be required by the regulatory authority.

Federal regulations do not currently require disinfection of groundwater unless the well has been designated by the state as vulnerable to contamination by surface water (termed "groundwater under the direct influence of surface water"). These are generally relatively shallow wells. Many states, though, have their own requirements for required disinfection of various sizes, types, or classes of well systems.

BIBLIOGRAPHY

APHA, AWWA, and WEF (American Public Health Association, American Water Works Association, and Water Environment Federation). 2012. *Standard Methods for the Examination of Water and Wastewater*, 22nd ed. Edited by E.W. Rice, R.B. Baird, A.D. Eaton, and L.S. Clesceri. Washington, D.C.: APHA.

AWWA. 1993. *Drinking Water Handbook for Public Officials*. Denver, Colo.: American Water Works Association.

AWWA. 2010. Principles and Practices of Water Supply Operations—*Water Transmission and Distribution*, 4th ed. Denver, Colo.: American Water Works Association.

Edzwald, J.K., ed. 2011. *Water Quality and Treatment*, 6th ed. New York: McGraw-Hill.

USEPA Final Implementation Guidance for the Stage 1 Disinfectants/Disinfection Byproducts Rule. http://water.epa.gov/lawsregs/rulesregs/sdwa/stage1/upload/s1dbprimplguid.pdf.

USEPA Guidance on Ground Water Rule (GWR). http://www.epa.gov/ogwdw/disinfection/gwr/pdfs/fs_gwr_finalrule.pdf.

USEPA Guidance on Total Coliform Rule. http://www.epa.gov/safewater/disinfection/tcr/pdfs/RTCR%20draft%20fact%20sheet%2061710.pdf.

USEPA Office of Ground Water and Drinking Water. http://water.epa.gov/drink/index.cfm.

USEPA. 1999. *Guidance Manual for Conducting Sanitary Surveys of Public Water Systems; Surface Water and Ground Water Under the Direct Influence (GWUDI)*. April 1999. US Environmental Protection Agency. Accessed June 22, 2009. http://www.epa.gov/OGWDW/mdbp/pdf/sansurv/sansurv.pdf.

USEPA. 2003. *Long Term 2 Enhanced Surface Water Treatment Rule: Toolbox Guidance Manual*. EPA 815-D-03-009. http://www.wqts.com/pdf/2003-01_LT2_ESWTR_Toolbox.pdf.

USEPA. 2004. *Assessing Capacity Through Sanitary Surveys*. US Environmental Protection Agency: Drinking Water Academy. Accessed June 22, 2009. http://www.epa.gov/safewater/dwa/electronic/presentations/pwss/assesscapacity.pdf.

Chapter 2

Managing System Operations

The objective of distribution system operations is to reliably and efficiently deliver high-quality drinking water to meet customers' needs and satisfy fire flow requirements. Achieving this objective can be difficult. To be successful, system operators must understand the purpose and consequence of the many operational options they can employ. Then they must develop an integrated operations management plan that can be implemented and the results tracked.

Three main distribution system performance categories are included in the distribution system operations objective: water quality, reliability, and efficiency. Achieving excellence in one category may adversely affect the performance in the other two. System operators must balance these sometimes-competing objectives to satisfy the priorities established by their customers, regulatory agencies, and system managers. These categories are further defined as follows:

- *Water quality.* Ideally, water delivered to customers would be the same quality as the water that entered the distribution system (from the treatment plant or other source). However, water quality is usually changed as it travels through miles of distribution piping and resides in storage facilities. These changes mostly do not improve the water quality. System operators can use many practices to mitigate the changes and deliver water of high quality to users.
- *Reliability.* The goal of system operators is to deliver an uninterrupted supply of water to customers and to meet fire-fighting requirements. Main breaks, power supply problems, and emergency failures (from natural disasters or intentional acts) can cause temporary interruption to water service. System operators can adopt practices and plans to reduce the frequency and extent of these service stoppages.
- *Efficiency.* Energy usage and water losses are the two main elements affecting efficient delivery of water. Reducing these factors can save water and reduce distribution system operating costs. System operators can participate in improving efficiency by employing practices and procedures as part of an overall plan.

Many operational practices can affect the performance of the distribution system. System operators are often overwhelmed with the choices and resort to a reactive approach by using practices only when a problem is apparent. This can be effective, but most often the effect is only temporary and is not documented, so when the problem reoccurs, the operators are again searching for a solution. To ensure lasting improvement in system performance, operators should use a management plan that is based on a thorough assessment of the system and selection of the operational practices that can make a difference.

The management plan should include performance measures and goals for individual practices and the three main performance categories (water quality, reliability, and efficiency). System operators should continuously review the performance results so the plan can be modified if needed.

Some of the performance improvement practices are necessarily only initiated by executive management because they may involve considerable capital expense. However, operators must be knowledgeable of these initiatives so they can operate the system effectively if they are adopted. Operators should also be included for input by management when considering system performance-enhancing options. This then requires that operators are knowledgeable of the merits of these options so they can provide educated advice to management.

DEVELOPING AN OPERATIONS MANAGEMENT PLAN

A distribution system operations management plan includes performance objectives, operational practices to achieve the objectives, and performance measures and numeric goals of the plan. Also, persons or departments responsible for leading implementation of certain operational practices are identified in the plan. Periodic plan reviews are scheduled to review the results and make modifications if needed.

System Performance Assessment

An initial step in developing an operations management plan is to assess the current status of the system performance. Evaluating performance for water quality, reliability, and efficiency can be an involved process. However, there are several good resources that can simplify this task.

The Partnership for Safe Water is supported by the six largest drinking water organizations, including the American Water Works Association (AWWA) and US Environmental Protection Agency (USEPA). The Partnership provides a comprehensive program to optimize distribution system performance. A key element of this program is a self-assessment of the current system status. The results of the self-assessment are then used to create an improvement plan that leads toward improved performance. The program provides extensive guidance on how to conduct the self-assessment. This includes software to calculate performance measures that are used to assess the degree of optimization. Performance goals (that define excellence) are provided so that systems can compare and track the results of their improvement action plan. For further information, the program website can be consulted: www.partnershipforsafewater.org.

AWWA has developed a standard for distribution system management, ANSI/AWWA G200. This standard includes the consensus management practices of the drinking water industry. Conformance with the standard is voluntary. A companion document for the standard is the *Operational Guide to AWWA Standard G200: Distribution Systems Operation and Management* (Oberoi 2009). This guide provides useful information on how to use the standard, including conducting a system self-assessment.

Another reference that may be helpful is *Water Distribution System Assessment Workbook* (Smith 2005). The workbook contains worksheets for conducting a self-assessment, and references are provided for additional information.

Performance Measures and Goals

An important element of an operations management plan is to use performance measures and track system results against established goals. The most appropriate measures are identified in the system performance assessment.

Several approaches are used when establishing goals. Goals can represent the ultimate performance level, reflect current operation, be short term or long term, be those recommended by others or internally developed, and may include financial limitations. No matter which approach is used, the goals must be recognized by all operations staff.

Goals should be achievable with effort. One approach is to have an interim or short-term goal that is slightly better than the current status. This provides for interim success while allowing time to reach a higher level. Good goals are those that are within the control of the person or group that is responsible. There does not need to be a goal for everything being measured. Some performance measures are just for tracking and do not need a goal attached.

Operational Practices

The distribution system operations management plan lists the practices used to meet the system performance objectives (a list of the main operational practices affecting performance is shown in Table 2-1). Systems may emphasize selected operational practices depending of the need and the performance objectives. Many systems may list *all* of the listed operational practices in their plan.

The operational practices listed in Table 2-1 are the most common. Site-specific conditions may require the use of additional practices to meet system performance objectives. Information on the operational considerations for each of the practices is included in chapter 3. Examples of

Table 2-1 Operational practices affecting distribution system performance

Operational Practices	Chapter	Performance		
		Water Quality	Reliability	Efficiency
Corrosion control—external	5		•	
Corrosion control—internal	5	••		•
Cross-connection control	13	••	•	
Customer complaint response	20	•	•	•
Energy management	3		•	••
Flushing program	3	••	•	•
Hydrant and valve maintenance	6 & 9	•	••	•
Main breaks management	3	•	••	•
Maintaining disinfectant residual	3	••	•	•
–Nitrification control (chloramines)	3	•		
–DBP control	3	•	•	
Pipe rehabilitation and replacement	7 & 2		••	•
Pressure management	3	•	•	••
Sediment control	3	•		•
Security and surveillance	17	•	•	•
Storage tank management	10	••	•	•
Water age management	3	••	•	•
Water loss control	3	•	•	••
Water sampling and response	3	•	•	•

• Performance affecting practice.
•• Major performance affecting practice.

performance measures commonly used by utility system operators are given in this chapter. The table indicates the performance criteria mainly affected by each operational practice. Exceptions to these relationships may occur because of unusual local conditions, but these should be included in the management plan.

Discussions of several operational practices are located in separate chapters of this book. The chapters listed in Table 2-1 indicate where this information can be found. Discussion regarding development of a management plan for pipeline rehabilitation and replacement is included in this chapter because these practices are often adopted by executive management and included in a capital improvement plan.

Management Plan Review and Revision

After the distribution system operations management plan has been developed and is in use, it should be periodically reviewed to evaluate the results. Most systems review the plan annually. This review occurs most often after the data from a calendar year is available and before budgets for the next year are due. For many systems, the best time is in the spring.

The review includes an assessment of the performance measures as compared to the established performance goals. Any major operational challenges are noted along with suggestions how these could be approached if they reoccur. Projections for operations during the upcoming year should identify any unusual situations that may affect system operating practices. Adjustments to the operations management plan are made when performance goals are not being met or when changing circumstances require a modified approach.

SYSTEM OPERATIONS TRAINING

Managing the operation of a distribution system requires a trained staff. Distribution system operators require certification to ensure a level of understanding needed to provide potable water service. System operators need to meet regulatory requirements and provide emergency service for fire fighting and in response to natural disasters.

Training can be provided in several forms: experienced op30erators, reference materials, training courses, and system operations simulations. Experienced operators provide invaluable information to those individuals who are new to the system. Reference materials (like this book) provide standardized information gained from the operating experience of industry leaders and researchers. Also, these materials are usually the basis for certification exam questions. AWWA and other providers offer periodic training courses aimed at system operators (contact your local AWWA for availability).

Technology is providing another valuable training tool: hydraulic model operations simulations. Network computer models can be used to experience operational situations and practice responses. This can be accomplished by using fixed information or real-time operating data. Extreme events can be simulated so that operators are able to develop confidence when faced with catastrophic situations. If this technology is not available, as a substitute, operators can talk through various operational scenarios so that they can develop confident responses.

EXAMPLE PERFORMANCE MEASURES

Performance measures are given in the following list for the operational practices listed in Table 2-1. These measures are commonly used by utility system operators to quantify performance and to compare with established goals. Measures are not listed for all of the operational practices. In some cases, this is because no common measures could be found. Performance goals are also not listed because each system must establish its own according to its needs. Some common goals are available in particular where regulatory requirements exist. Many systems have adopted the regulatory limit as their goal. Other systems have adopted goals that are more stringent than regulations require. The goal-setting approach is specific to each system and its operational objectives.

- **Customer Complaint Response**
  - Technical complaint rate = # technical complaints/#customer accounts
  - Technical complaint response time = minutes to respond after receipt

  Technical complaints are directly related to the core services of the utility. They include complaints associated with water quality, taste, odor, appearance, pressure, main breaks, and disruptions of water service.

- **Energy Management**
  - Electrical usage rate = kW·h/# customer accounts
  - Unit cost of water delivery = $ cost/mgd, or $ cost/# customer accounts

- **Hydrant and Valve Maintenance**
  - % Valves inspected and exercised annually = # inspected/total number in system
  - % Hydrants inspected and exercised annually = # inspected/total number in system
  - % Large valves inspected and exercised annually = # >10-in. valves inspected/total large valves
  - Hydrant repair time = time (hours) returned to service upon receipt of repair order

- **Main Breaks Management**
  - Main break frequency rate = # reported breaks per 100 miles of utility-controlled mains
  - Main break frequency trend = a 5- or 10-year main break frequency rate that is declining

  Reported breaks are those leaks or breaks that come to the attention of the water utility as reported by customers, traffic authorities, or any outside party due to their visible and/or disruptive nature. A nonsurfacing break initially reported as loss of pressure by a customer is an example of a nonsurfacing reported leak. Events that can be inferred from alerts by supervisory control and data acquisition (SCADA) systems can be labeled as reported breaks. Water utilities respond to reported breaks in a *reactive* mode, often under emergency conditions (AWWA Manual M36, *Water Audits and Loss Control Programs*).

- **Maintaining Disinfectant Residual**
  - System disinfectant residual = monthly 95th percentile of all routine measurements.
  - Consecutive residual measurements below minimum at individual routine sample locations

  Routine measurements are measurements taken from samples that are collected on a regular schedule.

- **Pipeline Rehabilitation and Replacement**
  - Pipeline annual renewal rate = miles of pipeline renewed ÷ total miles of distribution system pipeline mains
  - Pipeline annual breaks rate = # of breaks/100 miles of pipeline mains
  - Service interruption rate = # of 4-hour service interruptions/# of customer accounts

- **Pressure Management**
  - Minimum pressure (in each zone)
    - Normal water demand period, monthly average of daily minimums
    - Peak water demand and fire flow periods, hourly minimum during simultaneous peak water demand and fire flow conditions
    - Emergency conditions, during emergencies such as main breaks and power outages

- **Storage Tank Management**
  - Storage facility sampling frequency = number of water quality samples taken monthly
  - Storage facility inspection frequency = number of inspections conducted annually

- **Water Age Management**
  - Pipeline maximum water age = days resident in the system distribution mains (calculated by a calibrated hydraulic network model)
  - Storage facility maximum water age = depending on operating conditions (calculated by a calibrated hydraulic network model)

- **Water Loss Control**
  - Infrastructure condition factor = see AWWA Manual M36, *Water Audits and Loss Control Programs*, for details
  - Infrastructure leakage index = see AWWA Manual M36 for details
  - Annual real losses = see AWWA Manual M36 for details

PIPELINE REHABILITATION AND REPLACEMENT

Pipeline rehabilitation and replacement (R&R) has a significant effect on water distribution system infrastructure condition and performance. By targeting older or problem pipelines for renovation, utilities can reduce leaks and breaks, improve water quality, and enhance water delivery reliability. Management planning for major pipeline renovation programs is described in this section, and pipeline rehabilitation methods are discussed in chapter 7.

Pipeline R&R Plan

Infrastructure renovation (replacing and rehabilitating pipelines) is an expensive undertaking. Utilities usually cannot afford to replace aging system components at a rate they would prefer and must make choices regarding the amount of system improvements they can perform annually. A pipeline R&R plan should be developed to provide priorities so that the most needed improvements can be made while others must be delayed. The primary components of this plan are preparing a pipeline inventory, conducting a performance assessment and tracking, and determining priorities.

Pipeline Inventory

An inventory of pipelines with physical and location attributes is an essential element of a pipeline R&R plan. Data in the inventory should minimally include pipe size, material, class, lining, and year installed. Location data should include street name, municipality, and pressure zone. The effective management of physical asset and location information for pipelines can enable a utility to locate and repair mains and plan for rehabilitation or replacement.

Performance Assessment and Tracking

Performance information may include customer complaints, breaks and leaks, and water quality, flow, and pressure. The frequency and type of main breaks and leaks are often the most important performance information. Several methods are available to assess the pipe condition that may lead to failure. Some of the more common methods are soil corrosiveness testing, pipe coupon testing, and acoustic leak detection. Video inspection of pipelines is also common practice for large mains.

Planning and Prioritization

A pipeline renewal program should be coordinated with the utility capital improvement plan. The program must consider the appropriate annual funding and select the specific main segments to be renewed. Because of financial constraints, most utilities must limit the number of renewal projects to only those that have the highest priority.

AWWA Manual M28, *Rehabilitation of Water Mains*, describes several approaches for prioritizing renewal of mains, and some models are commercially available. These approaches include scoring methods (e.g., point systems), economic analysis (like break-even, cost-benefit), failure probability and regression analyses, and mechanistic models to predict pipe failure based on loading and condition.

Operational Considerations

Utility management is usually responsible for developing a pipeline R&R program and for the decisions to implement the plan due to the cost and impact of these programs. System operators, however, have an important role. Operators must often acquire the data needed for the system inventory to construct performance measures and that are used to set renewal priorities. Another critical operational consequence is that as projects are developed they present situations where there may be water supply interruptions or require unusual operating conditions. System operators should plan for these adjustments to minimize adverse customer impacts.

BIBLIOGRAPHY

ANSI/AWWA G200. *AWWA Standard for Distribution Systems Operation and Management.* Denver, Colo.: American Water Works Association.

AWWA. M28—*Rehabilitation of Water Mains.* Denver, Colo.: American Water Works Association.

AWWA. M36—*Water Audits and Loss Control Programs.* Denver, Colo.: American Water Works Association.

AWWA. 2010. Principles and Practices of Water Supply Operations—*Water Transmission and Distribution*, 4th ed. Denver, Colo.: American Water Works Association.

Friedman, M., G. Kirmeyer, J. Lemieux, M. LeChevallier, S. Seidl, and J. Routt. 2010. *Criteria for Optimized Distribution Systems.* Denver, Colo.: Water Research Foundation.

Great Lakes–Upper Mississippi River Board of State Public Health and Environmental Managers. 2012. *Recommended Standards for Water Works.* Albany, N.Y.: Health Education Services.

Grigg, N.S. 2004. *Assessment and Renewal of Water Distribution Systems.* Report #91025F. Denver, Colo.: Awwa Research Foundation.

Oberoi, K. *Operational Guide to AWWA Standard G200: Distribution Systems Operation and Management.* Denver, Colo.: American Water Works Association.

Smith, C., ed. 2005. *Water Distribution System Assessment Workbook.* Denver, Colo.: American Water Works Association.

Chapter 3

Operational Practices

Three major performance categories for drinking water distribution system operations are (1) maintain water quality from the point of entry into the distribution system to the point of use, (2) deliver sufficient water to reliably satisfy customer demands and meet fire-fighting requirements, and (3) efficiently perform system operations by controlling water losses and reducing energy use. Simultaneously satisfying all these objectives can be difficult.

Operators should engage in practices that improve or maintain distribution system performance in all three performance categories. Therefore, there must be an understanding about the purpose and consequence of the many operational practice options so that an integrated plan can be developed and implemented.

Some of the performance improvement practices are necessarily initiated only by management since they may involve considerable capital expense. However, operators must be knowledgeable of these initiatives so they can operate the system efficiently if they are adopted. Operators should also be included for input by management when considering system performance-enhancing options. This then requires that operators are knowledgeable of the advantages and disadvantages of these options so they can provide educated advice to management.

Many operational practices can be employed to affect changes in distribution system performance (listed in chapter 2). System operators should be aware of how these practices can be used and which ones may have the most effect on system performance. The distribution system operations management plan (discussed in chapter 2) includes performance measures and operational goals. When the desired results are not being achieved, adjusting the operational practices included in this chapter may result in the needed improvement.

Although the most common operational practices are listed that have the most influence on performance for each of the three main performance categories, operators should appreciate that many of the practices affect performance of more than one category. Also, there will be operational practices that are not listed which may have a large effect on system performance for specific local conditions. These practices should be continued and included in a distribution system operations management plan.

MAINTAINING WATER QUALITY

The operational practices included in this section have the greatest effect on this performance category. Where water quality performance goals are not being met, adjusting these practices may result in the needed improvement. Operators can use the information provided to help select the most effective practices and the adjustments that can lead to improved performance.

Sources of Water Quality Problems

It is the responsibility of distribution system operations personnel to be aware of possible sources of water quality problems and to utilize practices to improve or maintain water quality. Degraded water quality can result in customer dissatisfaction and may even present a threat to public health. Unsatisfactory water quality can be caused by water quality problems with the water entering the distribution system, changes in the water quality within the distribution system, or contamination of the distribution system from outside sources.

Water Quality Entering the Distribution System

Stringent regulatory requirements (US Environmental Protection Agency [USEPA], states, provinces, and other agencies) are in place for water produced by treatment plants and from well supplies used to provide drinking water for community water systems. These requirements include monitoring for a large number of contaminants and other water quality measurements. Unless there is a violation of the regulatory requirements, it is highly unlikely that these contaminants could be present in the water entering the distribution system.

There are, however, substances that can affect water quality that are not regulated. Some changes can cause disagreeable odors, tastes, and color without an impact on health. Odors from natural compounds in lakes and rivers can sometimes resist treatment and, thus, could cause customer dissatisfaction with their drinking water. Iron or manganese can occur, particularly in well water, in high amounts to discolor the water and even cause staining of plumbing fixtures or laundry.

Some water systems purchase water from another system. In a few cases, the purchased water has passed through several intermediate systems. Changes in water quality can occur due to the length of time in transport or other factors. The receiving system should take special samples (that may not be required by regulations) to ensure that the water is satisfactory. Regulatory agencies may need to be consulted if results indicate possible concern.

Changes in Water Quality Within the Distribution System

Many factors can contribute to changes in water quality in distribution systems. Water age (the time from production or source to the customer's tap) is one of the most important of these factors. Excessive water age can cause disinfection residuals to dissipate, and this can lead to the growth of microorganisms. Tastes and odors can develop, and metals (e.g., iron and manganese) and other substances can enter the water, causing color and increasing the potential for staining of plumbing fixtures and laundry. Increasing water temperatures can aggravate this problem. Also, the presence of nutrients from the water treatment process or from natural sources can accelerate the growth of problem organisms. Distribution system storage facilities are often the source of water quality problems, because they increase water age and provide surfaces for biofilm growth.

Corrosion of the interior surface of water mains can cause deterioration in water quality. The corroded areas can shield microorganisms from disinfection, and materials from these locations can enter the water and cause problems for customers. Lead and copper levels at customer taps can increase when corrosive water is left in contact with lead solder that joins copper pipe or plumbing fixtures that contain lead. Some older homes are connected to the water main with lead service lines that can be the source of increased lead levels in tap water. Lead levels in home tap water are regulated under USEPA's Lead and Copper Rule.

Contamination From Outside Sources

Cross-connections can be a source of water quality problems. Chapter 13 is devoted to cross-connection control and backflow prevention. Backflow through a cross-connection can be a very serious, perhaps life-threatening, water quality problem. Suspected cross-connection contamination must be dealt with immediately.

Construction and maintenance activities provide opportunity for contamination of the water supply. Proper procedures must be followed to disinfect new water mains and when making repairs.

Intentional contamination of a water supply is very rare. However, sabotage or terrorism could target water systems. The appropriate emergency response and security agencies must be contacted if intentional contamination is suspected.

PRACTICES THAT AFFECT WATER QUALITY

The best way to deal with water quality problems is to prevent them from occurring. Employing practices that can reduce or eliminate the development of water quality problems will reduce costs related to the necessary responses as well as improve customer satisfaction. Several operational practices can be used to maintain or improve water quality in distribution systems. Operators should consider employing some or all of these practices to reduce or eliminate water quality deterioration.

System Design

The development of water quality problems in the distribution system can often be avoided by using system designs that consider these issues. Long-range system plans can anticipate potential water quality changes and include features to mitigate the impacts. The use of hydraulic and water quality system network computer models (chapter 4) can reveal areas where water quality problems may develop. With this information, the system design can be modified or operational procedures employed that can reduce any potential problems.

Dead-end mains (e.g., dead ends caused by closed valves or pressure zone boundaries) can be the source of water quality problems. The system design should strive to eliminate or reduce these troublesome conditions (chapter 4).

Finished water storage facilities (chapter 10) are another major system component that is often the source of deteriorating water quality. The best design for storage facilities results in good mixing and frequent water exchange (Figure 3-1). The facility should also include access for sampling and inspection, and isolation capability for cleaning and maintenance.

System Operation and Maintenance Practices

Several operational practices can be used to help prevent the onset of water quality problems. The most effective procedures either reduce hydraulic detention time (decrease the time that the water is in the system before it is used, which also should help maintain a good disinfectant residual); remove potentially poor quality water from the system; or use materials or water quality characteristics to protect system components from corrosion.

Preventing Internal Corrosion

Corrosion on the inside of pipelines (internal corrosion) is caused by chemical, biological, and hydraulic characteristics of drinking water that decay internal metallic components (pipes, steel tanks, metal water service lines, and interior piping in buildings). This can result in elevated metals in the bulk water, metal deposits, pitting, tuberculation, or weakening of piping. Internal corrosion of distribution system components (illustrated in Figure 3-2) affects water quality, system reliability (by contributing to increased main break frequency), and efficiency (by affecting the flow characteristics of pipelines).

Internal corrosion can be serious on unlined cast-iron or steel mains, steel tanks, and other metal surfaces in the distribution system. As a result, unlined cast-iron and steel pipe are no longer installed, but there are still

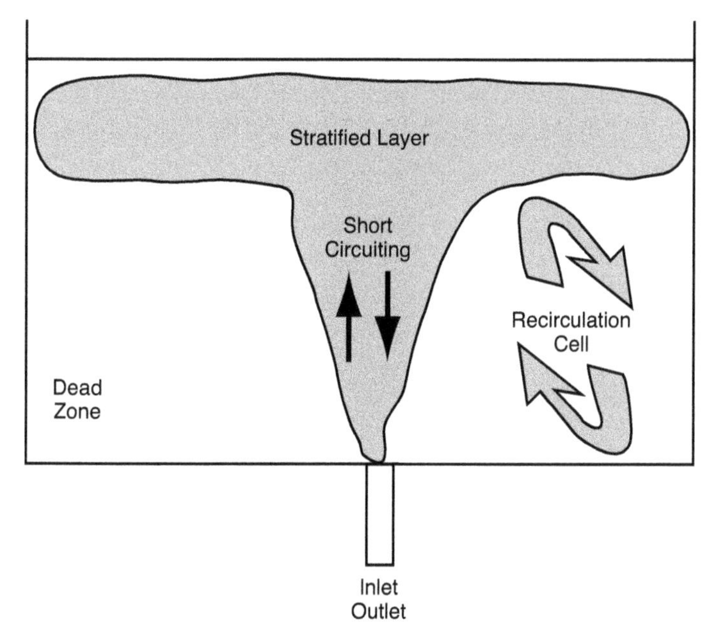

Figure 3-1 Nonideal flow patterns in a storage tank

thousands of miles of old, unlined pipe in use. Cathodic protection is often used to control corrosion in steel tanks.

USEPA's Lead and Copper Rule provides protection from the potential adverse health effects of lead and copper leached from water service lines and customer plumbing. All water systems must monitor for the presence of lead and copper in drinking water samples from customer taps, and, if excessive levels are found, the system must add treatment to reduce the corrosiveness of the water.

Strategies used to reduce lead and copper levels involve pH and alkalinity adjustment, the use of corrosion inhibitors, and calcium adjustment. Lead control in the pH range of 6–9 generally requires the dissolved inorganic carbon (DIC) to be greater than 2 mg C/L (carbon per liter). Carbon dioxide, soda ash, and sodium bicarbonate may be used to adjust the DIC. Orthophosphate (or other phosphate-based inhibitors) may be used as a corrosion inhibitor in the pH range of 7.4–7.8 at a typical dosage of 1–5 mg PO_4/L (phosphate per liter). Copper and iron corrosion are reduced by using orthophosphate or by adjusting pH and DIC. Lead service lines may require replacement in some cases.

A system-wide corrosion control program includes scheduled monitoring and testing to assess corrosion potential and occurrence in distribution systems. Monitoring for corrosion of metal pipelines and system components is also needed to verify the effectiveness of any control procedures. For example, the treatment of the water with corrosion inhibitors (zinc,

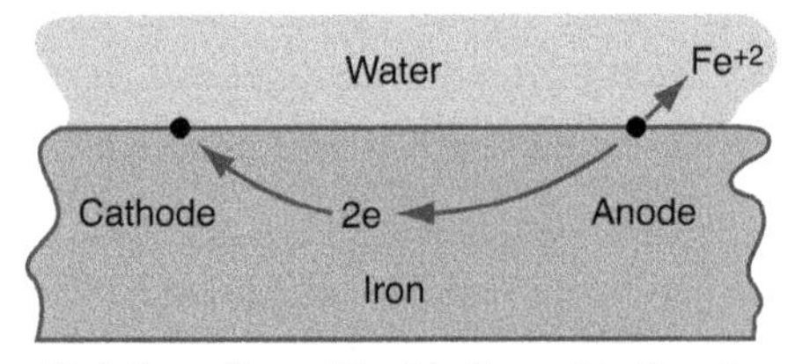

A. Minor Variations Cause Electric Current to Develop

$H_2O \rightleftarrows H^+ + OH^-$

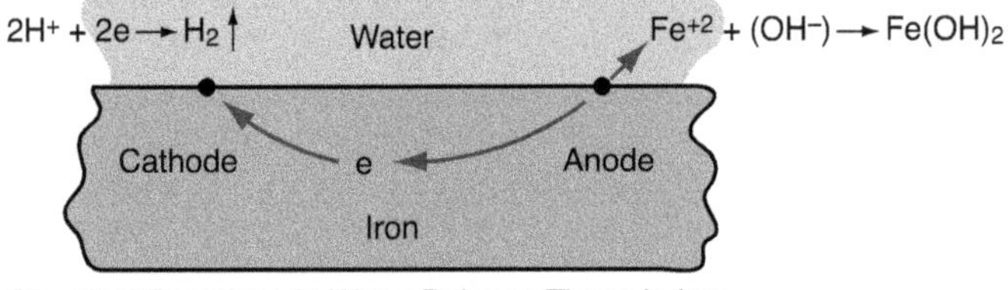

B. Chemical Reactions in Water Balance Those in Iron

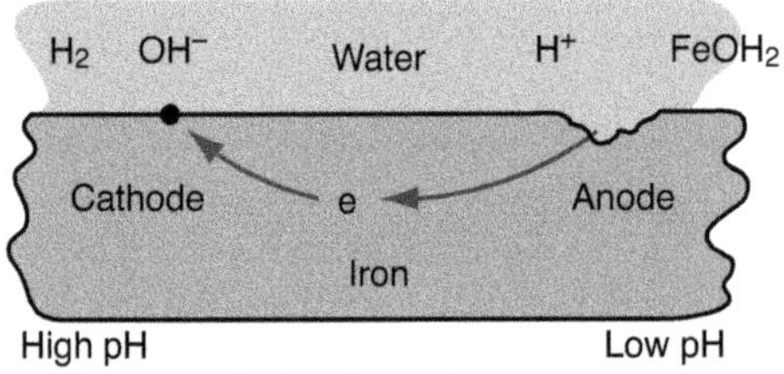

C. Rate of Corrosion is Accelerated

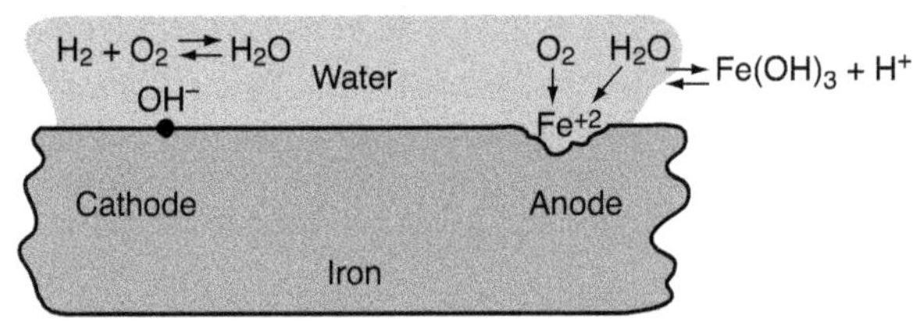

D. Rust Forms

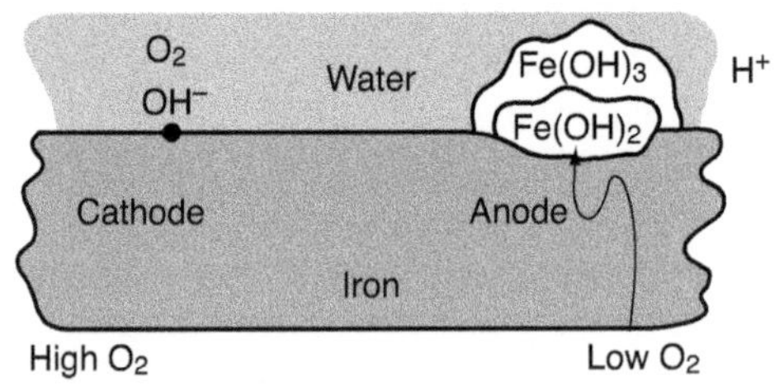

E. Rust Precipitates to Form Tubercules

Source: AWWA 2010.

Figure 3-2 Chemical and electrical reactions in a corrosion cell

phosphate, and silica) necessitates testing for excessive residuals. Water quality parameters (like pH, alkalinity, conductivity, calcium, and redox [oxidation reduction] potential) are also tested to ensure they are not changing in the distribution system.

In addition to monitoring, programs should include regular inspection of internal pipeline conditions by looking for tuberculation, pitting, holes, or scaling. These inspections may be coordinated with pipeline break or leak repairs. Transmission mains or key distribution mains should be flow tested periodically to determine *C* values for historical comparison. Reductions in *C* values may indicate a developing corrosion problem.

Corrosion control programs should include a periodic review that may lead to a revision of the corrosion control strategies. State regulatory agencies may need to approve any change in the control strategy, and they may require additional testing to ensure the effectiveness of the revised procedures.

Effective corrosion control programs include system surveillance practices to identify isolated pockets of increased corrosion potential. Areas of concern may have low water use, can be oxygen deficient, often have low disinfectant levels, and may promote biological growths. Galvanized, copper, and steel pipes are vulnerable to corrosion and should be monitored. Corrosion-related customer complaints can provide important information that can help identify developing corrosion locations.

System operators can contribute to improved corrosion control by reducing water age and maintaining optimum water quality within the distribution system. Operators should be alert to information from leak and break investigations since this data can reveal corrosion locations that may need special attention.

Water Main Flushing

Flushing water from pipelines is one of the main tools of system operators to clean and refresh water being delivered to customers. Several objectives can be achieved by a carefully designed flushing program, including removing sediment and other impurities, improving disinfectant residual levels, decreasing disinfection by-product (DBP) levels, resolving water quality complaints, or decreasing water age.

The extent of a systems problem with discolored water caused by sediment depends on water characteristics and the type of piping in the system. Although it may be bacteriologically safe, customers are reluctant to drink discolored water. The sediment may also discolor washed clothes and can be a source of annoying taste and odor. Slime growth is also a problem in some systems, but this can usually be controlled by maintaining an adequate chlorine residual in the distribution system.

Operators of systems with sediment problems usually find it best to set up a regular flushing program that will keep the mains relatively clean and minimize customer "dirty water" complaints. Flushing can also help prevent obstructions in the system caused by sediment buildup. Some water system operators find it most convenient to conduct main flushing (coordinated with hydrant testing) in the fall after the heavy water use of summer is over. Other system operators flush in spring to rid the system of sediment that has accumulated during the winter, before flow rates increase in the summer sprinkling season. Many water systems have particularly bad sediment accumulation in dead-end mains, and some system operators have found it necessary to flush these mains as often as weekly to avoid customer complaints.

It is not normal practice for a larger water system to flush the entire distribution system. Rather, selected areas that have initiated customer complaints may be placed on a regular flushing program. If flushing of a large area of the system is necessary, the conventional procedure is to start at the well or treatment plant and work outward, flushing each hydrant in turn until the water is relatively clear.

Flushing programs may be preventive or in response to a specific problem. Programs can also be system-wide or localized. In any case, a unidirectional flushing program is often preferred. With this method, all of the valves to connecting mains are closed so that a section of main about 2,000 ft (610 m) long draws from a relatively clean source and discharges from a single hydrant at the end. The unidirectional flushing method can increase the velocity of the water so sediment and biofilms can be removed. Many commercially available software packages can aid in developing a unidirectional flushing program.

It is often helpful to flush late at night to provide better flow rates, cause less traffic disruption from water on the street, and result in fewer customer complaints. In addition, customers will not see the condition of the water being flushed. Flushing should be preceded by various announcements advising customers of when flushing will take place and suggesting that they not wash clothes if the water is discolored. The announcements can be made in local media, and many utilities also place "water main flushing this week" signs on key roads in the area. Customer service and communications employees should be prepared to inform customers of the timing and duration of the program, what to expect during the flushing, and any actions needed to clear services lines.

The flushing velocity should be appropriate for the specific water quality objective. Lower velocities (1 ft/sec or less) may be used when the objective is to move water through the system (to improve disinfectant residual or reduce water age). Higher velocities (2.5–5 ft/sec) may be required to flush sediment or remove biofilms. System water pressures should be monitored during flushing to avoid creating low or negative pressure areas. Monitoring flows can ensure that fire flow requirements can be achieved. Monitoring of water quality parameters (such as disinfectant residual, turbidity, and color) can help to set performance goals to optimize a flushing program.

Water quality can be adversely affected by rapid changes in velocity or the direction of flow in pipelines. Flushing (and other uses of fire hydrants) can cause a loss in pressure and even a change in the normal flow

Courtesy of Pollardwater.com.

Figure 3-3 Flow diffuser

in adjacent pipelines. Training should be provided to personnel using fire hydrants so they operate the hydrants properly.

Methods of diffusing flow must also be considered for each site so as not to damage public or private property. Various types of flow diffusers are available to break up and divert the flow to prevent damage to lawns (Figure 3-3).

Systems should have performance measures (chapter 2) for assessing the water quality in the distribution system when conducting flushing. These could include the number or frequency of water quality complaints received, the disinfectant residual, dissolved oxygen levels, or color and odor values.

There is an opportunity, while flushing, to assess the condition of the distribution system. Exercising valves and hydrants often reveals broken or improperly functioning system components that can then be repaired to ensure they are in proper working condition.

The utility flushing plan should consider procedures to neutralize disinfectant residuals to protect waterways and reduce environmental impacts. Chloraminated systems may consider stopping ammonia addition and convert to free chlorine to minimize the impacts on nitrogen discharge to receiving streams (see discussion of nitrification control later in this chapter for procedures and precautions). Local regulations may require dechlorination to discharge water flushed from the system.

Automatic flushing systems are an option to address areas of deteriorating water quality that result from low water usage. Plans for automated flushing systems include consideration of the location, frequency of operation, and any environmental impacts. These systems should be routinely inspected and monitored for any impacts on system pressure. Regulations may require dechlorination of discharged water, and this may complicate automated flushing systems.

System operators are often those executing a flushing program. Therefore, they are well informed about the location and extent of the day's flushing activities. Monitoring pressure and flow during flushing ensures that adequate operating conditions are maintained. The results of the flushing program need to be recorded and reviewed since these may include important information concerning the condition of the distribution system.

Cross-Connection Control

An effective cross-connection control and backflow prevention program is essential to protect the distribution system from contamination by outside sources. Chapter 13 describes the practices and procedures for these programs. A more detailed description of cross-connection control programs is included in AWWA Manual M14, *Recommended Practice for Backflow Prevention and Cross-Connection Control.* When consulting these references water systems should recognize that they must comply with state or local cross-connection regulations.

In general, comprehensive programs include these common elements:

- Legal authority and administrative responsibility
- Adequate budget, funding, and staff resources
- Record-keeping and data management systems
- Qualified and trained personnel
- A clearly defined service policy
- Established minimum standards for acceptable backflow-prevention assemblies
- Training and certification for backflow device installers and testers
- Public education program (Explain the reasons for the program and its benefits.)
- Required testing and maintenance of backflow-prevention assemblies
- Enforced action for noncompliance
- A periodic backflow-prevention program report

System operators should be aware of the threat for contamination from cross-connections and the elements of the utility prevention program. An operator's primary weapon against backflow from cross-connections is to maintain adequate system pressure. If a situation develops where pressure is lost in an area (e.g., from a large main break), the area may need to be isolated until possible contamination can be investigated. An approved standard operating procedure (SOP) for this situation should be available for operators to follow.

Customer Complaint Response

Customer complaints contain important information about water quality, system reliability, and security that can be used to assess distribution system performance. Customer complaint response is important to water utilities, because this interaction often is the basis for a customer's perception about their satisfaction with service. In some locations, reporting procedures and content may be specified under state or local regulatory requirements. Chapter 20 contains information to formulate an effective customer complaint response program.

Distribution system operators may be assigned to investigate customer complaints. This provides an opportunity to solve a problem for the customer and, perhaps, discover a larger system problem. It is often a challenge to collect and share customer complaint information when several utility departments are involved. The complaint records should, therefore, be kept in a location accessible to all those who will use the information.

Maintaining Disinfectant Residual

Maintaining adequate disinfectant residual throughout the distribution system is needed to ensure drinking water quality. Disinfectant residual requirements are included in state and federal regulations. Most regulations require a minimum residual (the level varies) at all required monitoring sites. Some systems can obtain a variance to operate without a disinfectant residual. USEPA regulations include maximum residual disinfectant levels for water entering the distribution system for chlorine (4 mg/L), chloramine (4 mg/L), and chlorine dioxide (0.8 mg/L).

Loss of disinfectant residual may be an indication of many problems. Some possible causes of residual loss may include

- Sediment or biofilm accumulation,
- Cross connections,
- Nitrification (usually only in chloraminated systems) intrusion due to leaks or breaks,
- Poor-quality main installations,
- Uncontrolled blending of free chlorine and chloraminated waters,
- High water age (storage facilities and dead ends are often sources), or
- Pipeline corrosion.

Some distribution systems use booster chlorination to raise residuals in the system. Boosting residuals in the system can be an effective way to more evenly distribute. Continuous online analyzers can provide operators with necessary information to properly control these facilities.

Chloramine residuals undergo auto decomposition over time, releasing ammonia, which can promote nitrification. Once established, nitrifying bacteria can be difficult to eradicate and can cause many water quality problems, as summarized in the "Nitrification Control" section of this chapter. Nitrification is prevalent in high water age areas such as storage facilities and low-flow mains during warmer seasons. Therefore, utilities using chloramine should set up a program of monitoring and responding to nitrification indicator parameters (e.g., nitrite, free ammonia, heterotrophic plate counts [HPCs], and rapid chloramines loss). Areas prone to stagnation should be identified so that operations in these areas can be modified.

Chlorine residual measurements taken from storage facilities should be collected to ensure that water from storage is leaving the tank during sampling. This practice will provide residual values that represent water stored in the tank and will alert operators if these values are declining. Operators should use procedures for filling and withdrawing water from storage facilities that ensure the water age is minimized and all parts of the tanks are being used.

Chlorine is highly toxic to aquatic life. This requires special consideration where water from the system must be discharged to a river or lake (such as system flushing). In these situations, disinfectant residuals may need to be removed by treatment. Operators may have limitations due to the treatment procedures because these may affect the rate that operators can drain tanks or perform water main flushing.

Disinfectant residual performance goals. Goals should be established for distribution system disinfectant residuals. The goals must satisfy all regulatory requirements, but more stringent requirements can be used by systems to provide an early warning to avoid violating regulations. Water system operators often do not consider maintaining disinfectant residuals to be "their job." Other utility departments may collect and analyze water samples for this measurement. Still others may respond if values are not at an acceptable level. The system operators are often not informed of measurement results and remediation procedures. This is unfortunate, since operators can many times use operational procedures to mitigate many of these problems.

System operators should, therefore, be aware of disinfectant residual measurements, and they should be consulted about ways to reduce residual decay. This can be particularly effective where system operations can increase water use from problem storage facilities and pipeline dead ends. Employing preventive operating procedures can help avoid customer complaints and eliminate the need for emergency flushing to restore residual and meet regulatory requirements as well as performance goals.

Nitrification Control

Many utilities have converted from free chlorine to chloramine as a secondary disinfectant because of increasingly strict DBP regulations. Systems using chloramines can encounter excessive disinfectant residual loss due to nitrification in high water age areas. Thus, nitrification is associated with the "maintaining disinfectant residual" factor.

Nitrification is a microbial process where ammonia is oxidized to nitrite and nitrate. Ammonia may be present in drinking water naturally or by ammonia addition to form chloramines. Nitrification can lead to disinfectant residual decay and result in growth of bacteria, reduced pH, alkalinity, and dissolved oxygen.

One factor to control nitrification is to ensure that the chloramine formation process is optimized. When forming chloramines, the chlorine to ammonia-N ratio should be 4.5:1–5.0:1 in most cases. Free ammonia should be routinely monitored at key distribution system locations. Both total and free chlorine residual should also be tested at all routine sample locations. To retard nitrification, most systems strive to maintain a minimum total chloramine residual of at least 0.5 mg/L throughout the system.

Operational practices to prevent nitrification focus on reducing water age. Storage facilities are of special concern because design features (e.g., standpipes) can contribute to increasing water age. Some storage facilities can develop unmixed areas where water age can differ significantly from the bulk of the water in the facility.

The size and location of distribution mains (creating dead ends) can increase the probability of nitrification due to water age. As water temperature increases, operators should be particularly diligent to monitor possible dead-end areas because this has a major effect on the nitrification rate.

If nitrification must be eradicated, seasonal switches from chloramine to free chlorine may be helpful. Targeted breakpoint chlorination of storage facilities or mains has also been successful to combat nitrification in these areas. Residual changeovers require comprehensive efforts that include informing consumers (like those with special-use needs such as kidney dialysis and aquatic life keepers). Regulatory agencies typically require approval for disinfectant residual changes since such changeovers can result in increased DBP levels. Discharges to the environment may need special treatment (dechlorination) to protect aquatic life from chloramines. Lead concentration levels may also be affected by disinfectant changeovers.

Utilities that use chloramines as secondary disinfectants should establish action levels and have response plans to control nitrification. A carefully designed water quality monitoring program is needed to detect nitrification and verify the effectiveness of any actions. Monitoring programs should routinely sample storage facility outlets, pressure zone boundaries, large mains serving sparsely populated areas, and dead-end mains. Samples should be tested for chlorine residual (free and total), free and total ammonia, nitrite, heterotrophic bacteria, and pH. Nitrification detection is usually based on a combination of factors such as rapid loss of chloramine residual, free ammonia increases, increased nitrite, increased HPC, and decreased pH.

System operators must be aware of the signs of nitrification and the action levels established by the utility. Water age can be calculated by most hydraulic models (see chapter 4) and this information should be available for review. If other utility departments are responsible for water quality tests, they should communicate with operations personnel regarding the results and trends.

When faced with nitrification, operators should evaluate each situation to determine the most appropriate response that will resolve the problem while maintaining quality water services to customers. Actions taken in response to nitrification need to be quick and appropriate for the situation, ranging from adjusting the chloramine formation process, flushing to remove effected water and replacing with acceptable disinfectant residual water, improving water turnover in affected areas, or performing breakpoint chlorination.

Disinfectant By-product Control

Disinfectants used in water treatment, and in the distribution system, can combine with other substances to form undesirable (and regulated) by-products. Chlorine dioxide usage can result in chlorite levels of concern (the maximum contaminants level, or MCL, is 1.0 mg/L). Chlorine can produce regulated DBPs: total trihalomethane (TTHM) and haloacetic acids (HAA5).

It can be difficult to control DBP levels while providing an adequate disinfectant residual level to achieve water quality goals. Several factors influence DBP formation and composition, such as

- Point of application,
- Disinfectant,
- Disinfectant contact time,
- Disinfectant dosage and residuals,
- Organic matter precursors,
- Water age,
- pH,
- Temperature,
- Bromide concentration, and
- Booster chlorination.

A comprehensive DBP control program balances all of the competing factors to ensure meeting regulatory requirements and system operational goals. DBP levels can change, depending on conditions, in the distribution system. To develop a DBP control program, a system should first determine the levels entering the system. Comparing these values with distribution system levels can reveal formation patterns.

Once the DBP formation patterns are known, a control program can be developed. The effect of various factors on TTHM and HAA5 formation must be understood to construct a customized DBP control plan. Some of the more important relationships are as follows:

- The maximum TTHM and HAA5 levels often do not occur at the same locations. This is partly because HAA5 formation is typically faster than TTHM formation.
- Higher pH favors TTHM formation, and lower pH favors HAA5 formation.

- HAA5 can degrade with time in the distribution system. HPC bacteria can metabolize HAA5, and this often occurs farther from the entry point to the distribution system.
- Chloramines do not form high levels of TTHM or HAA5.
- Increasing water age often results in higher levels of DBPs.
- Lower-temperature water reduces formation of DBPs. Seasonal DBP variations should be examined for site-specific results.

Operators can usually execute the DBP control plan by using one or more of these practices:

- Using chloramine as a distribution system residual disinfectant;
- Reducing free chlorine dosage at entry points and adding chlorine boosters in the system;
- Reducing water age;
- Aerating storage tanks to reduce the levels of certain TTHM species such as chloroform, but not necessarily other DBPs;
- Flushing to remove high-DBP water and reduce water age; or
- Blending source waters or purchased finished water with low DBP levels.

Sediment Control

Sediment can cause a disinfectant demand and contribute to scale that can shield microorganisms. Accumulation of sediment usually occurs in areas of low flow like dead-end mains, large mains with low usage, and storage facilities. Sediment can affect water quality and in some cases lead to violations of drinking water regulations.

Sediment is usually inorganic, caused by several common conditions including post-treatment precipitation of aluminum or iron compounds from treatment coagulants; zinc or phosphate compounds used as corrosion inhibitors; carbonate scales and particulates from pH or alkalinity adjustment chemicals or source water; and manganese or iron from source water or treatment residuals. Control programs usually comprise several steps: assessing the existing conditions, removal of deposits, and adjusting conditions to prevent reoccurrence.

System operators may be called on to conduct high-velocity flushing to remove existing deposits. If flushing is not successful, more aggressive pipeline rehabilitation (such as pigging or mechanical scraping, discussed in chapter 7) may be needed. Operators may also be involved in adjustments to the water pH or other chemical parameters that may be needed to prevent future accumulation of scales or deposits.

Storage Facility Management

Storage facilities are one of the most important components of drinking water distribution systems. Storage is used to provide an adequate water supply for fire fighting and to more easily adjust to the variability in customer demand. However, poor design, use of ineffective operating procedures, and lack of preventive maintenance can cause these assets to become the source of system problems. Storage facility management practices affect water quality. Additionally, these practices can affect system reliability and contribute to the efficient delivery of water to meet customer needs. Chapter 10 contains extensive information of storage facility design, operation, and maintenance.

Operators should understand the factors that can lead to developing water quality problems in storage facilities. Procedures should then be used to operate storage facilities to minimize conditions that can lead to water quality deterioration. Using storage to ensure an uninterrupted water supply during emergencies like main breaks enhances system reliability. Also, operators can anticipate procedures that may require additional pumping and, with sufficient planning, employ storage-use strategies that can save energy and lower operating costs.

Water Age Management

Water quality deterioration within the distribution system is often the consequence of excessive water age. Chemical, microbiological, and physical changes in water quality take time. The time needed to make some changes is affected by flow and temperature. If water resides in the system long enough, disinfectant residual can decline, and this allows biological growths to accelerate, creates corrosive conditions, and provides opportunities for other adverse water quality changes.

Water age is an important factor that has a great impact on water quality. It also affects system reliability and efficiency. For example, corrosion within pipelines can accelerate because of conditions that develop in high water age areas. This can weaken pipes, main breaks can occur, and these often temporarily interrupt water service. This same condition can reduce the flow characteristics of pipelines and cause an increase in pumping and energy use (affecting efficiency).

It is not easy to monitor water age. Hydraulic models, tracer studies, and other technical methods can be used to estimate water age. Most water system operators use hydraulic models to estimate water age (many with integrated water quality modeling capability to predict concentrations of chlorine, DBPs, and other constituents in a distribution system).

System operators can control water age by the following methods:

1. Incorporate water age evaluation as a system design consideration.
2. Monitor key locations in the system for water age and water quality indicator parameters such as
   - Chlorine residual,
   - pH,
   - Temperature,

- HPCs, and
- DBPs,

Key monitoring locations should include:

- Downstream of storage facilities (when drawing out),
- Areas where large-volume piping serves low-demand consumers, and
- System extremities, including dead ends and pressure zone boundaries.

3. Use preventive maintenance practices (such as targeted unidirectional flushing programs) to combat water age in isolated locations.
4. Operate storage facilities (filling and draining) to enhance mixing and decrease water age.
5. Use customer complaints as a tool to detect water age–related issues.

Water Sampling and Response

An effective water sampling plan obtains water for testing that reflects the conditions in the distribution system and also acquires information used to meet performance goals. These results are primarily used to evaluate and respond to water quality issues but also provide data about circumstances that affect reliability and efficiency.

Drinking water quality regulations require monitoring and reporting for disinfectant residual, bacteria, and other water quality parameters within the distribution system according to approved monitoring plans including the Surface Water Treatment Rule, Total Coliform Rule, Disinfectants and Disinfection By-products Rule, Groundwater Rule, and the Lead and Copper Rule. An effective sampling plan should establish a sufficient number of locations for the specific compliance sampling or special evaluation purpose.

In addition to sample locations required by regulations, samples should be collected from known problem areas. Routine samples from storage facilities should be representative of water leaving the storage facility. Chemical treatment points (like chlorine booster stations) in the distribution system should be monitored. Nonroutine sampling locations may include areas upstream and downstream of main breaks and repairs, or other system changes (such as maintenance, cleaning) and customer complaint locations.

System operators may be responsible for collecting water samples used for system monitoring. Therefore, they should receive training on collection techniques and test methods. The training should be based on the SOPs for these topics.

The results from water quality tests should be promptly reviewed. Operators should respond immediately to any irregular results. Samples taken for regulatory compliance must follow response requirements. Test results of all sampling should be communicated to consumers according to the response plan of the system. Consumers should be notified quickly and appropriately if sampling results suggest possible hazards.

MANAGING SYSTEM RELIABILITY

The operational practices included in this section have the greatest effect on this performance category. Where system reliability performance goals are not being met, adjusting these practices may result in the needed improvement. Operators can use the information provided to help select the most effective practices and the adjustments that can lead to improved performance.

Reliability Factors

Every distribution system occasionally experiences interruptions in water service. These are often caused by main breaks and associated repair procedures, power supply interruptions, equipment failure, insufficient water storage, or a serious water quality issue. Operators can anticipate most of these conditions and use operational practices that will reduce the frequency and severity of future interruptions in water service, thus improving system reliability.

System Design

Reliability can be built into the distribution system. Features should be included in the distribution network to reduce service interruptions even during emergencies. Some of the more common design elements include the following:

- *Adequate storage.* The system should have sufficient storage volume for emergency service and to provide an operating tool to mitigate fluctuations in daily and hourly demand. Special consideration, however, should be given to the possibility of deteriorating water quality if water age in storage facilities is too great.
- *Integrated main network.* The distribution network should provide supplies from several directions. Most service areas should have the ability to direct water from more than one source (valves may need to be operated to make this possible).
- *Construction standards.* The system should have standards for installation and repair of distribution system components. Improper installation practices can lead to premature system failures.
- *Looping water mains.* Continuous flow through water mains reduces water age and provides redundancy in the event of a break. Looped mains are less likely to freeze, and carrying capacity is often improved.
- *Right-sizing mains.* Mains are often sized to provide improved fire flows. This can lead to oversizing and create areas where water can become stagnant. Corrosion and biological growths can then develop and contribute to circumstances that can increase the frequency of main breaks. Sizing mains to provide needed fire flow while considering water quality and system reliability is termed *right-sizing.*

Alternate Power Sources

Power outages can cause interruptions in service due to inoperable pumps, disabled booster disinfection facilities, and nonfunctional telemetry and control systems. Natural disasters, power grid problems, excessive power demand, and many other possible situations can cause a power outage. Systems should develop power interruption contingency plans that consider the impact of an extended power outage and the risk the utility wishes to accept.

There are several common approaches to provide water service during an outage. Many systems use more than one of these alternatives to increase reliability.

- *Emergency interconnections.* Many systems have agreements with neighboring utilities to interconnect the systems in an emergency.
- *Multiple power feeds.* If more than one power feed into the system is available, this is often a form of redundancy.
- *Standby generators.* Diesel- or gas-powered generators can provide emergency power to critical system components during a power outage. Battery backup systems are also used for this purpose.

Redundant Equipment

Reliability is increased if redundant equipment is available. Most systems have spare critical pumps. These may be duplicate units in size and capacity or various sizes to provide added flexibility. Systems should evaluate and identify critical facilities and equipment and provide redundancy where needed.

Another often-forgotten factor to improve reliability is the availability of spare parts and system components. Utilities should keep readily available an adequate supply of pipes, valves, hydrants, meters, and other system repair parts. Having timely access to the needed parts may mean that they are nearby rather than on the utility premises.

SYSTEM OPERATION AND MAINTENANCE PRACTICES

A number of operational practices can be used to affect system reliability. The most effective procedures either reduce main break frequency and duration or enhance system security. Pressure and pressure variability can contribute to main breaks and leaks; this important factor is a primary operational consideration. Pressure control is discussed later in this chapter in the “Maintaining System Efficiency” section, but it has an important effect on reliability as well.

Main Breaks Management

Main break frequency has a major impact on the reliability of the distribution system to deliver an uninterrupted supply (system reliability).

Some other consequences of main breaks are water losses, customer complaints, and effects on water quality. Water losses occur from the break but also from water used in the repair procedures. Customer complaints and judgments about system performance are influenced by the frequency, duration, and extent of reported main breaks. System contamination can be caused by main breaks either at the break site or indirectly by depressurizing the system and causing backflow at other locations.

Important factors that affect the frequency and severity of main breaks are hydrant and valve maintenance, pipeline rehabilitation and replacement rate, storage tank maintenance and operation, pressure management, and internal and external corrosion control. Operators can, therefore, use adjustments to these practices to reduce main break frequency.

Reducing the main break frequency requires a long-term management program. Conditions that contribute to the frequency of main breaks often develop over decades. The effect of changes in practices (described previously) on the main break frequency, therefore, may take years before system performance goals can be completely realized. An effective main break management program has several common elements: monitoring and reporting, break history review and analysis, standard response and repair procedures, repair crew and responder training requirements, communications procedures, performance measures and goals, and system operator main break scenario training.

Most utilities monitor reported main breaks and the associated direct costs. A comprehensive management program also documents and analyzes the *indirect* costs and the causes before planning for improvements.

Field data collected at the break site should include location, size, material, visible condition, and failure type. Data from system records is added to these field observations (such as the year the pipe was installed, typical pressure and flow rates, and weather data). Soil and material testing data is added to the record after the break event. Cost data completes the break documentation when labor, material, and contractor costs are processed and after indirect costs, such as damage claims and customer service interruptions, are available.

The objective of data analysis is to find mains with high failure rates. Break clusters often have similar causes. Some common analysis topics are pipe materials, location characteristics, seasonal trends, cost options, and technical issues.

The main break management program uses written operating procedures and provides training to employees for investigation, locating, isolation, and repair of leaks and breaks. Procedures ensure that other utilities like gas or electric are notified and their facilities marked before excavation begins. Other standard procedures are directed at safety and water quality.

Specific procedures are needed to deal with the situation where the main was depressurized for repair. Local regulatory requirements are noted and followed before placing the main back into service. If the main was depressurized and there is a potential for contamination, precautions should be taken to ensure that the pipeline is disinfected or otherwise restored to convey safe drinking water. State drinking water regulations have requirements for issuing and rescinding boil-water advisories or boil-water orders and for bacteriological sampling and testing. A utility must conform to these requirements where they apply to a main break.

The main break management program identifies employees who notify and interact with customers, the community, other utilities, public officials, regulators, and the media. When a leak occurs on a customer-controlled service line, the utility must have a process in place to notify the customer. Contractors or other utilities may also cause main breaks, and these situations require specific response procedures.

The management program establishes targets to ensure timely investigation and repair of breaks. Response times can be affected by many factors, including dispatch efficiency, labor and equipment availability and expertise, accessibility and condition of valves, traffic conditions, and location. A first responder is often dispatched to investigate the event and secure the area. A system main break program establishes processes and procedures to achieve the target response time within cost and resource limits.

Systems should establish performance goals for main break frequency. Most systems base their performance goals for main breaks on reported breaks. Definitions can vary and this can affect the results. The following definition is from AWWA Manual M36, *Water Audits and Loss Control Programs.*

> *Reported breaks*—Those leaks or breaks that come to the attention of the water utility as reported by customers, traffic authorities, or any outside party due to their visible and/or disruptive nature. A nonsurfacing break initially reported as loss of pressure by a customer is an example of a nonsurfacing reported leak. Events that can be inferred by alerts by supervisory control and data acquisition (SCADA) systems can be labeled as reported breaks. Water utilities respond to reported breaks in a *reactive* mode, often under emergency conditions.

Operators may institute practices included in the main break management program to reduce the number of breaks and improve system reliability, including managing pressure to avoid unnecessary extremes, increasing rehabilitation and replacement of selected mains; and conducting improved corrosion control measures.

When responding to a major main break, system operators are often challenged to provide uninterrupted service. By operating the correct valves, operators can isolate the area and reduce the impact. Specific training is needed to respond correctly to these events. Some systems use the network computer hydraulic model to simulate breaks and practice responses. When this tool is not available, operators should discuss situations and responses with supervisors and trainers before they occur so that they are adequately prepared. Practicing the correct operator response to a major main break emergency can make a difference in the outcome.

Emergency Repairs

The level of emergency in making distribution repairs varies by how serious the problem is, where the problem is, and the weather conditions. A large main break must obviously be taken care of as quickly as possible. A small service leak can often be guarded and repair deferred a day or two if necessary in summer, but in winter it may have to be repaired immediately to prevent dangerous icing conditions. Various other levels of emergency also exist. The main point is that the distribution repair crew must be prepared and organized to quickly, efficiently, and safely respond to the situation.

Emergency Repair Preparations

In the event of a major emergency, every minute counts, so the better organized the work crew is, the sooner workers will arrive at the site and start repairs. The starting point is to ensure that trained and experienced workers are available to respond at all times. Most water utilities have a system where at least two key personnel are always on call at night and on weekends. The duty is rotated, and the persons on duty must be accessible by phone. In many cases, the person on call is allowed to take the emergency service truck home so that a response can be made without delay. It is that person's responsibility to be sure the truck has a full fuel tank and all equipment is ready for use at the end of each day.

The best method of achieving immediate reaction is to have at least one service truck equipped with the essential tools and equipment required for emergency response. Among the items that should be included are valve keys, picks and shovels, hand tools and special wrenches, commonly used repair sleeves, barricades and warning signs, ladder, bucket, boots, rope, flashlights, hardhats, and other safety gear. A dewatering pump, generator with floodlights, and can of gasoline for the pump and generator should also be carried on the truck. Other important items are copies of maps and records showing main, valve, hydrant, and service locations. A valve-box locator will also facilitate finding valve boxes, particularly if there is snow on the ground.

Identification procedures must be established to make sure the items to be carried on the truck are not borrowed and inadvertently not returned. Among the methods used is to have a special place on the truck for each item, or to paint all of the tools for the emergency truck a special color.

Having a regular program of valve and hydrant exercising and inspection is also part of preparing for an emergency. Being able to quickly locate valves and having them operate easily can greatly facilitate isolating a main break. Systems without a valve inspection program often have to shut several valves before they can find a combination that will work properly and shut tightly. This delays the repair work and increases the water lost.

It is best to consider various emergency scenarios and to plan equipment and procedures for each one. If, for instance, repair parts are not kept in stock for all sizes and types of water pipe in the system, a supply house or other water utility must have the parts and be accessible at any time. It is also wise to have interconnections in place with adjoining water systems and agreements on when water may be used.

Special thought and planning should be given to providing employees with proper personal safety equipment and gear that might be required for confined space entry, trench wall stabilization, and public safety. Water quality testing and sampling as part of an emergency repair procedure needs special attention and training.

Repair Procedures

The first three tasks in repairing a main break are (1) locate the valves that must be closed to isolate the section, (2) notify the customers that their water service will be off for a period of time, and (3) call the local "dig safe" system to have the other utilities locate their pipes and cables at the location to be excavated. If enough workers are available, one crew can start closing valves while another crew is notifying customers. The customers should be advised to draw some drinking water, provided with an estimate of how long the water will be off, and warned not to wash clothes if the water is discolored after it is turned back on.

Common practice is to close all but one valve as quickly as possible and leave the last one partially open until the last minute before repair is to be made. If possible, a leak should be repaired without completely shutting off pressure to reduce the possibility of dirty water flowing into the main through the break. Maintaining some pressure in the main also reduces the possibility of water draining (backflow) from customer services. After all preparations have been made for installing the repair sleeve, the last valve that has been kept partially open can be closed further so that water continues to flow from the break at a nominal rate.

The excavation for repairing a water main or service leak should normally be parallel to the pipeline and located to one side to allow a worker to stand next to the pipe while making the repair. The excavation should not be made too small in an attempt to save time. If the pipe is buried 5 ft (1.5 m) or more deep, the opening must be large enough for a person to work safely. It usually helps to dig a sump hole a little deeper than the bottom of the rest of the excavation, and locate it in a corner as far away from the main break as possible. The pump suction hose can then be placed in the sump to keep the excavation as free of water as possible.

Repair sleeves of the type shown in Figure 3-4 are most commonly used to repair beam breaks or to plug holes in corporation stops. Sleeves should be kept on hand to fit the outside diameter of each type of pipe installed on the system.

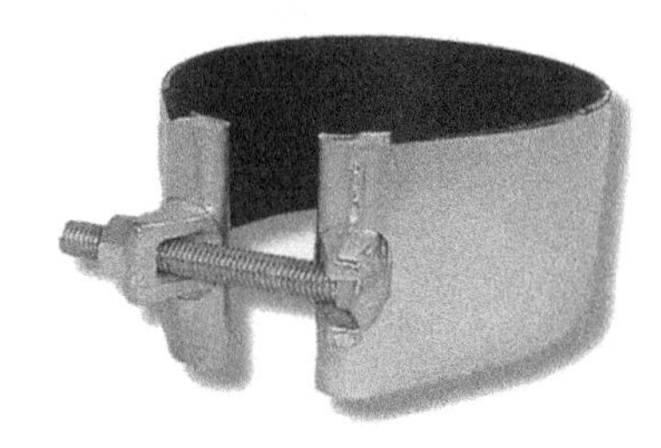

Courtesy of The Skinner Company.

Figure 3-4 Pipe repair devices

Installation of a sleeve includes first scraping and wire brushing the exterior of the main to remove as much corrosion and dirt as possible. The pipe and repair sleeve should then be disinfected by spraying or daubing with 5 percent sodium hypochlorite (ordinary liquid bleach). The sleeve is then installed around the pipe adjacent to the break, slid into position, and rotated to make sure the skived edge is not folded under. Particular care must be taken to prevent foreign matter from being caught between the pipe and gasket.

Although the sleeve nuts can be tightened with a large pair of pliers, this requires an unnecessarily long time, especially because it is prudent to do the job and leave the ditch as quickly as possible. Several types of long sockets and special wrenches are available for use to quickly tighten the sleeve bolts.

After the sleeve is tightened, pressure should be restored by partially opening one valve to make sure there is no leakage before the excavation is backfilled. The line should then be flushed to remove any air and dirt that may have entered the pipe during the repair, and the line should be chlorinated to reduce the danger of contamination. If pressure has been completely shut off at any time during the repair, bacteriological samples should be collected for analysis to make sure the main has not been contaminated.

Record Keeping

A detailed record should be kept of every break or leak that is repaired. If it can be determined, the cause should be recorded, such as beam break, pipe split, blowout, joint leak, or failure due to corrosion. If any old pipe is removed during the repair, a piece of it should be tagged and kept for future reference on the condition of the pipe interior and exterior.

Emergency Repair Safety

Safety for both workers and the public must not be ignored in the haste to stop the flow of water and restore service to customers. Time must be taken to direct traffic and keep onlookers, especially children, away from the excavation. Time must also be taken to properly install sheeting or shoring if necessary to protect workers from excavation wall cave-in.

Hydrant and Valve Maintenance

It is common to neglect the many valves and hydrants in a distribution system since they are not often needed to deliver water to customers. These system components become critical, however, during emergencies or when making system changes. It is important to institute an effective preventive maintenance program aimed at ensuring that these mechanisms will operate as intended when needed and thus enhance system reliability. Chapter 6 contains extensive information on valve types, application, and preventive maintenance. Hydrant types, application, inspection, repair, and maintenance are the subject of chapter 9.

Hydrant and valve maintenance can also influence water quality and system efficiency. Unidirectional flushing, for example, requires the operation of specific valves and hydrants. Malfunctioning valves can prevent performing this procedure and cause degrading water quality. Leaking valves, particularly at pressure boundaries, can affect pumping requirements and, thus, system efficiency. An effective maintenance program includes an inventory, condition assessment plan, maintenance management process (tracking, prioritization, and coordination), planning process (analysis and budgeting), and communication plan.

The valve and hydrant maintenance program should establish performance measures and goals (chapter 2). Inspecting valves and hydrants is expensive, so the performance targets should be coordinated with operating budgets and integrated into plans for capital replacement and repair.

System operators need to know the location and operable status of valves and hydrants. This is crucial when responding to main breaks and to direct needed flows for fire fighting. Routine system operations also benefit from this information, so flows can be routed to provide uninterrupted service and avoid causing water quality disturbances (e.g., flushing mains).

Pipeline Rehabilitation and Replacement

Pipeline rehabilitation (chapter 7) and replacement programs (chapter 2) can have a significant impact on system reliability and to a lesser extent on efficiency. Effective programs renovate problem pipeline segments to reduce main break frequency and improve flow characteristics. Water quality can also be enhanced where these pipelines are interjecting substances from pipe wall deposits. System operators should help identify problem pipelines for rehabilitation or replacement consideration.

Pressure Management

Managing distribution system pressure is the most fundamental activity of operators. Pressure is a key factor for system design and a major element considered when expanding distribution systems. Operators monitor pressure continuously, and unusual pressure measurements are often the first sign of an emergency situation. Maintaining pressure throughout the distribution system is a primary practice that affects system reliability. Although pressure management is included under the heading "Reliability," it also contributes to maintaining water quality (by preventing backflow) and system efficiency (since pumping requires a majority of system energy use).

Maintaining a positive pressure continuously and throughout the distribution system is an important barrier for protecting water quality. Most states have requirements for minimum pressure although there are no federal pressure regulations. The "Ten States Standards" (adopted by many states; Great Lakes–Upper Mississippi River Board of State Public Health and Environmental Managers 2012) stipulates that water systems "shall be designed to maintain a minimum pressure of 20 psi (140 kPa) at ground level at all points in the distribution system under all conditions of flow." Additionally, these standards specify that normal working pressure in the distribution system should be approximately 60–80 psi (410–550 kPa) and not less than 35 psi (240 kPa).

Distribution system hydraulic models (chapter 4) are used to design the system to maintain pressures within targeted ranges. Calibration data for models describe the physical characteristics of the system, water demands, and system control status (initial and boundary conditions). Pressures and flows are primary model outputs. EPANET (free from USEPA) is the basis for the many software programs available for utility use.

Pressure should be monitored at various locations around the distribution system. Important monitoring sites include storage facilities, pump stations, major metering sites, valve chambers, and critical customer services. Monitoring at the location of the lowest and highest pressure within each zone is needed for a complete system assessment. These locations can be determined with a calibrated hydraulic model or by a system survey.

Many factors influence system pressure and its variability. Adequate, stable pressure throughout the distribution network should be an operational goal. Common practices and situations that involve pressure and pressure variability include

- Pipeline maintenance,
- Operation and maintenance of storage facilities,
- Main breaks (which often have a temporary but major impact on system pressure),
- Energy management programs (which may also involve pressure limits),
- Water loss control programs,
- Backflow from cross-connections, and
- Flushing procedures.

Operators should be aware of practices (e.g., internal corrosion or incorrectly sized pipelines) that influence system pressure. With this knowledge, adjustments can often be made to reduce the impact. Customer complaints can result from changes in pressure and pressure variability. These can be reduced when operators take necessary actions.

External Corrosion Control

Pipeline (and other system components) external corrosion can cause loss of pipe material and generally deteriorates the metal. Weakened pipe walls are vulnerable to leakage and failure. This condition can affect system reliability.

External corrosion is caused by either chemical or electrical conditions in the soil surrounding the pipe. The corrosiveness of soil can vary widely from one area of the country to another, sometimes even within the same community. In general, concrete pipe will suffer harmful corrosion only under very corrosive soil conditions. Ductile-iron pipe does not require protection under normal soil conditions, but if the pipe is to be installed in corrosive soil, a polyethylene wrap should be loosely installed around the pipe to prevent corrosion. Steel pipe must have a heavy protective coating for all soil conditions and is usually provided with cathodic protection.

Some conditions that are likely to increase the corrosiveness of soil include the following:

- High moisture content
- Poor aeration
- Fine soil texture
- Low electrical resistivity
- High organic content, such as in a swamp
- High chloride or sulfate content
- High acidity or high alkalinity
- Presence of sulfide
- Presence of anaerobic bacteria

Corrosion cells are typically created in metallic pipe by surface imperfections, such as nicks, or impurities in the metal. The corrosion usually takes the form of pits in an otherwise relatively undisturbed pipe surface. The pits may eventually penetrate the pipe wall.

Another type of external corrosion is caused by direct current that leaves its intended circuit, collects on a pipeline, and discharges into the soil. The problem can be caused by operation of light-rail cars, especially in many older cities (Figure 3-5). This type of stray-current corrosion may occur from some other sources, such as cathodic current being applied to other nearby structures—for example, natural gas pipelines—as well as current from subway trains.

Corrosion caused by stray current often appears as deep pits concentrated in a relatively small area of the pipe. Stray-current corrosion is rather complicated, so if it is suspected as affecting water distribution piping, it is best to seek professional assistance.

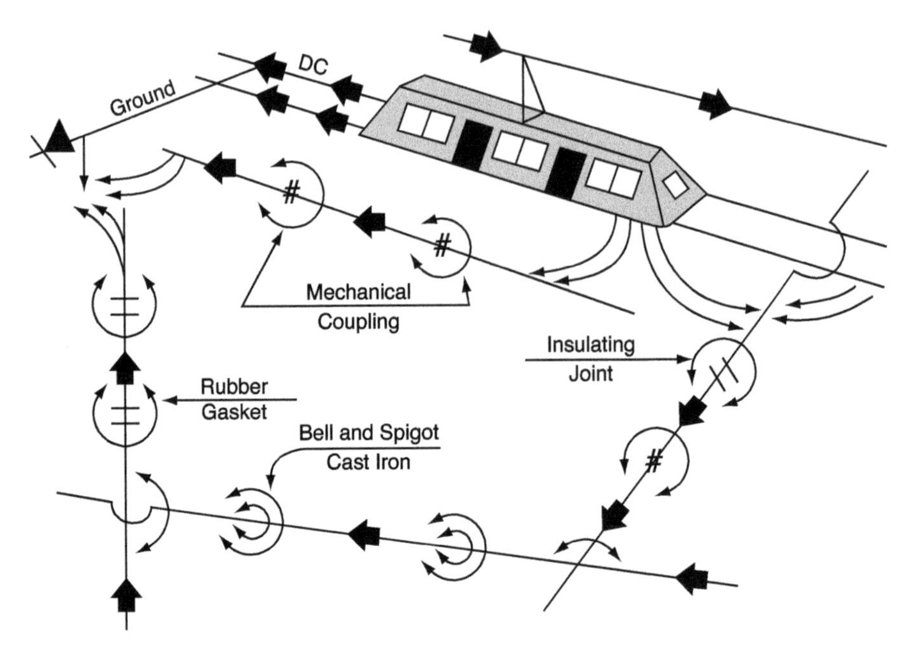

Figure 3-5 Stray-current flow on underground pipelines

Bimetallic Corrosion

Bimetallic corrosion is most often seen in the plumbing systems of buildings. It is caused by the connection of two electrochemically different metals, such as a brass fitting to a galvanized iron pipe. The two metals form a corrosion cell (galvanic cell) that results in loss of the anodic metal (the iron pipe) and protection of the cathodic metal (the brass fitting).

The galvanic series for common waterworks metals is shown in Table 3-1. Each metal may be corroded by any metal below it in the series. The greater the separation between the metals, the more potential and rapid the corrosion process will be. If dissimilar metals must be connected together, an insulating coupling should be installed between them.

Security and Surveillance Practices

Continuous and widespread distribution system surveillance (chapter 17) is an effective method not only to detect but also to prevent system contamination. Effective security surveillance programs use a combination of online sensors and trained observers to identify possible contamination events. The preservation of water quality is often the main goal of these programs.

The benefits of these programs also extend to enhancing system reliability and efficiency. Vandalism is one of the most often encountered security breaches. Alert observers can prevent disruptions in service due to

Table 3-1 Galvanic series for metals used in water systems

Corroded End (Anode)	
Magnesium	Most Active
Magnesium alloys	+
Zinc	
Aluminum	Corrosion Potential
Cadmium	
Mild steel	
Wrought iron	
Cast iron	
Lead–tin solders	
Lead	
Tin	
Brass	
Copper	
Stainless steel	-
Protected end (cathode)	Least Active

vandalism. Security systems at pumping stations, for example, can deter unauthorized entry and prevent damage to these facilities.

Operators are critical members of the system security team. Field observations coupled with online sensors and system data are effective tools to provide improved system security.

MAINTAINING SYSTEM EFFICIENCY

The operational practices included in this section have the greatest effect on this performance category. Where system efficiency performance goals are not being met, adjusting these practices may result in the needed improvement. Operators can use the information provided to help select the most effective practices and the adjustments that can lead to improved performance.

System efficiency can be improved by reducing energy use and controlling water losses. Both of these practices can result in lower operating costs and, thus, contribute to the financial stability of the water system. Operators can contribute to the financial strength of the system by employing these operational practices.

Several operational practices can be used to affect system efficiency. System efficiency primarily focuses on managing energy use and reducing water loss. Water conservation (reducing water demand) may be an important efficiency practice for some utilities. This is not considered an operational practice but a system management policy; therefore, it is not included in this discussion, except for the infrequent practice where pressure limitation is used to reduce water use. Pressure limitation is mentioned in the previous discussion of pressure management.

Energy Management

Water utility energy (mostly electricity) use varies by the source of supply and the physical layout of the system. Regardless of these differences, the typical system uses approximately 90 percent of its energy for distribution system pumping. Therefore, reducing distribution system pumping can be an effective way to lower energy costs. Effective energy management is a key factor to improve distribution system performance efficiency.

Implementing an effective energy management program can reduce operating costs considerably. Developing an energy management program involves several steps:

1. Consider seeking expert advice.
2. Perform an energy audit (energy assessment).
3. Analyze audit results.
4. Develop a customized energy management program:
   a. Track results,
   b. Communicate to customers and management, and
   c. Perform continuous improvement (repeat these steps as needed).

A good understanding of energy usage and cost is essential to identify areas for improvement and to monitor usage trends. At least 12 months of energy and cost data is needed to establish a baseline. Most of the necessary data is found in energy bills, equipment energy information, and lighting fixture data. USEPA provides a free tool to help with this analysis, the Energy Use Assessment Tool (USEPA 2012).

Utility operators should establish a pumping efficiency goal. To meet a goal that targets lower energy use, system operators can

- Replace inefficient pumps and motors;
- Replace pump impellers and bowls;
- Trim pump impellers;
- Refurbish existing pumps with new low-friction coatings;
- Eliminate pump discharge throttling and bypassing;
- Add variable speed drives, where appropriate; and
- Replace inefficient adjustable speed drives.

Other practices that reduce energy usage include the following:

- *Cleaning and lining mains* (chapter 7). This practice improves pipeline flow characteristics and reduces friction losses.
- *Reducing or optimizing system pressures.* Implementing improved pressure management can reduce energy by creating less demanding pumping performance.
- *Pump operation scheduling.* For example, in areas where there is "time-of-day" pricing, shifting pumping to off-peak times and using system storage during peak hours can reduce energy costs; however, doing so doesn't necessarily use less energy.

- *Enhancing water loss control efforts.* A reduction in system leakage results in less water produced for the same customer population, thereby reducing water production energy.

System operators need to be aware of the energy management program and goals. Implementing elements of the strategy may require special training (pump use timing is critical) and equipment. Energy usage results should be readily available and goals established.

Water Loss Control

Water utilities experience a variety of water losses. Leakage from pipelines and storage facilities are recognized by operators as sources of real losses. Other significant losses result from poor accounting, inaccurate meters, and unauthorized uses. These apparent losses may approach or surpass the magnitude of the real losses. It is, therefore, essential to understand the volume and type of water losses so that an effective control strategy can be developed. Water loss control is a major contributing factor affecting the efficient delivery of water through distribution systems. Reducing water losses can significantly reduce operating costs.

Water Audit

A comprehensive water audit accounts for the water supplied and consumed within the delivery system. A *water balance* is achieved when all of the water entering the system is accounted for. After the magnitude and type of water losses are identified, a plan can be developed to reduce these over time. The various water supply and consumption categories contained in a water balance are shown in Figure 3-6.

Loss Control Measures

When the water audit is complete, a loss control plan can be developed. The plan concentrates in two main areas: apparent losses and real losses.

Apparent loss reduction. These losses are caused by meter inaccuracies, systematic data consumption errors (usually from billing systems), and unauthorized consumption. Reducing these losses requires improvement in meter accuracy and may necessitate more frequent meter replacement. Data handling errors require detailed analysis to uncover. Often these are the result of meter reading mistakes or billing system data entry errors. Reducing unauthorized water use can require public education and restricted access procedures.

Reducing real losses. Most real losses come from leaks or breaks in pipelines or service lines, reservoir walls, and tank overflows. Factors contributing to pipeline leaks are temperature stresses, fittings and joints, pipe breaks, defective materials, corrosion, operational errors, and accidental damage. Operators can reduce the volume lost through leaks by better

<table>
<tr><td rowspan="9">Water From Own Sources (corrected for known errors)</td><td rowspan="11">System Input Volume</td><td>Water Exported</td><td rowspan="5">Authorized Consumption</td><td rowspan="3">Billed Authorized Consumption</td><td>Billed Water Exported</td><td rowspan="3">Revenue Water</td></tr>
<tr><td rowspan="10">Water Supplied</td><td>Billed Metered Consumption</td></tr>
<tr><td>Billed Unmetered Consumption</td></tr>
<tr><td rowspan="2">Unbilled Authorized Consumption</td><td>Unbilled Metered Consumption</td><td rowspan="8">Non-revenue Water</td></tr>
<tr><td>Unbilled Unmetered Consumption</td></tr>
<tr><td rowspan="6">Water Losses</td><td rowspan="3">Apparent Losses</td><td>Unauthorized Consumption</td></tr>
<tr><td>Customer Metering Inaccuracies</td></tr>
<tr><td>Systematic Data Handling Errors</td></tr>
<tr><td rowspan="3">Real Losses</td><td>Leakage on Transmission and Distribution Mains</td></tr>
<tr><td rowspan="2">Water Imported</td><td>Leakage and Overflows at Utility's Storage Tanks</td></tr>
<tr><td>Leakage on Service Connections Up to Point of Customer Metering</td></tr>
</table>

NOTE: All data in volume for the period of reference, typically one year.

Figure 3-6 Water balance

controlling pressure, improving leak detection and repair response, and increasing the frequency of pipeline rehabilitation.

An effective leak control program includes a system to locate pipe, detect leaks, and conduct targeted leak surveys. To make significant real-loss reductions requires a plan; time to execute the plan; and ongoing, accurate water accounting to monitor the effects of the plan.

Locating Pipe

Electronic locators are used for locating metallic water mains, service pipes, valve boxes, and access covers. The units will also locate metallic gas pipe, and telephone and television cables, but it is generally best to let other utilities locate their own pipes and cables to ensure accuracy and avoid liability.

Ground-probing radar. Radar is now used for many purposes and would be very useful for all utilities if it could be easily used to detect all underground pipes and cables. Unfortunately, at the present time, the units that are available work only under certain soil conditions, require special expertise to interpret the results, and are expensive. Perhaps radar equipment will eventually be developed for use by all water systems.

Metal detectors. Units similar to military mine detectors have flat detection coils on the ends of their handles (Figure 3-7). When the coil is near a metal object that is relatively close to the ground surface, there is a change in audible tone or meter reading. Relatively inexpensive units can save a lot of time locating metal access covers, valve boxes, and meter-pit covers that have been paved over, have grass growing over them, or are covered by snow.

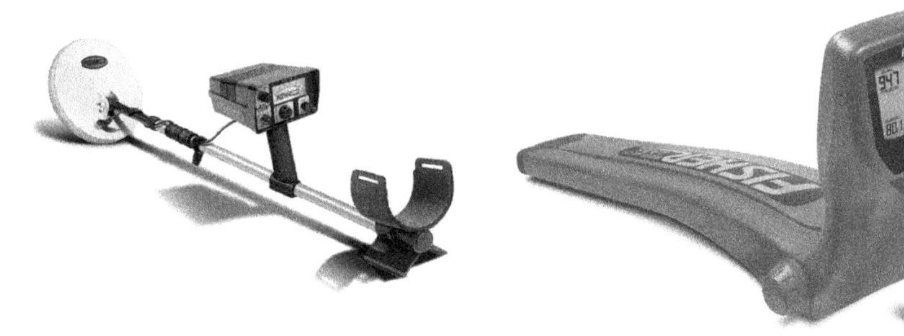

Courtesy of Pollardwater.com.

Figure 3-7 Valve and box locator

Courtesy of Pollardwater.com.

Figure 3-8 Electronic pipe detector

Some valve boxes and meter pits are now made entirely of plastic. To make one of these detectable by an electronic locator (Figure 3-8), a small piece of metal or a small magnet can be fastened to the underside of the cover.

Magnetic locators. A magnetic locator consists of a single unit that monitors the earth's magnetic field. When it is brought near any object containing iron or steel, there is an imbalance in the magnetic field, which the locator translates into a change in sound or meter reading. A unit will generally detect an 8-in. (200-mm) ductile-iron pipe 8 ft (2.5 m) deep. It will not detect noniron metals, such as aluminum cans, copper water service pipe, or cables.

Radio transmission units. Another type of locator uses a radio transmitter and receiver. Commonly called a line tracer, continuous metal locator, or pipe and cable locator, it consists of two units: a radio transmitter that sends out a signal and a receiver with a loop antenna that receives a maximum signal only in the plane of the loop. It will locate any continuous pipe or cable made of any type of metal. The transmitter introduces a signal into the metal, either by a direct connection or by placing the transmitter above the line (Figure 3-8).

Pipe and cable locators will locate copper and galvanized service pipes over a considerable distance. But when used for new ductile-iron pipe with rubber joints, the signal may travel for only a few pipe lengths because the signal does not conduct well from one pipe to the next.

Nonmetallic pipe locators. The best way of locating nonmetallic pipe is to have metallic tape or tracer wire buried in the ditch at the time the pipe is installed. If metal tape is used, it is usually buried about 1 ft (0.3 m) below the surface so it will be easily detectable and will act as a visual warning to anyone excavating in the vicinity of the pipe. The tape or wire can be easily located with a pipe and cable locator by either direct or inductive signal.

Unfortunately, few installers have had the foresight to install tracers, so there are many water systems with nonmetallic pipe and no location

Table 3-2 Water loss versus pipe leak size

	Water Loss*			
	Per Day		Per Month*	
Pipe Leak Size	gal	(L)	gal	(L)
●	360	(1,360)	11,160	(42,241)
●	3,096	(11,720)	95,976	(363,270)
●	8,424	(31,890)	261,144	(988,430)
●	14,952	(56,593)	463,512	(1,755,392)

*Based on approximately 60-psi (410-kPa) pressure.

records. One way of locating nonmetallic mains and services is to employ a unit that uses a transmitter to send small shock waves through the water. The pipe is then located using a receiver that detects the vibration in the soil above the pipe. In most soils, pipe can be located at least 250 ft (76 m) from the transmitter and may work over as long as a mile (1.5 km) under ideal conditions. This type of unit will usually not work well in dry, loose soil or very wet ground.

Locating Leaks

Besides wasting water, leaks from mains and service lines can cause a number of problems, including property damage and personal injury. Numerous cases have involved large sections of pavement that caved in after being undermined by hidden water leaks. Large leaks occasionally cause flooding that flows into basements of adjacent buildings. Also, icing caused by water leaks in freezing weather can cause very serious traffic problems. Relatively small leaks can waste a surprisingly large quantity of water, as indicated by Table 3-2.

In areas with clay or other dense soil, leaks generally come to the surface or find their way into a sewer. A leak detection program, therefore, includes asking police and other municipal employees to watch for unusual water on the street or parkway, and requesting sewer maintenance workers to watch for and report sewers with unusually high flow. If unusually high sewer flow is suspected to be due to a water leak, the flow can sometimes be traced upstream until a section is found with heavy flow in one manhole and low flow in the next, indicating that the leak must be entering somewhere between them. Leak locations can also occasionally be identified by a sunken piece of pavement or particularly lush growth of grass in a particular area.

If the ground is sand or gravel, locating small leaks can be difficult because the water is directly dissipated. One method that may work under some conditions is to use an infrared thermometer gun. The gun can instantly read the temperature of the ground surface at a distance of about

3 ft (1 m), so by reading the temperature of the ground over a water line, a subtle change in temperature caused by the leaking water may be noted.

Listening devices. Listening devices can be used to locate leaks by detecting the sound caused by escaping water. In general, the smaller the leak, the higher pitch the sound will be. One mechanical device that has been used by water system operators for many years is the aquaphone, which is like the receiver of an old-fashioned telephone with a metal spike protruding from where the telephone wire should go. The spike can be placed against a metal pipe, meter, or fire hydrant and sounds are amplified surprisingly well. Aquaphones continue to be used because they are inexpensive, trouble free, and fit in a worker's pocket.

The problem with simple amplifiers is that they amplify all sounds equally, including traffic, wind, and any other noises in the area. More accurate sound detection of leaks can be accomplished using an amplifier that can reduce the band of unwanted sound frequencies and enhance the frequency band of escaping water. More complete leak detection kits include equipment for listening for sounds on the ground surface as well as probes and direct contact devices to connect to hydrants, valves, and service boxes.

Firms specializing in contractual leak detection use even more sophisticated equipment. A leak technician checks for leaks on a section of main by feeding sound picked up by transducers into a computer (called a *correlator*) for analysis. The computer takes into account pipe size, pipe material, and other factors that affect the speed at which sound travels through the pipe. As illustrated in Figure 3-9, sound information obtained from two listening points can pinpoint a leak location quite accurately.

Leak Surveys

The simplest type of leak detection survey is to just listen for sounds of leaks on selected hydrants, and then follow up with a more complete survey in suspected areas. A very complete survey collects sound information from all hydrants, valves, and water services.

Leak surveys are best conducted at night when there is less traffic noise and lower flow in the mains. If it appears that a leak may be on a water service, the service should be shut off at the curb stop to determine whether there is a change in the sound. If there are wide gaps in the system with no hydrants or valves, metal rods may be driven down to contact the main and a listening device connected to the top of the rod.

If the suspected location of a leak is under pavement, it is often prudent to check further before completely opening the roadway. A small hole is drilled down to the main, and if there really is a leak at the location, the water will usually come to the surface through the hole.

Copper pipe transmits sound best, followed in order by steel, cast or ductile iron, plastic, asbestos–cement, and concrete. Unfortunately, mains with rubber gaskets will often not transmit sound much beyond the pipe length having the leak.

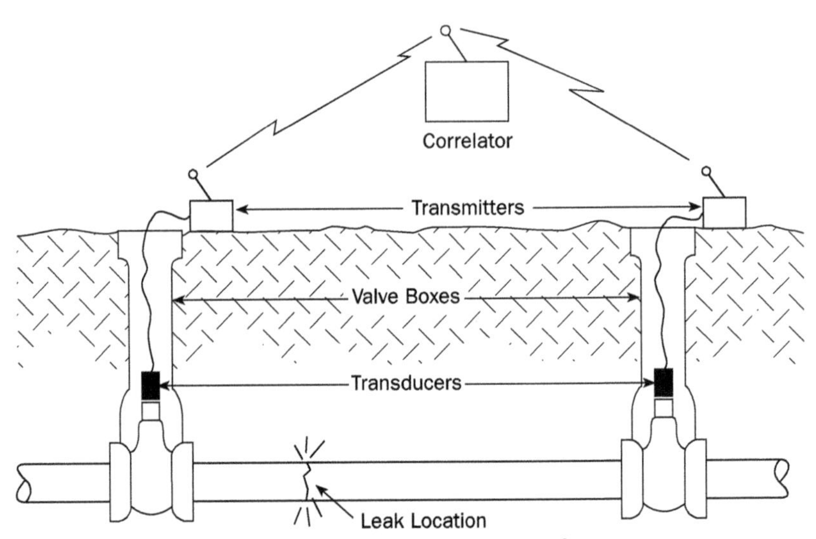

Figure 3-9 Pinpointing a leak from data provided by two transducers

System operators can participate in water loss control and, thus, help reduce water waste and operating costs. Alert operators can detect possible leaks and initiate investigations that may lead to more rapid repairs. When major leaks are found, operators can reduce the volume lost by operating appropriate valves and isolating the area during repair procedures.

BIBLIOGRAPHY

ANSI/AWWA C510. *AWWA Standard for Double-Check Valve Backflow-Prevention Assembly.* Denver, Colo.: American Water Works Association.

ANSI/AWWA C652. *AWWA Standard for Disinfection of Water-Storage Facilities.* Denver, Colo.: American Water Works Association.

ANSI/AWWA D102. *AWWA Standard for Coating Steel Water-Storage Tanks.* Denver, Colo.: American Water Works Association.

ANSI/AWWA D103. *AWWA Standard for Factory-Coated Bolted Carbon Steel Tanks for Water Storage.* Denver, Colo.: American Water Works Association.

ANSI/AWWA D104. *AWWA Standard for Automatically Controlled, Impressed-Current Cathodic Protection for the Interior Submerged Surfaces of Steel Water Storage Tanks.* Denver, Colo.: American Water Works Association.

ANSI/AWWA G200. *AWWA Standard for Distribution Systems Operation and Management.* Denver, Colo.: American Water Works Association.

APHA, AWWA, and WEF (American Public Health Association, American Water Works Association, and Water Environment Federation). 2012. *Standard Methods for the Examination of Water and Wastewater*, 22nd ed. Edited by E.W. Rice, R.B. Baird, A.D. Eaton, and L.S. Clesceri. Washington, D.C.: APHA.

AWWA. 1985. *Maintaining Distribution System Water Quality.* Denver, Colo.: American Water Works Association.

AWWA. 1986. *Corrosion Control for Operators.* Denver, Colo.: American Water Works Association.

AWWA. 1987. *Distribution System Maintenance Techniques.* Denver, Colo.: American Water Works Association.

AWWA. 1987. *Leaks in Water Distribution Systems.* Denver, Colo.: American Water Works Association.

AWWA. 1991. *Fire Hydrant and Water Main Maintenance. Training package.* Denver, Colo.: American Water Works Association.

AWWA. 2010. *AWWA Video Field Guide: Hydrant Flow Tests DVD.* Denver, Colo.: American Water Works Association.

AWWA. 2010. Principles and Practices of Water Supply Operations—*Water Transmission and Distribution*, 4th ed. Denver, Colo.: American Water Works Association.

AWWA. M14—*Recommended Practice for Backflow Prevention and Cross-Connection Control*, 3rd ed. Denver, Colo.: American Water Works Association.

AWWA. M17—*Installation, Field Testing, and Maintenance of Fire Hydrants.* Denver, Colo.: American Water Works Association.

AWWA. M19—*Emergency Planning for Water Utilities.* Denver, Colo.: American Water Works Association.

AWWA. M20—*Water Chlorination and Chloramination Practices and Principles.* Denver, Colo.: American Water Works Association.

AWWA. M27—*External Corrosion—Introduction to Chemistry and Control.* Denver, Colo.: American Water Works Association.

AWWA. M28— *Rehabilitation of Water Mains.* 2001. Denver, Colo.: American Water Works Association.

AWWA. M36—*Water Audits and Loss Control Programs.* Denver, Colo.: American Water Works Association.

AWWA. M42—*Steel Water-Storage Tanks.* Denver, Colo.: American Water Works Association.

AWWA. M44—*Distribution Valves: Selection, Installation, Field Testing, and Maintenance.* Denver, Colo.: American Water Works Association.

AWWA. M56—*Fundamentals and Control of Nitrification in Chloraminated Drinking Water Distribution Systems.* Denver, Colo.: American Water Works Association.

AWWA. M58—*Assessment and Control of Internal Corrosion in Drinking Water Distribution Systems.* Denver, Colo.: American Water Works Association.

Black and Veatch. 2010. *White's Handbook of Chlorination and Alternate Disinfectants*, 5th ed. New York: John Wiley & Sons.

Brandt, M., J. Clement, and J. Powell. 2004. *Managing Distribution Retention Time to Improve Water Quality, Phase I and Phase II.* Report #91006F. Denver, Colo.: Awwa Research Foundation.

Burton, F. 1996. *Water and Wastewater Industries Characteristics and Energy Management Opportunities.* Report CR-106941. Palo Alto, Calif.: Electric Power Research Institute.

Carlson, S.W., and A. Walburger. 2007. *Energy Index Development for Benchmarking Water and Wastewater Utilities.* Report #91201. Denver, Colo.: Awwa Research Foundation.

Chang, Y., D.J. Reardon, P. Kwan, G. Boyd, J. Brant, K.L. Rakness, and D. Furukawa. 2008. *Evaluation of Dynamic Energy Consumption of Advanced Water and Wastewater Treatment Systems.* Denver, Colo.: Awwa Research Foundation.

Clark, R.M. 1998. *Modeling Water Quality in Drinking Water Distribution Systems.* Denver, Colo.: American Water Works Association.

Friedman, M., G. Kirmeyer, G. Pierson, S. Harrison, K. Martel, A. Sandvig, and A. Hanson. 2005. *Development of Distribution System Water Quality Optimization Plans.* Report #91069. Denver, Colo.: Awwa Research Foundation.

Friedman, M., G. Kirmeyer, J. Lemieux, M. LeChevallier, S. Seidl, and J. Routt. 2010. *Criteria for Optimized Distribution Systems.* Denver, Colo.: Water Research Foundation and Partnership for Safe Water.

Friedman, M., K. Martel, A. Hill, D. Holt, S. Smith, T. Ta, C. Sherwin, D. Hiltebrand, P. Pommerenk, Z. Hinedi, and A. Camper. 2003. Establishing Site-Specific Flushing Velocities. Project #2606. Denver, Colo.: Awwa Research Foundation.

Friedman, M.J., A. Hill, G. Korshin, R. Valentine, and S. Reiber. 2010. *Assessment of Inorganics Accumulation in Drinking Water System Scales and Sediments.* Denver, Colo.: Water Research Foundation.

Grayman, W.M., L.A. Rossman, C. Arnold, R.A. Deininger, C. Smith, J.F. Smith, and R. Schnipke. 2000. *Water Quality Modeling of Distribution System Storage Facilities.* Denver, Colo.: Awwa Research Foundation.

Great Lakes–Upper Mississippi River Board of State Public Health and Environmental Managers. 2012. *Recommended Standards for Water Works.* Albany, N.Y.: Health Education Services.

Jordan, J.K. 1990. *Maintenance Management.* Denver, Colo.: American Water Works Association.

Kirmeyer, G.J., M. Friedman, and J. Clement. 2000. *Guidance Manual for Maintaining Distribution System Water Quality.* Denver, Colo.: Awwa Research Foundation and American Water Works Association.

Kirmeyer, G.J., M. Friedman, K. Martel, and A. Sandvig. 2002. *Guidance Manual for Monitoring Distribution System Water Quality.* Denver, Colo.: Awwa Research Foundation and American Water Works Association.

Kirmeyer, G.K., L. Kirby, B.M. Murphy, P.F. Noran, K.D. Martel, T.W. Lund, J.L. Anderson, and R. Medhurst. 1999. *Maintaining Water Quality in Finished Water Storage Facilities.* Denver, Colo.: Awwa Research Foundation.

Rishel, J.B. 2002. *Water Pumps and Pumping Systems.* New York: McGraw-Hill Professional.

Roberts, P.J.W., X. Tian, F. Sotiropoulos, and M. Duer. 2006. *Physical Modeling of Mixing in Water Storage Tanks.* Denver, Colo.: Awwa Research Foundation.

USEPA. 2003. *Revised Guidance Manual for Selecting Lead and Copper Control Strategies.* EPA-816-R-03-001. Cincinnati, Ohio: US Environmental Protection Agency.

USEPA. March 2006. *Interactive Sampling Guide for Drinking Water System Operators.* EPA 816-F-03-016. Cincinnati, Ohio: US Environmental Protection Agency.

USEPA. March 2007. *Simultaneous Compliance Guidance Manual for the Long Term 2 and Stage 2 DBP Rules.* EPA 815-R-07-017. Cincinnati, Ohio: US Environmental Protection Agency. http://www.epa.gov/safewater/disinfection/stage2/pdfs/guide_st2_pws_simultaneous-compliance.pdf

USEPA. 2008. *Water Quality in Small Community Distribution Systems: Reference Guide for Operators.* EPA/600/R-08/039CD. Cincinnati, Ohio: US Environmental Protection Agency. www.epa.gov.

USEPA. 2012. Energy Use Assessment Tool available for download at USEPA Determining Energy Usage Web page: http://water.epa.gov/infrastructure/sustain/energy_use.cfm.

USEPA. EPANET. www.epa.gov/nrmrl/wswrd/dw/epanet.html.

Walski, T.M., D.V. Chase, D.A. Savic, W. Grayman, S. Beckwith, and E. Koelle. 2003. *Advanced Water Distribution Modeling and Management.* Waterbury, Conn.: Haestad Press.

Chapter 4

Distribution System Design

Many considerations are involved with planning and designing a water distribution system. Some of the factors that may affect the design, in addition to those discussed in chapters 2 and 3, are the source or sources of water, population density, economic conditions of the community, geographical location, and history and practices of the water system.

WATER SOURCE EFFECTS ON SYSTEM DESIGN

The type and location of the water source have considerable effect on the design, construction, and operation of a water distribution system. The general types of systems classified by source are

- Surface water,
- Groundwater,
- Purchased water, and
- Rural water.

Surface Water Systems

It is rare for groundwater to be available in large enough quantities to support a large community, so many medium-size and essentially all large water systems use surface water sources (Figure 4-1). One of the prime features of a surface water system is that the water often enters from one side of the distribution system. The situation this creates is that large-diameter transmission mains are usually required to carry water to the far sides of the distribution system.

Some exceptions exist, but in general, surface water is of good quality and plentiful. This in turn attracts industries that require process water for cooling, cleaning, and incorporation into a product. The availability of good-quality water at a reasonable price generally promotes rapid growth of the community, which in turn causes frequent expansion of the water distribution system.

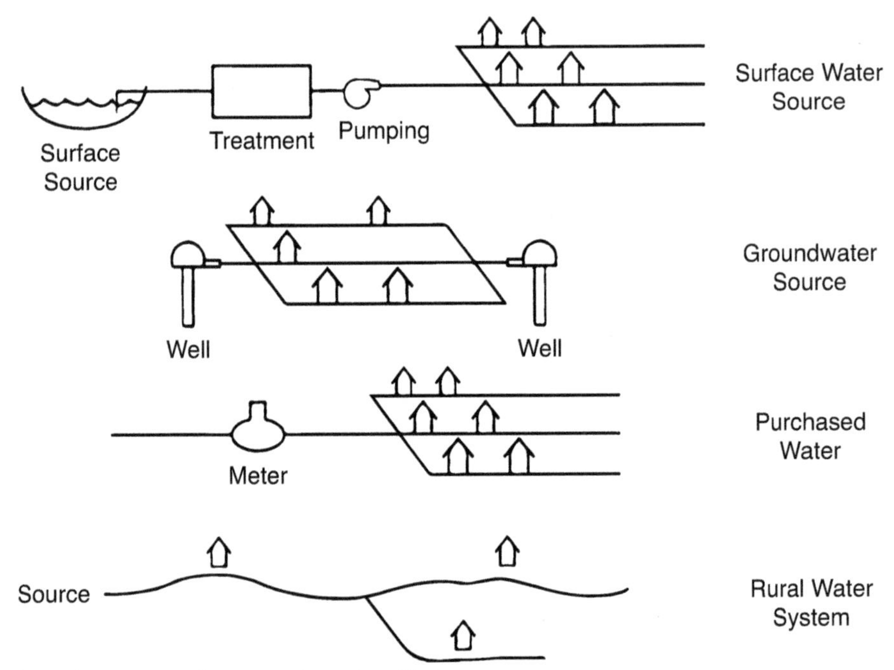

Figure 4-1 Types of water systems classified by source

Groundwater Systems

Although groundwater is generally available in most areas of the United States, the amount available for withdrawal at most locations is limited. If groundwater is generally available and the water requires no special treatment, some water systems are able to install several wells at various locations in the distribution system. These water systems may require few if any transmission mains because water flows from several directions through the piping grid.

If groundwater is available at only one location, all of the wells may be located at one side of the distribution system, so the piping design is similar to that of a surface water system. A similar situation arises if the groundwater must receive treatment for contaminant removal or aesthetic improvement. In this case, water from several wells is piped to a central treatment facility and then pumped to the distribution system.

Purchased Water

Many small water systems that started out using water from their own wells eventually had to change to either a surface water source or purchased water when their community outgrew the capacity of the groundwater source. Other communities have switched to purchased water when it was discovered that their groundwater source was contaminated and treatment for contaminant removal was not economically practical.

Examples of large numbers of purchased water systems are in the Chicago, Ill., and Detroit, Mich., areas where hundreds of surrounding communities draw water from a few large treatment plants using water from the Great Lakes. Some of these systems rechlorinate the water as it enters their systems, but otherwise no treatment is necessary.

Purchased water systems must usually provide a large amount of water storage because they depend on a single connection. If the connection should break, they could be without water for hours, or even days. Purchased water systems must maintain particularly tight water accountability because they are paying for all water metered to them, including unmetered uses and water wasted in leaks.

Rural Water Systems

Another class of water system has developed in recent years that has distribution system design and operating problems somewhat different than other water utilities. Rural water systems have developed in areas where both groundwater and surface water are nonexistent or of extremely poor quality. Many rural systems have been funded by government programs. The systems obtain water from remote sources, treat it if necessary, and then run long mains across the countryside to provide water to individual farms, homes, and small communities.

In most cases, the water mains are plastic pipes installed by plowing them into the ground, and, in most systems, there is no intent to provide fire protection. The water main capacity is sufficient only to provide domestic water in limited quantities. Operators of these types of rural water systems face many unique problems in operating and maintaining their systems.

TYPES OF WATER SYSTEM LAYOUT

The three general ways in which distribution systems are laid out include an arterial-loop system, a grid system, and a tree system.

Arterial-loop systems (Figure 4-2) are designed to have large-diameter mains around the water service area. Flow will be good at any point within the grid, because water can be supplied from four directions.

Grid systems (Figure 4-3) have most of the water mains that serve homes and businesses interconnected, and they are reinforced with larger arterial mains that feedwater to the area. If the grid mains are all at least 6-in. (150-mm) diameter, flow is usually good at most locations because water can be drawn from two or three directions.

Tree systems (Figure 4-4) have transmission mains that supply water into an area, but the distribution mains that branch off are generally not connected, and many are dead ends. This is usually considered poor design because flow to many locations is through only one pipeline. Flow near

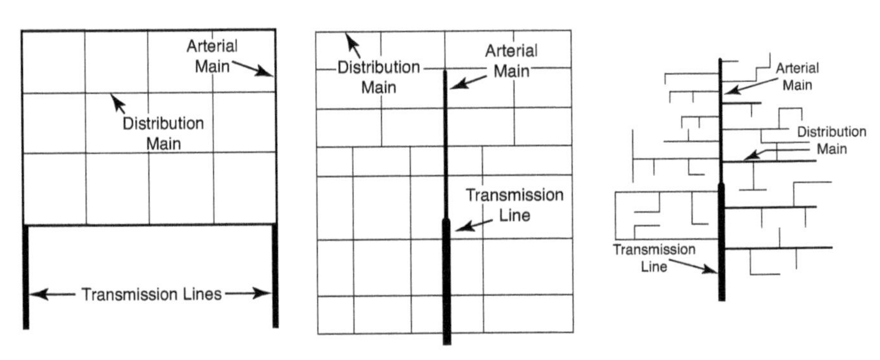

Figure 4-2 Arterial-loop system **Figure 4-3 Grid system** **Figure 4-4 Tree system**

the end of a long "branch" may be relatively poor. This design is poor also because a relatively large number of customers may be without water while repairs are made at a point near the connection of the branch to the transmission main. Customers at the ends of long branches may complain of poor water quality. This is caused by poor circulation of water in the system.

Unfortunately, few distribution systems are completely laid out in an ideal pattern. Most systems have been added onto as new housing or industrial areas have been developed, and the original systems required reinforcement to carry the additional loads. Most distribution systems actually combine grid and tree systems.

Dead-End Mains

Residents usually like to live on dead-end streets because there is very little traffic.

From a water supply standpoint, though, dead-end streets are undesirable because it is difficult to "loop" the piping system. A dead-end main can cause the following types of problems:

- The fire hydrants on the dead-end main draw from only one direction, so they may not provide adequate flow.
- Domestic use on the main provides a very low flow rate, so water quality in the dead-end main often degrades to the point of prompting customer complaints about taste, odor, or rusty water. Many water systems have had to set up regular schedules of flushing dead-end mains to avoid these customer complaints.

If a dead-end main is at least 6 in. (150 mm) in diameter and has sufficient flow and pressure, a fire hydrant should be installed at the end of the pipe. If the main does not have sufficient capacity to provide minimum fire flow, an approved flushing hydrant or blowoff should be installed for flushing purposes. Flushing devices should be sized to provide a flow velocity of at least 2.5 ft/sec (0.76 m/sec) in the water main.

Valves and Hydrants

Shutoff valves should be installed at frequent intervals in the distribution system so that areas may be isolated for repair without having to shut off too many customers. As a rule, at least two valves should be installed at each main intersection, and three is much better (Figure 4-5; map symbols are shown later in Chapter 19, Figure 19-5). Where there are long sections of main with no intersections, intermediate valves should also be installed at what would normally be one-block intervals. The Ten States Standards (Great Lakes–Upper Mississippi River Board of State Public Health and Environmental Managers 2012) recommend a maximum spacing of 500 ft (152 m) in business districts, 800 ft (244 m) in residential areas, and 1 mi (1.6 km) in rural areas where future development is not expected.

Unless there are special circumstances, the best location for valves at a street intersection is opposite the right-of-way line for the intersecting street. This usually keeps them beyond the street paving and makes them easy to find. Mid-block valves should be located opposite an extended lot line, which is least likely to place them in a residential or business driveway.

Fire hydrants are best located at street intersections, and if the blocks are long, additional hydrants should be located near the middle of each block. Hydrant spacing should generally be between 350 and 600 ft (107 and 183 m), depending on the density and valuation of the area being served.

WATER MAIN SIZING

Water mains must carry water for domestic use in residential homes and apartments, and irrigation of plants and lawns; commercial uses, such as those at stores, public buildings, and industries; and fire flow.

Water Use Terms

The following terms are frequently used in determining pipe sizing for distribution system design:

- *Average day demand* is the total system water use for 1 year divided by 365 days in a year.
- *Per-capita water use* is the average day demand divided by the number of residents connected to the water system. This figure varies widely from system to system, depending primarily on the quantity of water used for commercial and irrigation purposes. Nationally, the figure is estimated to be about 105 gpcd (gallons per capita per day) (397 L/d per capita).
- *Maximum day demand* is the water use during the 24 hours of highest demand during the year.
- *Peak hour demand* is the water use during the highest 1-hour period during the year or sometimes during the history of the system.

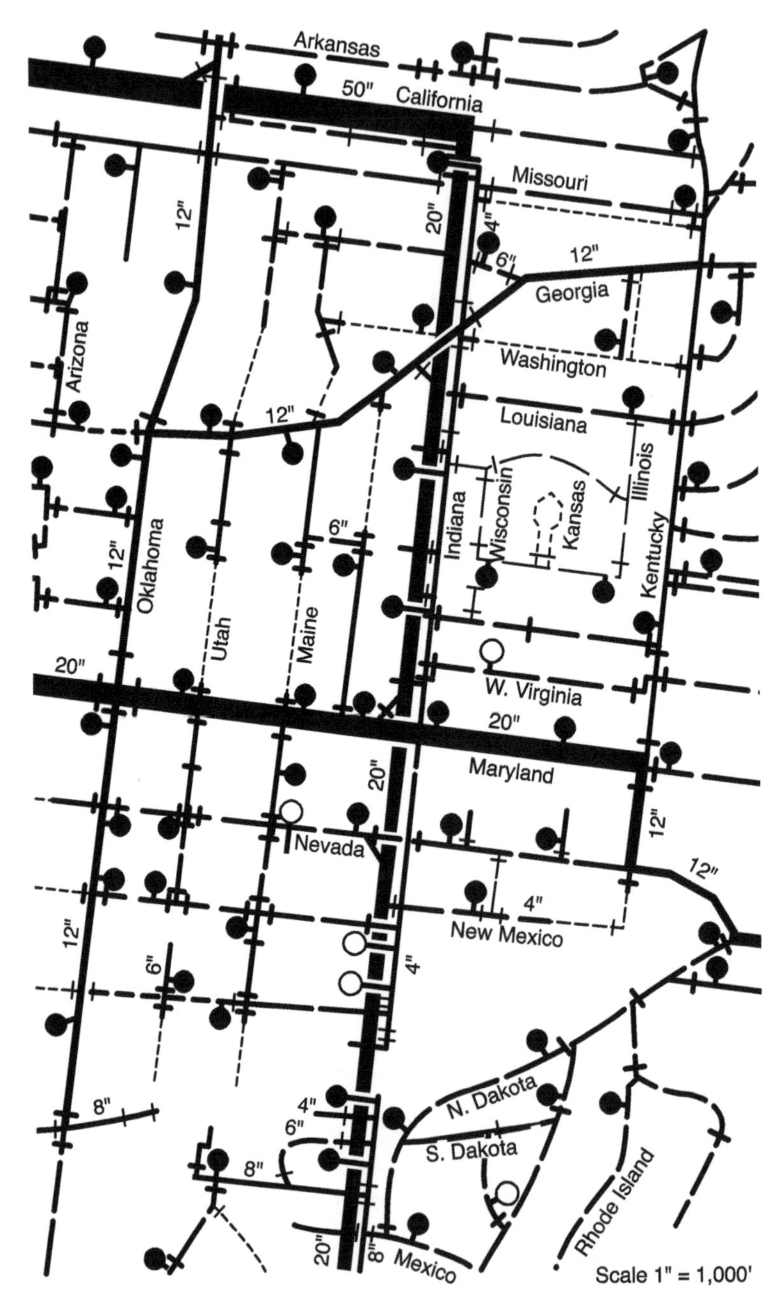

Figure 4-5 Portion of a typical distribution system map

Residential and Commercial Water Use

Residential flow is very small in comparison to the amount of water that can be carried in water mains of the size required for fire flow. In areas where there is no restriction on water use for sprinkling, the maximum hourly water use on a hot summer day may be two or three times the average rate, but this is still small in comparison to the capacity of mains required for fire flow.

Some industries and businesses, such as hospitals, incorporate large quantities of water in their products. Except under extreme emergencies, normal water service must be maintained to key facilities, such as hospitals, at the same time as fire flow is maintained in the area. In this case, the average water use of key facilities must be added to the fire flow requirements when designing mains.

Fire Flow Requirements

The primary factor in determining the size of water mains, storage tanks, and pumping stations for water systems serving a population of fewer than 50,000 is usually the fire protection requirements. The requirements for each community are set by the Insurance Services Office (ISO), which represents the fire insurance underwriters in the United States. The ISO determines the minimum flow that the water system must be able to maintain for a specified period of time in order to receive a specified fire protection rating. The fire insurance rates in the community are then based, in part, on this classification.

Many small, older water systems were originally built with 4-in. (100-mm) diameter and smaller mains. These small lines will often provide adequate capacity for domestic flow, but fire hydrants located on them will not yield adequate flow to meet ISO requirements. The general requirements for a well-designed system meeting fire flow requirements are as follows:

- Mains in residential areas should have a minimum size of 6 or 8 in. (150 or 200 mm).
- Mains serving business, industrial, or other high-value areas should be at least 8 in. (200 mm) and may need to be larger to provide adequate fire flow.
- Mains smaller than 6 in. (150 mm) should be installed only when they are to be used to provide circulation in a grid system. The general policy today is that water mains smaller than 6 in. (150 mm) should not have fire hydrants connected to them.

Additional information on fire flow requirements can be found in AWWA Manual M31, *Distribution System Requirements for Fire Protection.*

Water Pressure Requirements

The normal working pressure in the distribution system should preferably be between 35 and 65 psi (240 and 414 kPa). Higher pressures will significantly increase main and service leaks and will hasten the failure of water heaters and water-using appliances. In addition, customers do not like very high pressure because it will blow dishes out of their hands if faucets are opened quickly. The Uniform Plumbing Code requires that water pressure not exceed 80 psi (552 kPa) at service connections, unless the service is provided with a pressure-reducing valve.

The minimum pressure at ground level, at all points in the distribution system, *under all flow conditions*, must be 20 psi (140 kPa). In other words, the pressure should not drop below this pressure during fire flow conditions.

A water system supplying an area that has varying elevation must usually be divided into pressure zones to maintain reasonable pressure to customers. Water is furnished initially to the highest zone and then admitted to the lower zones through pressure-reducing valves.

Water Velocity Limitations

The velocity of water flow in pipes is also a consideration in determining pipe sizes. The limit for normal operations should be about 5 ft/sec (1.5 m/sec). When the velocity is higher, friction loss becomes excessive. Higher velocities can usually be tolerated under fire flow conditions.

Sizing to Maintain Water Quality

When sizing water mains, there is a temptation to add a "safety" factor, enlarging the mains an additional size. This practice should be avoided because long detention times may result, which degrades water quality. The minimum size that can satisfy the fire and domestic demand should be selected.

Network Analysis

The sizing of water mains depends on a combination of factors, including system pressure, flow velocity, head loss resulting from friction, and the size of all mains that lead to a particular location. The calculation used for many years for analyzing flow in a distribution system involves use of the Hazen–Williams formula.

When a distribution system is to be expanded, it is usually necessary to analyze the entire system and to determine whether and how the system must be modified to properly handle the new loading. This analysis is often performed by computer modeling techniques.

Hydraulic computer models are an important tool used by most systems to help provide information as a basis for decisions regarding planning, design, and operation of the distribution network. The models use extensive and complex calculations to construct computer simulations of system performance under various operating conditions, thus providing insight for managers, engineers, and operators regarding the best options for their use. Distribution system network computer models are used for

- Planning applications including
  - Capital improvement program,
  - Conservation impact studies,
  - Water main rehabilitation programs,
  - Reservoir siting;
- Engineer design applications including
  - Fire flow studies,
  - Valve sizing,
  - Reservoir sizing,
  - Pump station location and sizing,
  - Calculation of pressure and flow at various locations,
  - Pressure zone boundary locating;
- Operating applications including
  - Personnel training,
  - Troubleshooting,
  - Water loss calculations,
  - Emergency operations simulations,
  - Source management,
  - Model calibration,
  - Main flushing programs,
  - Area isolation assistance,
  - Energy cost control;
- Water quality improvements including
  - Constituent tracking,
  - Water source and water age tracking,
  - Disinfectant residual level prediction, and
  - Water quality monitoring location selection.

Distribution system models may use proprietary software or public domain programs like EPANET. Most programs include a feature to export into the EPANET file format if desired. This USEPA program is available for public use. It is a program used mostly for network analysis and is the basis for most proprietary computer programs. It can perform steady-state and extended-period simulation of hydraulic and water quality in pipe networks.

Regardless of the computer network program used, it is critical to properly "calibrate" the model so that it will provide useful information that reproduces the operation of the physical distribution network.

BIBLIOGRAPHY

AWWA. 2010. Principles and Practices of Water Supply Operations—*Water Sources*, 4th ed. Denver, Colo.: American Water Works Association.

AWWA. M31—*Distribution System Requirements for Fire Protection*. 2008. Denver, Colo.: American Water Works Association.

AWWA. M32—*Computer Modeling of Water Distribution Systems*. Denver, Colo.: American Water Works Association.

Boulos, P.F., K.E. Lansey, and B.W. Karney. 2006. *Comprehensive Water Distribution Systems Analysis Handbook for Engineers and Planners*. Denver, Colo.: American Water Works Association.

Great Lakes–Upper Mississippi River Board of State Public Health and Environmental Managers. 2012. *Recommended Standards for Water Works*. Albany, N.Y.: Health Education Services.

Mays, L.W. 2000. *Water Distribution Systems Handbook*. Denver, Colo.: American Water Works Association; New York: McGraw-Hill.

USEPA. EPANET. US Environmental Protection Agency. www.epa.gov/nrmrl/wswrd/dw/epanet.html.

Chapter 5

Water Main Pipe

Water main pipe must have sufficient strength to resist a variety of internal and external forces. Internal pressure includes not only the static pressure, which may be 100 psi (690 kPa) or higher, but also surge pressures. Surge, also called *water hammer*, is a sudden change in pressure caused by rapidly opening or closing a hydrant or valve, or starting or stopping a pump. The shock wave that travels through the pipe can amount to several times normal pressure.

External pressures include the weight of the earth fill over the pipe and the loading of traffic that may drive over it. The pipe must be capable of resisting crushing or excessive deflection due to the external loads.

Standards for all types of water system pipe have been established and are published by the American Water Works Association (AWWA) to ensure adequate and consistent quality.

PIPE SELECTION

Several factors other than pipe strength should be considered when selecting the type of pipe to be installed in a distribution system. Some factors may make a difference in the type of pipe to be used and others may not.

Corrosion Resistance

The potential for both internal and external corrosion of the pipe should be carefully considered. One of the advantages of plastic pipe is that it will not corrode. Concrete pressure pipe is also corrosion resistant for most soil conditions. Ductile-iron and steel pipe both carry internal and external coatings that will resist corrosion under normal water and soil conditions. Where the soil is very corrosive, a plastic wrap material is available to cover ductile-iron pipe to improve exterior corrosion resistance.

Smoothness of Pipe Interior

Pipe with smooth walls ensures the maximum flow possible. The measure of pipe smoothness is called the *C* value or factor. Typical values are listed

Table 5-1 *C* values of various pipe materials

Pipe Material	*C* Value
Asbestos–cement	140+
Cast-iron (old)	100
Cast-iron, badly tuberculated	<100+
Concrete pressure	140+
Ductile iron, cement lined	140+
Plastic	140+
Steel	140+
Wood stove pipe	±120

in Table 5-1. Old, tuberculated, cast-iron pipe is the worst, with a *C* value of 100 or even lower. With internal coatings now being used, the value for all types of new pipe is about the same, and the value should not change much over the years provided aggressive waters are not being transported.

Compatibility With Existing Materials

If extensions or replacements are being made to an existing water distribution system, a change in the type of piping material should be made only after careful consideration. In addition to the other considerations listed here, the following types of questions should be asked to determine whether the system should continue to use the existing type material or should change to a new type of pipe:

- Has the existing pipe been satisfactory from a maintenance standpoint? Will the new type being considered be better?
- Will different types of fittings, repair parts, and tapping equipment be necessary for the new type of pipe? In other words, if a new type of pipe is added to the system, will it be necessary to stock two types of parts and fittings?
- Will the new type of pipe be significantly easier or harder to install, repair, tap, or locate?
- Will any disadvantages be offset by significant savings to the water system?
- If the type of pipe is changed, will the pipe and fittings, such as tees, elbows, and saddles, be readily available?

Economy

Installation cost is a major part of the total cost of a project, whether the work is done by contractors or utility personnel. In other words, if one type of pipe is slightly less expensive to purchase than another, but the excavation and installation costs are the same for either one, the difference in pipe cost will not make a very big difference in the end cost of the project.

On the other hand, if there is a difference in the cost of installing one type of pipe over another, it could make a significant difference in the overall project cost. Some of the factors that should be considered are

- Weight of the pipe—pipe that is lightweight can be handled more easily and move quickly;
- Ease of assembling joints—push-on joints can be assembled much faster than bolted joints; and
- Pipe strength—if one type of pipe requires special bedding to withstand external pressures while another pipe does not, it could make a significant difference in the installation cost.

In many cases, the design engineer will specify the bedding and installation process required for each type of pipe that will be allowed, taking into consideration the local conditions on the project. The contractors will then determine the type of pipe that is most economical to install and will submit their bids accordingly.

Local Conditions

Water mains must often be installed in less-than-ideal soil conditions, and careful selection of the type of pipe and joints can result in both a more satisfactory and less expensive job.

Winding or Uneven Terrain

Installing a pipeline along a winding road or through hills and valleys will often require use of many elbows to accommodate the deflection. In addition, each of these elbows must be restrained, both horizontally and vertically. Under these conditions, using a type of pipe that will allow greater deflection and has restrained joints should be considered. This type may save on installation costs and provide a more trouble-free line in the future. As an example, some utilities install river-crossing pipe in very rough terrain.

High Groundwater Level

Where groundwater is high and it is difficult to completely dewater the excavation, some types of pipe and pipe joints may be easier to satisfactorily install than other types. For example, under wet, muddy conditions, mechanical joints can generally be made up with a greater certainty of holding than push-on joints.

Other Considerations

Some other considerations that may have a bearing on the type of pipe selected for use in the water system or on a specific job are

- State and local regulations,
- Weather conditions at the time of installation,
- Likelihood of earthquakes in the area,

- Whether or not fire protection is to be provided, and
- Whether or not the pipe is to be exposed to the weather or sunlight.

TYPES OF PIPE SERVICE

Water system piping can generally be divided into two classes, for both purposes of service and type of material used: transmission mains and distribution system mains.

Transmission Mains

Transmission mains are designed to carry large quantities of water from the source of supply, such as a well or treatment plant, and provide water to the distribution mains. They usually run in a nearly straight line from point to point and have only a few side connections.

Service lines to homes and businesses are not normally connected to a transmission main, so ease of making service taps is not a major consideration in selecting the type of pipe to be used.

Distribution Mains

Distribution mains are the pipelines that carry water from the transmission mains and distribute it to the customers and fire hydrants throughout the water system. They have many side connections and are frequently tapped for customer connections.

TYPES OF PIPE MATERIALS

Some of the very early water systems in the United States had wooden water mains made by boring or burning a hole down the center of logs. The logs were usually reinforced but did not withstand much pressure, so they were primarily used for gravity flow to supply local fountains and wells where residents could obtain water to carry to their homes.

The principal types of pipe in use in water distribution systems today fall into the following categories:

- Gray cast-iron pipe
- Ductile-iron pipe
- Steel pipe
- Asbestos–cement pipe
- Plastic pipe
- Fiberglass pipe
- Concrete pipe
- Other materials

A summary of the features of currently used pipe materials and the type of joints available is provided in Table 5-2.

Table 5-2 Comparison of pipe materials and joints

Type	Common Sizes	Normal Maximum Working Pressure	Advantages	Disadvantages	Types of Joints
Asbestos–cement (A–C) pipe	4 to 35 in. (100 to 890 mm) diameter	200 psi (1,380 kPa)	• Good flow characteristics • Light weight, easy to handle • Small sizes have low flexural strength—requires special care in bedding	• Easily damaged by impact • Difficult to locate underground • Requires special care in tapping • Asbestos fibers are weakened during field modifications	• Individual coupling with ring gaskets
Ductile-iron pipe (DIP; cement lined)	4 to 64 in. (76 to 1,625 mm) diameter	350 psi (2,413 kPa)	• Durable and strong • High flexural strength (resists breaking if bent) • Good corrosion resistance in all except very corrosive soil • Thinner walls than A–C or PVC, so more carrying capacity for the same outside diameter • Taps easily made directly into the pipe	• Quite heavy—must be handled with mechanical equipment • Must be covered with plastic or cathodically protected for protection in corrosive soil	• Bell and spigot (lead)—no longer being used • Push-on and mechanical—general use for a flexible joint • Flanged—used where pipe is to be rigid in pumping stations • Flexible ball—used for river crossings • Restrained—used to resist thrust and in unstable ground
Concrete pipe	Reinforced: 12 to 168 in. (305 to 4,267 mm) diameter Prestressed: 16 to 144 in. (406 to 3,658 mm) diameter	350 psi (2,413 kPa)	• Durable with low maintenance • Good corrosion resistance, internal and external • High external load capacity • Minimal requirements for bedding and backfilling	• Very heavy weight • May require external protection in high-chloride soil	• Bell and spigot with elastomeric gaskets • Other joint types available for fittings or thrust restraint

(Table continues)

Table 5-2 Comparison of pipe materials and joints (Continued)

Type	Common Sizes	Normal Maximum Working Pressure	Advantages	Disadvantages	Types of Joints
Steel pipe	4 to 120 in. (100 to 3,048 mm) diameter	Just about any pressure by making the walls thick enough	• Light weight makes it easy to install with mechanical equipment • High tensile strength; good choice where some movement may occur	• Very flexible—requires significant bedding support • Poor internal and external corrosion resistance unless properly lined and protected on the outside • Pipe can collapse if a vacuum should occur so air-and-vacuum relief valves are imperative on larger sizes	• Mechanical sleeve—mostly for smaller pipe • Rubber gasket joints—for low-pressure applications • Welded joints—for high-pressure or large-diameter applications • Flanged joints—where valves and fittings are to be used • Expansion joints—allows movement due to pipe expansion
Polyvinyl chloride (PVC) pipe	4 to 36 in. (100 to 914 mm) diameter	200 psi (1,379 kPa)	• Light weight makes it easy to install • Excellent resistance to corrosion • High impact strength	• Difficult to locate underground • Requires special care during tapping • Susceptible to damage during handling • Requires special care in bedding • Damaged by exposure to sunlight (ultraviolet radiation)	• Push-on joints
High-density polyethylene (HDPE) pipe	4 to 63 in. (100 to 1,600 mm) diameter	250 psi (1,750 kPa)	• Light weight makes it easy to install • Excellent resistance to corrosion • Can use ductile-iron fittings	• Difficult to locate underground • Requires special care during tapping • Requires special care in bedding • Requires higher level installation skills	• Thermal butt-fusion

Gray Cast-Iron Pipe

The oldest known installation of cast-iron pipe (CIP) was in 1664 in Versailles, France. Cast iron was by far the predominant type of pipe installed in the United States and Canada for many years, and most of it is still in use.

The earliest pipe was made of gray iron that was cast in sand molds, so it is commonly called *sand-cast pipe*. It was usually made in 12-ft (3.7-m) lengths, with a bell on one end and spigot on the other, designed for a lead joint. The outside diameter of the pipe is larger than centrifugally cast pipe and often is not completely round. Sand-cast pipe can usually be identified by the rough texture of the metal.

In the 1920s, the centrifugal process of manufacturing CIP was developed. With this method, molten iron is poured into a rotating horizontal mold, and centrifugal force maintains the iron around the walls of the mold until it solidifies. The outside walls of this pipe are smooth and uniform in size, and the desired strength can be developed with thinner pipe walls than sand-cast pipe because the iron is more homogeneous. Although gray CIP is strong and provides a long service life, it is brittle and can break. Most main breaks are straight-across cracks known as *beam breaks* and are quite often caused by stones or other hard objects improperly left in the bottom of the excavations when the pipe was laid.

Until the 1920s, distribution system pipe joints were made of poured lead (Figure 5-1). The process of pouring a lead joint was slow and required special skills. It was not unusual for the joints to leak, but because lead is rather flexible, the joints had some flexibility. The process of pouring a lead joint consisted of first tamping oakum into the bell, then pouring molten lead into the joint, and finally caulking the joint with a hammer and flat-end caulking iron. Gray CIP is no longer manufactured in the United States.

Ductile-Iron Pipe

Ductile-iron pipe (DIP) outwardly looks the same as centrifugal gray CIP, and it is produced in the same type of mold. But by adding an inoculant, usually magnesium, to the molten iron, the molecular structure is changed so that the material is much stronger and tougher. DIP can actually be bent to some degree without breaking, so it is not subject to the beam breaks that are common to CIP.

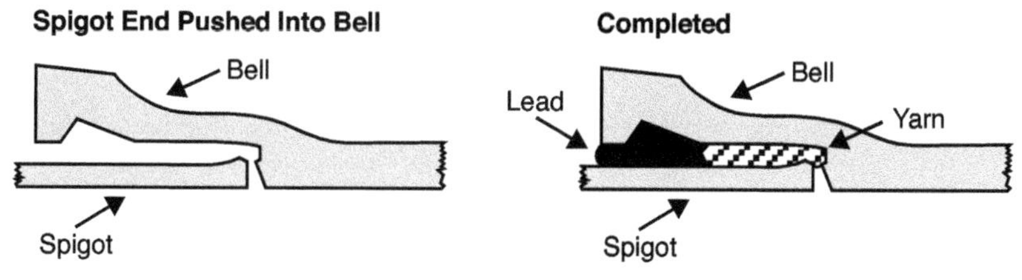

Figure 5-1 Leaded bell-and-spigot joint

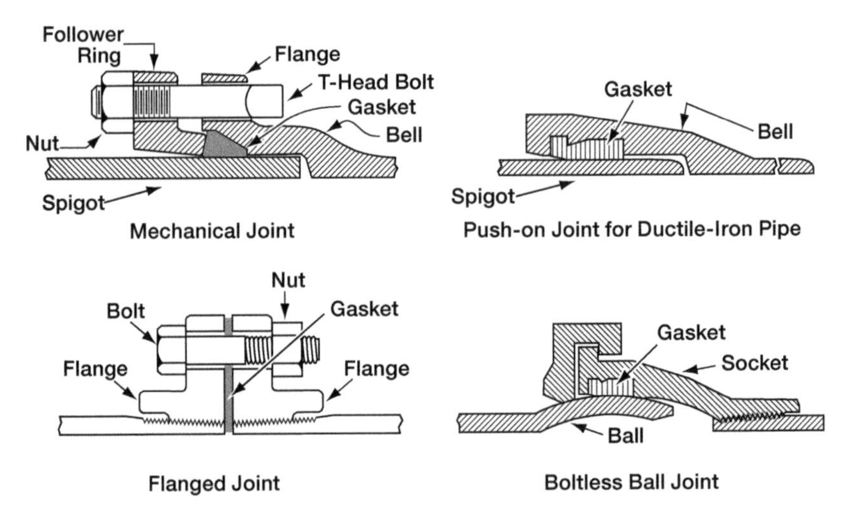

Source: Mechanical joint and push-on joint adapted from ANSI/AWWA C111/A21.11.

Figure 5-2 Common types of ductile-iron pipe joints

Bare DIP is subject to internal corrosion by aggressive water, the same as CIP, so most pipe is now lined with a thin cement–mortar coating that protects the metal and provides a smooth inner surface. The lining is between 1⁄16 and 1⁄8 in. (1.5 and 3 mm) thick, depending on the pipe size, and adheres to the pipe wall even when the pipe is cut or tapped. Cement–mortar is normally very durable, but it may be slowly attacked by very soft water, high-sulfate waters, or waters undersaturated in calcium carbonate. If the pipe is subject to any of these conditions, one should consider purchasing pipe with an epoxy or other lining.

The standard exterior coating used for DIP provides only moderate corrosion protection. In corrosive soils, supplemental corrosion protection should be provided.

Standard lengths for DIP are 18 and 20 ft (5.5 and 6.1 m). Standard pressure classes range from 150 to 350 psi (1,034 to 2,413 kPa).

Additional information on DIP may be obtained from the Ductile Iron Pipe Research Association.

Ductile-Iron Pipe Joints

Several types of joints are used today for connecting DIP and fittings. Details of more common types are illustrated in Figure 5-2 and described in the following paragraphs.

Flanged joints are often used in exposed locations, such as treatment plants and pumping stations, where the pipe installation is to be rigid and the pipe must be restrained from movement. Each of the matching flanges has a machined surface, and they are tightly bolted together with a gasket between them. Flanged joints should not be used underground because they have no means of flexing if there is ground movement.

Mechanical joints have special bells fitted with rubber gaskets. The joint is made tight by a bolted follower ring that squeezes the gasket into the bell. The joints are more expensive than push-on joints but make a very positive seal and allow a fair degree of deflection of the pipe.

Push-on joints are now the most popular type of joint for most main installations because they are the least expensive and can be assembled quickly in the field. The joint consists of a special bell fitted with a greased gasket that is inserted in the field just before the pipe is joined. The spigot end of the joining pipe must be beveled so it will slip into the gasket without catching or tearing. The spigot end slips into the gasket quite easily and makes a very tight seal as internal water pressure compresses the gasket.

The design of push-on joints varies with different pipe manufacturers, and although they are similar, they are not necessarily interchangeable. The joints must be made up by pushing the pipes in a straight line; but after the spigot has been inserted the full distance, the pipe can be deflected from about 2° to 5°, depending on pipe size.

Ball-and-socket joints are made to be restrained and deflect as much as 15°. The pipe is principally used for underwater intakes and river crossings, but it is also occasionally used for underground burial in very rough terrain. Both bolted and boltless ball-and-socket joint pipe are available.

Restrained joints of various types are also available from different manufacturers for use where it is necessary to ensure that the joints do not separate. Details of two types of restrained joints are illustrated in Figure 5-3.

Ductile-Iron Pipe Fittings

Before World War II, a wide variety of special fittings were available, such as reducing elbows, and tees with different sized outlets. But as manufacturing costs increased, manufacturers have cut back to making only a limited number of the most commonly used styles. Figure 5-4 illustrates some common styles that are available. Because of the limited choice of fittings, special connections must now be made up of a combination of fittings.

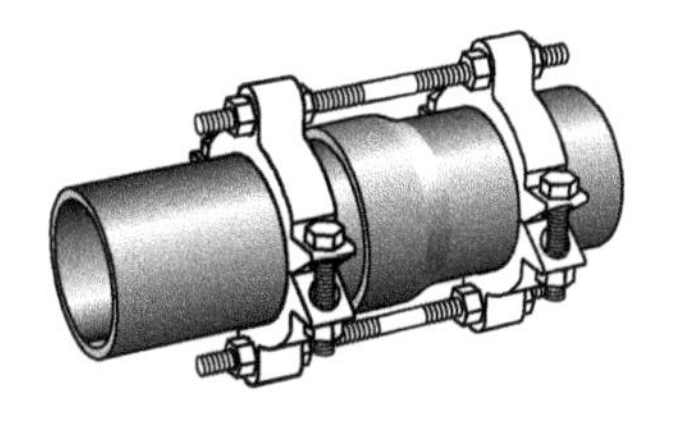

Pipe Bell Joint Restraint

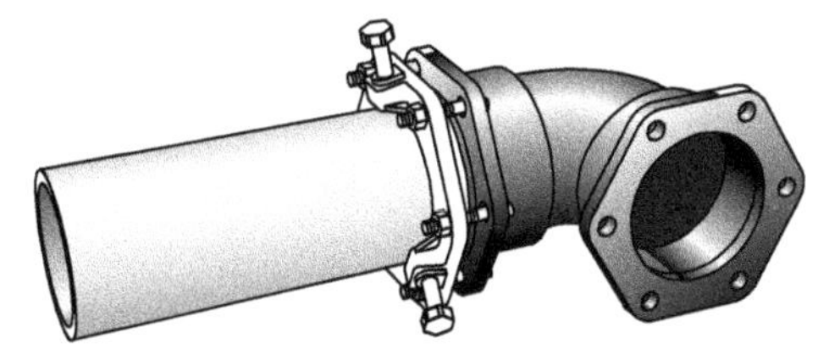

Mechanical Joint Pipe Restraint

Courtesy of The Ford Meter Box Company, Inc.

Figure 5-3 Restrained joints

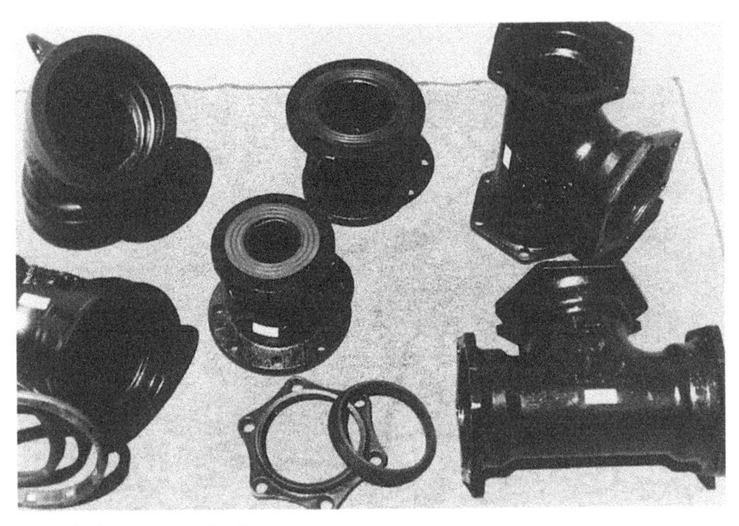

Courtesy of the US Pipe and Foundry Company.

Figure 5-4 Typical iron pipe fittings

In addition to use on DIP, iron fittings are also commonly used on types of polyvinyl chloride (PVC) pipe having the same outside diameter as DIP.

Steel Pipe

Steel pipe has been used in water systems in the United States since 1852. It can be made to withstand almost any pressure by making the walls thick enough, so it is frequently used for high-pressure pipelines.

Steel pipe is fabricated by three methods. In the mill process, a long, flat sheet is progressively rounded by a series of pressure rollers until a tube is formed, then the seam is welded shut. Large-diameter pipe is made by rolling flat plates to make short pieces of tube, and then several pieces are welded together to make lengths of pipe. In the spiral-weld method, a narrow piece of steel is fed from a coil to a machine that automatically welds it in a continuous spiral fashion. Steel pipe has been fabricated in diameters up to 30 ft (9 m).

The price of steel pipe is generally more competitive in sizes greater than 16 in. (400 mm). Unprotected steel is subject to corrosion, so both the interior and exterior must be protected. The interior is usually coated with cement mortar or epoxy as specified in AWWA Standards C205 and C210. AWWA standards provide for several types of plastic coatings, bituminous materials, and polyethylene (PE) tape that may be applied to the pipe exterior for protection against both abrasion and corrosion.

Cast-iron or ductile-iron fittings may be used on smaller sizes of steel pipe. For larger pipe, fabricated steel fittings are used. Various types of O-ring bell-and-spigot joints may be used with steel pipe, or the pipe lengths can be joined by welding, as illustrated in Figure 5-5.

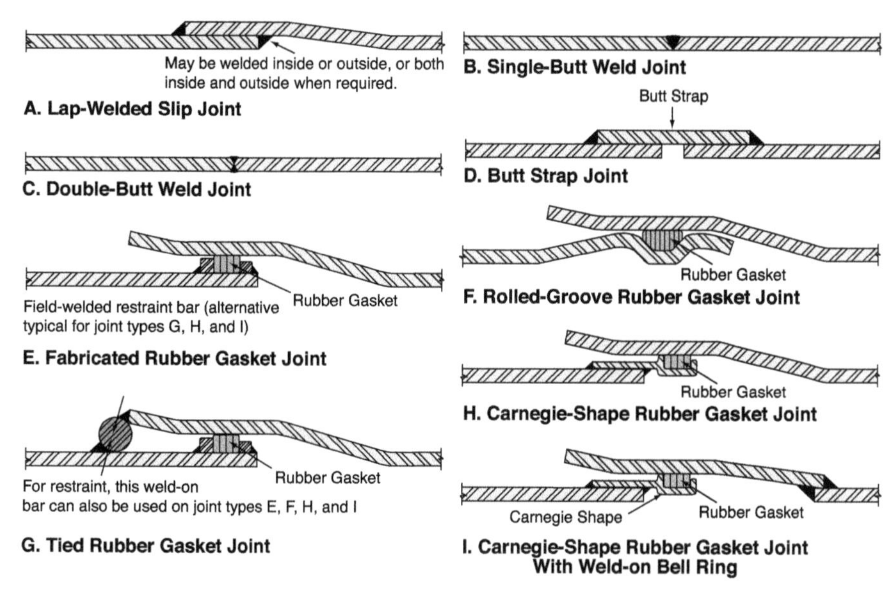

Figure 5-5 Common welded and rubber-gasketed joints used for connecting steel pipe

Additional information on steel pipe may be obtained from the Steel Tube Institute of North America.

Asbestos–Cement Pipe

Asbestos–cement (A–C) pipe was introduced in the United States around 1930. It is made of asbestos fibers, silica sand, and portland cement and is available in working pressures of 100, 150, and 200 psi (690, 1,030, and 1,380 kPa). Standard sizes are 4 to 42 in. (100 to 1,070 mm), and section lengths are 10 to 13 ft (3 to 4 m). Pipe sections are furnished with two plain ends and are coupled with an O-ring sleeve.

Although A–C pipe resists mild corrosion, it should not be used in very aggressive soils or for carrying very soft water. The advantages of the pipe include low initial cost, smooth interior walls, and light weight, which makes smaller sizes very easy to install.

Principal disadvantages are that it breaks easily if not handled and installed properly, it needs special care in tapping, and it requires special safety precautions during installation to prevent inhalation of asbestos dust. In recent years, the US Environmental Protection Agency (USEPA) has discouraged most uses of asbestos because of evidence that exposure to airborne asbestos fibers can greatly increase the chances of a person developing lung cancer. Studies by USEPA, AWWA, and other groups (USEPA 2012) have concluded that the asbestos in water mains does not generally constitute a health threat to customers. But A–C pipe is rarely installed in the United States today, both because of the fear of working with asbestos materials and because PVC pipe has taken its place as an inexpensive, lightweight piping material.

Plastic Pipe

Plastic pipe was first introduced in the United States around 1940 and has seen rapid acceptance since the 1960s. Many types of plastic materials can be manufactured to have various properties. The two important properties for plastic water main pipe are the ability to withstand both internal and external pressures, and the absence of any harmful substances that would cause taste, odor, or potential adverse health effects to customers.

Plastic pipe is commonly made from PVC, PE, and composite plastic in the form of plastic material reinforced with fiberglass. The interior surfaces of all types of plastic pipe are very smooth, so it has a very good flow factor, and the material is almost completely corrosion free.

All plastic pipe used for potable water must be certified for conformance with ANSI/NSF Standard 61 for potable water use. A marking must exist along the entire length of each section of pipe stating conformance with the standard.

Polyvinyl Chloride Pipe

PVC is by far the most widely used type of plastic pipe. In sizes from 4 to 12 in. (100 to 300 mm), the pipe is manufactured with the same outside diameter as DIP, so ductile-iron tees, bends, and other fittings made of that material may be used for connections. PVC fittings are also available. PVC transmission pipe in sizes 14 in. (356 mm) and larger is available in outside-diameter sizes based on either iron pipe size or DIP size.

The light weight of PVC pipe makes it considerably less expensive to ship, handle, and install than DIP. The pipe is also easy to cut and has moderate flexibility to adapt to ground settlement.

The pipe is susceptible to the ultraviolet radiation in sunlight, so it should not be stored outside for prolonged periods. The pipe is also very susceptible to damage during shipping and installation. Gouges deeper than 10 percent of the wall thickness can cause pipe failure later. Although the pipe may be directly tapped in the same manner as DIP, it must be done with extreme care. Many water system operators find it safer to make taps through a saddle.

Additional information on PVC pipe is available from the Uni-Bell PVC Pipe Association. Additional information on other types of plastic pipe is available from the Plastic Pipe Institute.

Polyethylene Pipe

PE pipe is available in nominal sizes through 63 in. (1,600 mm) in diameter, typically with the same outside diameter as iron pipe sizes. Pressure class ratings range from 40 to 198 psi (276 to 1,365 kPa), except that the available maximum pressure rating is reduced for larger pipe sizes. PE pipe is not quite as strong as PVC at ambient temperatures, but it is very tough, ductile, and flexible, even at low temperatures. PE pipe will not fracture under the expansive action of freezing water.

PE pipe can be joined by thermal butt-fusion, flange assemblies, or mechanical methods. It cannot be joined by solvent cements, adhesives, or threaded connections. The most widely used method is butt-fusion, which uses a portable tool to hold the ends of the pipe or fittings in close alignment while they are heated and fused together, and then allowed to cool. The fusion temperature, interface pressure, and cooling time are rather critical, so thermal fusion should only be performed by persons who have received training in the use of this type of equipment, according to the manufacturer's recommendations. Thermal fusion is covered in ASTM Standard D2657, *Standard Practice for Heat Fusion Jointing of Polyolefin Pipe and Fittings.*

Fittings are also available for joining by thermal fusion. Special flanged and mechanical joint adapters are available for connecting PE pipe to other types of pipe and fittings.

PE pipe is covered in AWWA Standard C906, *Polyethylene (PE) Pressure Pipe and Fittings, 4 in. (100 mm) Through 63 in. (1,575 mm), for Water Distribution and Transmission.*

Fiberglass Pipe

Fiberglass pipe is available for potable water use in sizes from 1 in. through 144 in. (25 mm through 3,600 mm) in diameter, and in five pressure classes ranging from 50 psi through 250 psi (345 kPa through 1,724 kPa). Several stiffness classes are also available that are incorporated into the design, depending on the exterior loading that will be applied to the pipe. Advantages of the pipe include corrosion resistance, light weight, low installation cost, ease of repair, and hydraulic smoothness. Disadvantages include susceptibility to mechanical damage, low modulus of elasticity, and lack of a standard jointing system.

One method of manufacturing the pipe is called *filament winding.* A continuous glass-fiber roving saturated with resin is wound around a mandrel in a carefully controlled pattern and tension. The inside diameter (ID) of the pipe is fixed by the mandrel diameter, and the wall thickness is governed by the pressure and stiffness class desired.

The other manufacturing method is centrifugal casting. The resin and fiberglass reinforcement are applied to the inside of a mold that is rotated and heated. The outside diameter of the pipe is determined by the mold and the ID varies depending on the wall thickness. Fittings are made by filament winding, by spraying chopped fiberglass and resin on a mold, or by joining cut pieces of pipe to make mitered fittings.

Several methods are used by the various manufacturers for joining pipe sections and fittings. One method is to butt the sections together and wrap the joint with fiberglass material and resin. The other methods include a variety of tapered bell-and-spigot joints that are bonded with adhesives. Pipe, fittings, and adhesive for joining are usually not interchangeable between pipe from different manufacturers.

Fiberglass pipe is covered in AWWA Standard C950, *Fiberglass Pressure Pipe.*

Permeation of Organic Chemicals Into Plastic Pipe

If gasoline, fuel oil, or other organic compounds have saturated the soil around plastic pipe, some molecules of the compounds can pass through the pipe walls in a process called permeation. This can give the water a disagreeable taste, but more important, it could pose a significant health threat to customers using the water. Continued exposure to the organic compounds can also soften the pipe and eventually lead to pipe failure.

Plastic pipe should not be installed in locations known to have soil contamination or where soil contamination is likely, such as near old petroleum storage tanks.

Concrete Pipe

Concrete pipe combines the high tensile strength of steel and the high compressive strength and corrosion resistance of concrete to produce pipe that is very durable at reasonable cost.

The pipe is generally available in diameters ranging from 10 to 252 in. (250 to 6,400 mm) and in lengths from 12 to 40 ft (3.7 to 12.2 m).

The four types of concrete pipe construction commonly used in the United States and Canada are

1. Prestressed concrete cylinder pipe,
2. Bar-wrapped steel-cylinder concrete pipe (formerly called pretensioned concrete cylinder pipe),
3. Reinforced concrete cylinder pipe, and
4. Reinforced concrete noncylinder pipe.

Prestressed Concrete Cylinder Pipe

Manufacture of the pipe starts with the assembly of a steel cylinder with the bell-and-spigot ends welded to it. The cylinder is first hydrostatically tested to verify watertightness, and then a cement–mortar core is placed on the interior by a centrifugal process or vertical casting.

After the core has cured on lined-cylinder pipe, hard-drawn steel wire is helically wrapped around the cylinder. On embedded-cylinder pipe, a concrete coating is placed around the steel cylinder, and after it has cured, it is wrapped with steel wire. Wrapping the cylinder with wire produces a compression of the concrete and steel core, which greatly strengthens the pipe. The pipe is finally coated with cement mortar for corrosion protection and is ready for use after the coating has cured.

Bar-Wrapped Steel-Cylinder Concrete Pipe

Bar-wrapped pipe is manufactured similarly to prestressed pipe except that, after the mortar lining has cured, the steel cylinder is wrapped with hot-rolled steel bar. The exterior is then protected with a cement–mortar coating (see Figure 5-6).

Reinforced Concrete Cylinder Pipe

Reinforced concrete cylinder pipe differs in that instead of wrapping the reinforcing wire around the steel cylinder, the reinforcing is cast into the wall of the pipe (see Figure 5-7). Use of this design is gradually declining in favor of prestressed or bar-wrapped pipe.

Reinforced Concrete Noncylinder Pipe

Noncylinder pipe has reinforcing but does not have a steel cylinder. Pressure is limited to about 55 psi (380 kPa), so use of this type of pipe in water systems is generally limited to intake lines.

Uses of Concrete Pipe

As shown in Figures 5-6 through 5-8, concrete pipe joints are made up with an O-ring that is compressed as the spigot is pushed into the bell. The metal parts are then protected from corrosion with a covering of mortar.

Some of the advantages of concrete pipe are that it can be designed to withstand high internal pressure and external loads, and it is resistant to both internal and external corrosion under normal water quality conditions. Among the principal disadvantages are that the relatively great weight makes shipping expensive and that exact pipe lengths and fittings required for an installation must be carefully laid out in advance. It is necessary for the concrete pipe supplier to provide a detailed layout of the pipe project to minimize field adjustments.

Additional information on concrete pipe is available from the American Concrete Pipe Association and the American Concrete Pressure Pipe Association.

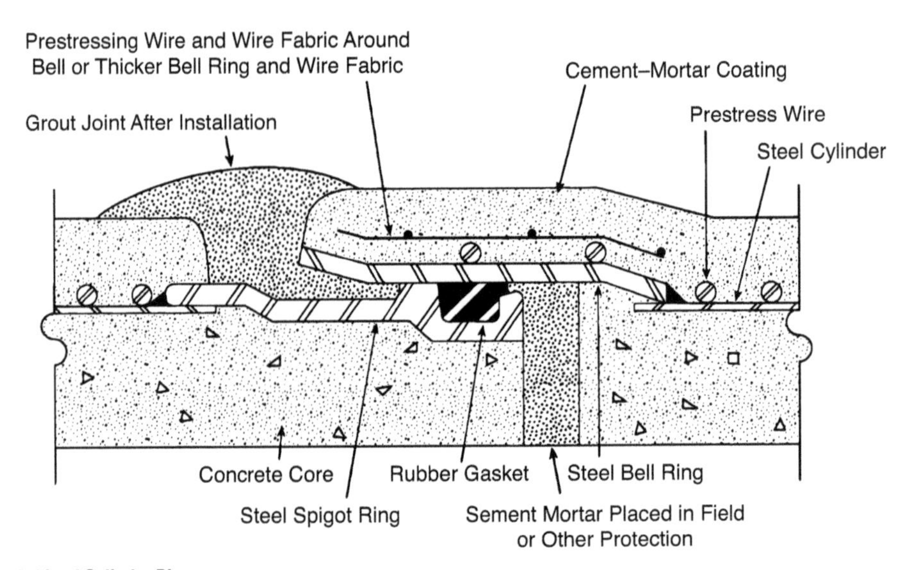

A. Lined Cylinder Pipe

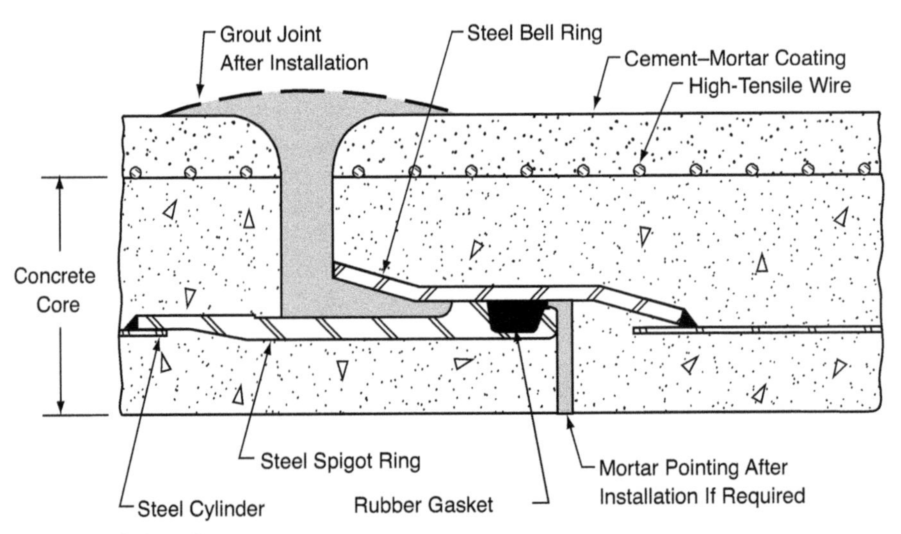

B. Embedded Cylinder Pipe

Courtesy of American Concrete Pressure Pipe Association.

Figure 5-6 Prestressed concrete embedded-cylinder pipe

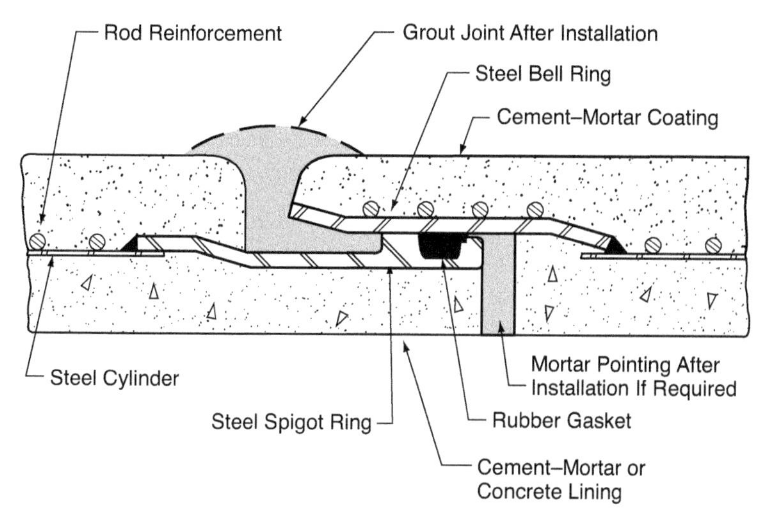

Courtesy of American Concrete Pressure Pipe Association.

Figure 5-7 Pretensioned concrete cylinder pipe

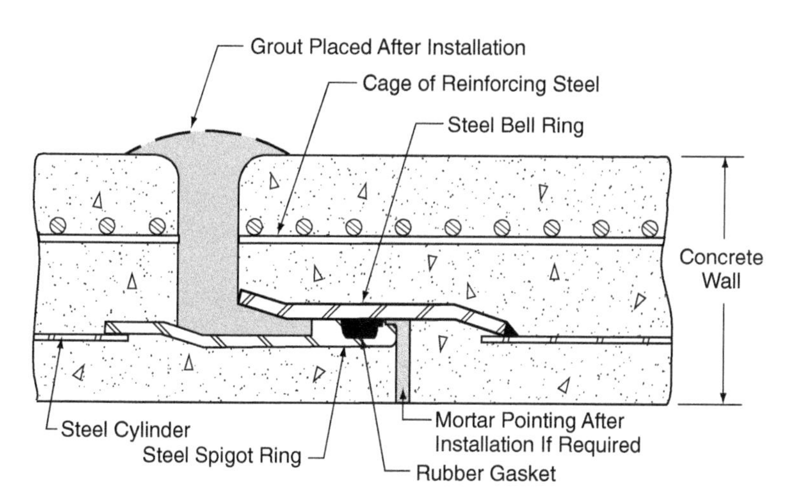

Courtesy of American Concrete Pressure Pipe Association.

Figure 5-8 Reinforced concrete cylinder pipe

BIBLIOGRAPHY

ANSI/AWWA C105. *American National Standard for Polyethylene Encasement for Ductile-Iron Pipe Systems*. Denver, Colo.: American Water Works Association.

ANSI/AWWA C111/A21.11. *American National Standard for Rubber-Gasket Joints for Ductile-Iron Pressure Pipe and Fittings*. Denver, Colo.: American Water Works Association.

ANSI/AWWA C205. *AWWA Standard for Cement–Mortar Protective Lining and Coating for Steel Water Pipe—4 In. (100 mm) and Larger—Shop Applied*. Denver, Colo.: American Water Works Association.

ANSI/AWWA C210. *AWWA Standard for Liquid Epoxy Coating Systems for the Interior and Exterior of Steel Water Pipelines*. Denver, Colo.: American Water Works Association.

ANSI/AWWA C300. *AWWA Standard for Reinforced Concrete Pressure Pipe, Steel-Cylinder Type*. Denver, Colo.: American Water Works Association.

ANSI/AWWA C301. *AWWA Standard for Prestressed Concrete Pressure Pipe, Steel-Cylinder Type*. Denver, Colo.: American Water Works Association.

ANSI/AWWA C302. *AWWA Standard for Reinforced Concrete Pressure Pipe, Noncylinder Type*. Denver, Colo.: American Water Works Association.

ANSI/AWWA C303. AWWA *Standard for Concrete Pressure Pipe, Bar-Wrapped, Steel-Cylinder Type*. Denver, Colo.: American Water Works Association.

ANSI/AWWA C304. AWWA *Standard for Design of Prestressed Concrete Cylinder Pipe*. Denver, Colo.: American Water Works Association.

ANSI/AWWA C606. *AWWA Standard for Grooved and Shouldered Joints*. Denver, Colo.: American Water Works Association.

ANSI/AWWA C906. *AWWA Standard for Polyethylene (PE) Pressure Pipe and Fittings, 4 In. (100 mm) Through 63 In. (1,575 mm), for Water Distribution and Transmission*. Denver, Colo.: American Water Works Association.

ANSI/AWWA C950. *AWWA Standard for Fiberglass Pressure Pipe*. Denver, Colo.: American Water Works Association.

ANSI/NSF 61. *Drinking Water System Components—Health Effects*. Ann Arbor, Mich.: NSF International.

ASTM Standard D2657, *Standard Practice for Heat Fusion Jointing of Polyolefin Pipe and Fittings*. West Conshohocken, Pa.: ASTM International.

AWWA. M9—*Concrete Pressure Pipe*. Denver, Colo.: American Water Works Association.

AWWA. M11—*Steel Pipe—A Guide for Design and Installation*. Denver, Colo.: American Water Works Association.

AWWA. M23—*PVC Pipe—Design and Installation*. Denver, Colo.: American Water Works Association.

AWWA. M41—*Ductile-Iron Pipe and Fittings*. Denver, Colo.: American Water Works Association.

AWWA. M45—*Fiberglass Pipe Design*. Denver, Colo.: American Water Works Association.

USEPA. 2012. Basic Information About Asbestos in Drinking Water. http://water.epa.gov/drink/contaminants/basicinformation/asbestos.cfm.

Chapter 6

Distribution System Valves

TYPES OF VALVES

Most of the valves used in water system operations fall into one of the categories illustrated in Figure 6-1. Many of these types are used only for special purposes. The valves most commonly found in a water distribution system are gate valves, butterfly valves, check valves, and control valves. Valves used in conjunction with water service lines are discussed in chapter 11.

Gate Valves

Gate valves are designed only to start or stop flow. They should not be used for throttling flow for prolonged periods because vibration of the gates will eventually wear and damage the valve.

The gate, or disk, of the valve is raised and lowered by a screw, which is operated by a handwheel or valve key. When fully open, the gates are pulled up into the bonnet, which provides almost unrestricted flow and very little head loss. When closed, the gates seat against two faces of the valve body. The principal causes of a gate valve not seating tightly are that the faces have become worn or an object has lodged under the gate.

Types of Gate Valves

Gate valves used in water systems are of two general types, rising stem and nonrising stem (NRS). Rising stem valves (Figure 6-2) are commonly called *outside screw and yoke* (OS&Y) because they have an exposed screw extending above the valve bonnet. This type of valve is commonly used in pump stations where it is an advantage to immediately tell whether a valve is open or shut by looking at the position of the screw.

OS&Y valves cannot be used where dirt might get into the screw, so all valves that are to be buried must be of the NRS type. The screw on NRS valves threads down through the gate mechanism, and the operating shaft is sealed at the top of the valve bonnet. New designs of gate valves are available that use a resilient seat of rubber or synthetic material. These valves

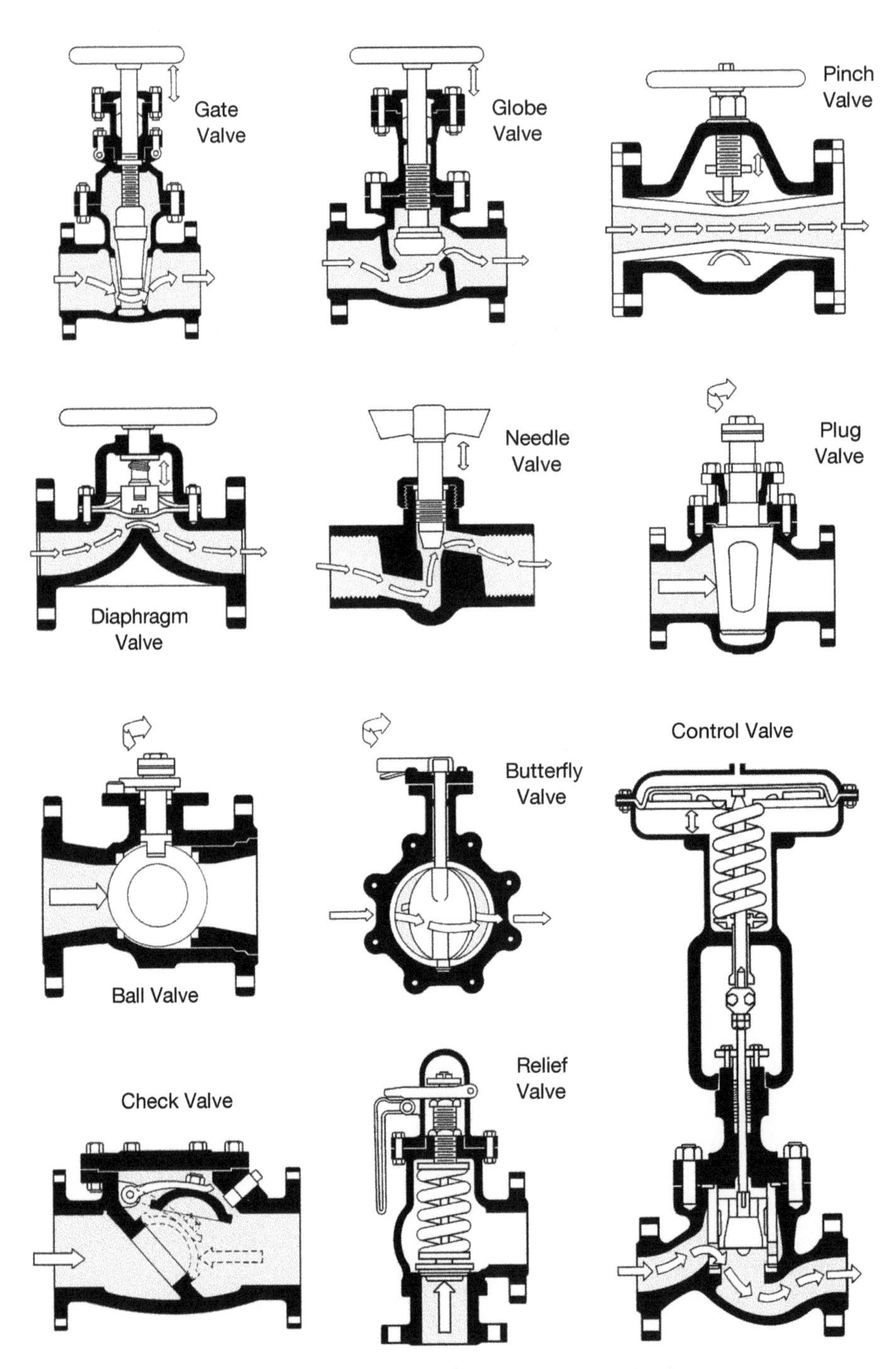

Courtesy of the Valve Manufacturers Association of America, Washington, D.C.

Figure 6-1 Types of water utility valves

Courtesy of Mueller Company, Decatur, Ill.

Figure 6-2 Valve with an outside screw and yoke

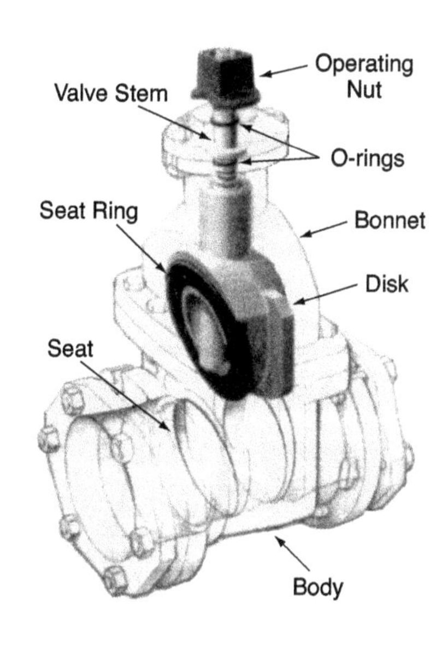

Courtesy of Mueller Company, Decatur, Ill.

Figure 6-3 Parts of resilient-seated gate valve

operate more easily and have more positive shutoff than older style valves. Figure 6-3 shows a cutaway of a resilient-seated NRS valve.

The stem seals on older valves used packing much the same as the packing used on some pump shafts. Over a period of years, the packing may have to be lubricated or adjusted to keep the seal from leaking at the stem opening. Newer valves use an O-ring seal that should not normally require maintenance.

Gate Valves Used in the Distribution System

Most of the gate valves used in a water distribution system are to isolate sections of the system, but some other types of specialty valves are also frequently used.

Hydrant auxiliary valves are internally the same as other gate valves, but the valve body has a special flange connection on one side for direct connection to a fire hydrant. This type of valve is shown in chapter 9.

Tapping valves are also internally the same but have a special flange for connection to a tapping tee and for connection of tapping equipment as discussed further in chapter 7.

Horizontal gate valves are much like other gate valves except they are designed to lie on one side. They are generally available in sizes greater than 16 in. (400 mm). One advantage to this design is that the operating mechanism does not have to lift the weight of the gate. Horizontal gate

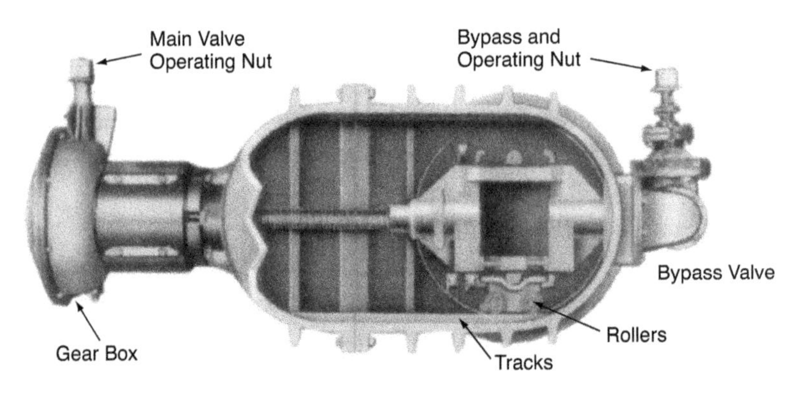

Courtesy of Mueller Company, Decatur, Ill.

Figure 6-4 Horizontal gate valve

valves must also be used where a large-diameter main is buried with relatively little cover because the operating mechanism of a vertical valve would otherwise extend above the ground surface.

Some designs of horizontal valves are equipped with special tracks in the valve body. Rollers on the disk ride in the tracks to support the weight of the disks (Figure 6-4). Bronze scrapers are provided to move ahead of the rollers to remove any foreign matter from the tracks. Another style is the rolling-disk valve, in which the disks themselves serve as rollers.

Bypass valves are often included within large gate valves. If a large gate valve has pressure on only one side, the pressure against the gates may make it difficult, if not impossible, to open the valve. A bypass valve is used to let water into the unpressurized main to equalize the pressure on the two sides of the gates so the valve may be operated.

Gate Valve Connections

Valve connections commonly used for distribution system gate valves are illustrated in Figure 6-5. Screw end connections are most often used in sizes up to 2 in. (50 mm). Flanged connections are most often used in exposed pumping station piping. Mechanical joints and push-type joints are used for buried water mains.

Gate valves are covered in AWWA Standard C500, *Metal-Seated Gate Valves for Water Supply Service*, and C509, *Resilient-Seated Gate Valves for Water Supply Service.*

Butterfly Valves

A butterfly valve has a disk that is rotated on a shaft. When in the open position, the disk is parallel with flow, and when closed, it seals against a rubber or synthetic elastomer bonded either on the valve body or on the edge of the disk. Because the disk of a butterfly valve remains in the flow path when the valve is open, the valve creates a somewhat greater pressure

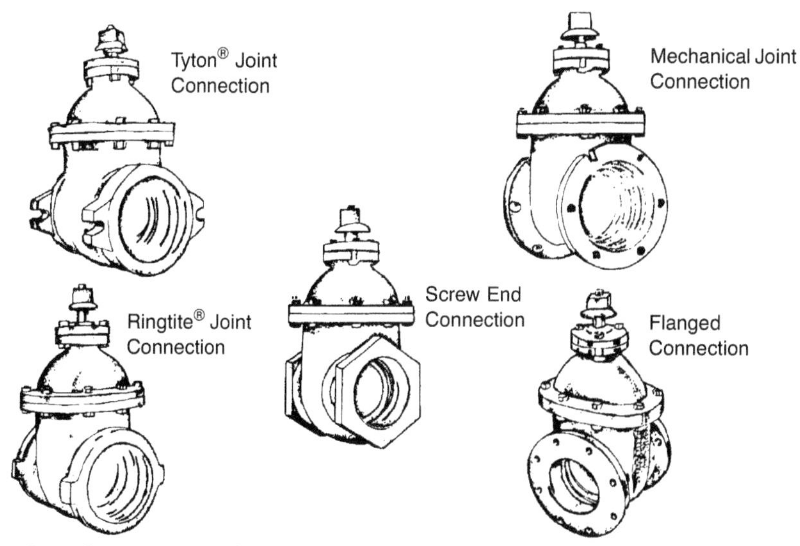

Figure 6-5 Common types of valve connections

loss than a gate valve, but it is generally not significant in comparison to other line losses.

Butterfly valves have the advantage of operating easily and quickly because the pressure on the two halves of the disk is relatively balanced and only one-quarter turn is necessary to fully open or shut them. On the other hand, care must be taken not to close a valve too quickly because it could produce a serious water hammer.

For use as water distribution system isolation valves, butterfly valves are less expensive than gate valves, particularly in larger sizes. The one consideration that should be made in considering their use is that, if the water main should ever have to be cleaned using pigs or swabs, the valves would present a serious obstacle.

Butterfly valves are primarily designed for on–off service. They can be used for occasional throttling, but if used for prolonged high-pressure throttling, the disk will vibrate until it is eventually damaged.

The wafer-type valves are installed by "sandwiching" them between two flanges. The short laying length of valves installed in this manner is often an advantage in pumping station piping systems. The mechanical joint and push-joint connections are intended for burial in water distribution systems.

Butterfly valves are covered in AWWA Standard C504, *Rubber-Seated Butterfly Valves.*

Check Valves

Check valves are designed to allow flow in only one direction. The most common use in a water system is on the discharge of pumps to prevent

Slanting Disk Check Valve

Cushioned Swing Check

Rubber Flapper Swing Check Valve

Double Door Check Valve

Foot Valve

Reprinted with permission of APCO/Valve & Primer Corp., 1420 S. Wright Blvd., Schaumberg, IL 60193 from APCO Valve Index by Ralph DiLorenzo, Exec. V.P., Copyrighted 1993.

Figure 6-6 Five types of check valves

backflow when the pump is shut down. Several types of check valves are illustrated in Figure 6-6. A foot valve is a special type of check valve installed at the bottom of a pump suction so the pump will not lose its prime when the power is turned off.

Depending on how and where a check valve is installed, there could be a problem with the valve slamming shut, potentially creating a serious water hammer. A variety of devices are available on some check valves to dampen the closing, including external weights, springs, and automatic slow-closing motorized drives.

Old check valves should be periodically inspected for wear. A worn valve on a pump discharge might stick in the open position, which could conceivably drain a whole water system backward through the pump if not noticed immediately. And worse, if the valve should finally slam shut, the water hammer force could be sufficient to break piping and move the pump off its foundation.

Check valves are covered in AWWA Standard C508, *Swing-Check Valves for Waterworks Service, 2 In. (50 mm) Through 24 In. (600 mm) NPS.*

Pressure-Regulating and Relief Valves

Throttling the flow of water requires a design that will withstand the high pressure forces that are created. Pressure-reducing valves, altitude valves, and pressure relief valves are all of similar globe-type valve design.

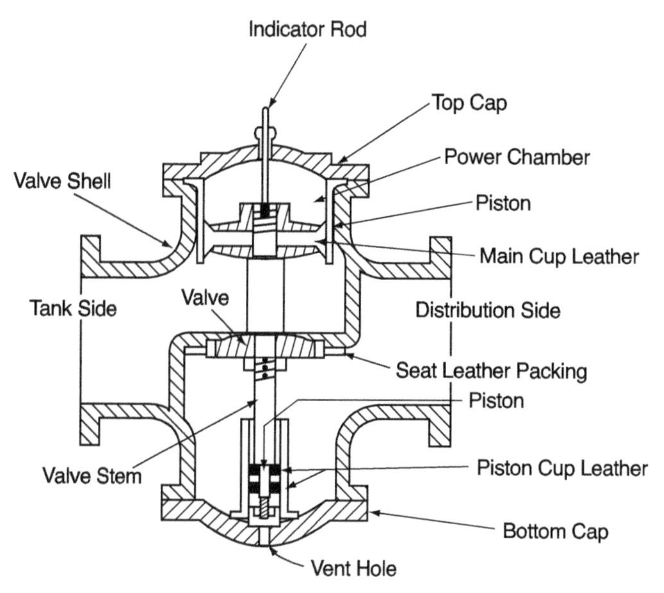

Figure 6-7 Principal parts of an altitude valve

Pressure-Reducing Valves

It is generally desirable to maintain the pressure in residences and businesses at less than 60 psi (414 kPa). Higher pressures cause more main failure, more leakage, and reduced life of appliances. A water system located in hilly country must usually establish several system pressure zones. The pressure-reducing valves operate automatically to throttle flow and maintain a lower pressure in the lower distribution system zones.

A pressure-reducing valve has two upper operating chambers sealed from each other by a flexible, reinforced diaphragm (Figure 6-7). The operating chambers receive pressure from the system and are adjusted to modulate the valve stem up and down to maintain the desired discharge pressure. The valve in the lower chamber constantly moves up and down to pass the correct quantity of water into the lower pressure system.

Altitude Valves

A ground-level reservoir is usually filled through an altitude valve. An altitude valve is similar in design to a pressure-reducing valve and can be adjusted to allow water to fill a reservoir at a controlled rate and is activated by the water pressure from the reservoir to close automatically when the reservoir is full.

Altitude valves are also used to control flow to an elevated tank when the tank is not high enough and would overflow under full system pressure. As illustrated in Figure 6-8, a double-acting valve restricts flow in both directions and automatically closes when the tank is full. A single-acting

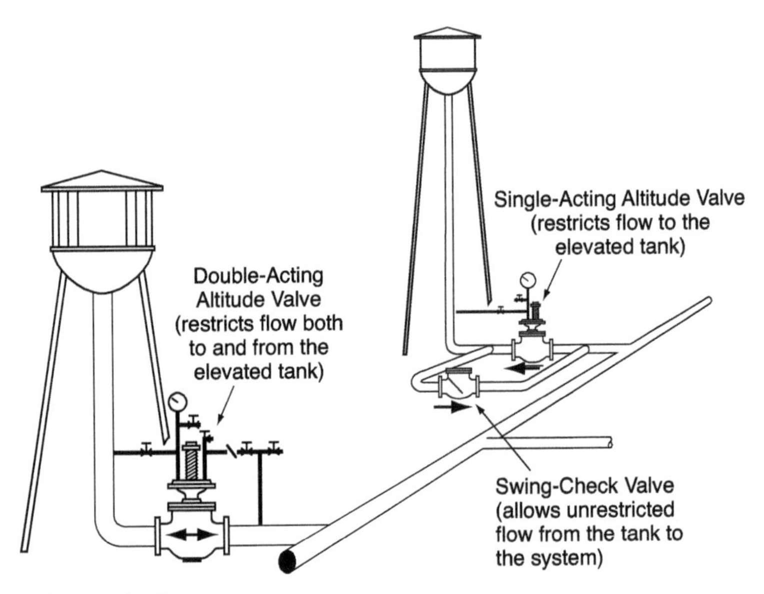

Courtesy of GA Industries, Inc.

Figure 6-8 Altitude control valves

valve restricts flow into the tank and closes when the tank is full, but unrestricted flow is allowed from the tank through a check valve.

Pressure Relief Valves

A rapid increase in pressure in a water system is called surge or water hammer. The pressure wave that moves rapidly down a pipe can damage valves, burst pipes, or blow pipe joints apart. Water hammer can also damage customer water services and plumbing fixtures. Common causes of water hammer include starting or stopping a large pump or opening or closing a valve or fire hydrant too quickly. Pressure relief valves can be installed at points on a water system to release some of the high-pressure water created by the water hammer.

Plumbing codes also require a small pressure relief valve to be installed on all water heaters and boilers to vent excessive pressure so the tank will not burst. A pressure relief valve is essentially just a globe valve with an adjustable spring to maintain pressure on the valve seat to keep the valve closed under normal pressure conditions.

Air-and-Vacuum Relief Valves

As illustrated in Figure 6-9, an air-and-vacuum relief valve consists of a float-operated valve that allows air to escape when the float is down. When water enters the container, the air vent valve is closed. If a vacuum occurs in the connection to the container, the relief valve will admit air to the system.

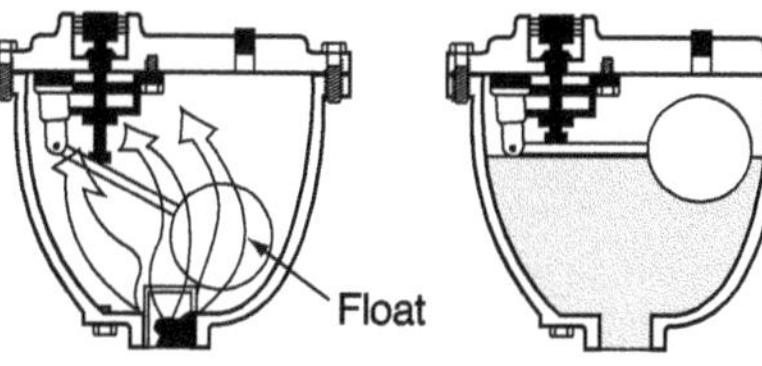

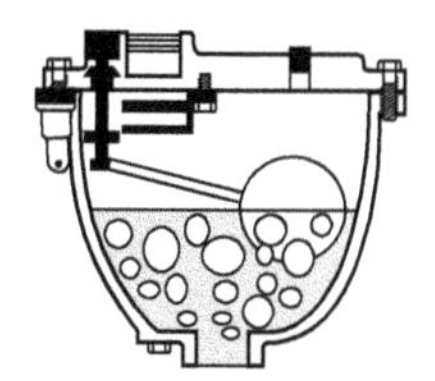

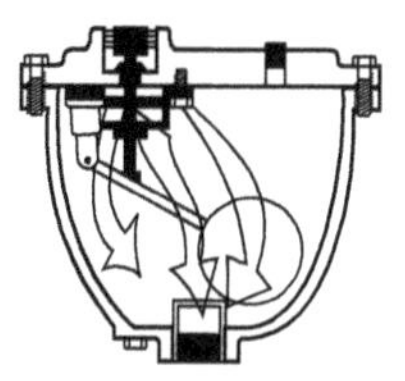

During the filling of the line, air entering the valve body will be exhausted to atmosphere. When the air is expelled and water enters the valve, the float will rise and cause the orifices to be closed.

The large and small orifices of the air-and-vacuum valve are normally held closed by the buoyant force of the float.

While the line is working under pressure, small amounts of trapped or entrained air are exhausted to atmosphere through the small orifice.

Air is permitted to enter the valve and replace the water while the line is being emptied.

Courtesy of GA Industries, Inc.

Figure 6-9 Operation of an air-and-vacuum relief valve

One common use of air relief valves is on the discharge of a well pump. The valve is installed on the discharge of the pump to vent air that has accumulated in the well column while the well is idle; otherwise, the air would be blown into the distribution system.

Air relief valves should also be installed at the high points in transmission pipelines. If pockets of air are allowed to accumulate at high points, the effective area of flow is greatly restricted, resulting in restricted flow and increased pumping costs.

VALVE OPERATION AND INSTALLATION

Because they are operated frequently, many of the valves located in treatment plants and pumping stations are power operated. Distribution system valves are usually operated infrequently, so they are manually operated.

Valve Operators

Several types of power operators are available for valves. Each has some advantages and disadvantages in terms of speed of operation, reliability, and ease of control.

Electric Operators

An electric valve operator uses a small electric motor to rotate the valve stem through a gear box. The motor is usually activated by a switch that turns on the motor, and the motor is turned off by a limit switch when the valve is fully open or closed. One advantage of electric control is that some units can be set to intermediate positions for throttling flow. A major disadvantage is that if there is a loss of electrical power, the valve cannot be

Table 6-1 Approximate number of turns required to operate most water system valves

Valve Size	Number of Turns	Valve Size	Number of Turns
3 in.	7.5	12 in.	38.5
4 in.	14.5	14 in.	46
6 in.	20.5	16 in.	53
8 in.	27	18 in.	59
10 in.	33.5	20 in.	65

power operated unless standby electric service is provided. Some units have handwheels for manual operation in the event of power failure.

Hydraulic Actuators

Hydraulic valve actuators are frequently used in treatment plants and pumping stations. They may be operated by either water pressure or hydraulic fluid. The fluid is admitted to the operating cylinder by electric solenoid valves to operate the valve in each direction. One advantage of hydraulic operators is that the speed of operation can be varied from very fast to very slow by throttling the flow of fluid to the cylinder. In installations where it is absolutely essential to have the valves operable under all conditions, including complete loss of water pressure, a separate hydraulic system is provided from a tank that is maintained partially full of compressed air.

Pneumatic Actuators

Pneumatic actuators are similar to hydraulic actuators, except that they are operated with compressed air.

Distribution System Valves

Small- and medium-sized distribution system gate valves are generally direct drive. Larger valves require a geared operator, so several turns of the valve key are required to make one turn of the valve stem. A gear-driven 20-in. (500-mm) valve may, for instance, require about 20 minutes for two workers to close by hand. The approximate number of turns required to operate most water system valves is shown in Table 6-1, but this may vary by make or model.

Operators are urged to count and record the number of turns for each valve in their distribution systems. In this way, when the valves are exercised, it is possible to determine when they are fully opened and fully closed.

Manual Operation

A distribution system valve has a 2-in. (50-mm) square operating nut that is operated by hand with a valve key (Figure 6-10). A valve that has not been operated for several years will often have to be closed using a series of up-and-down motions to clear the stem threads and accumulation of

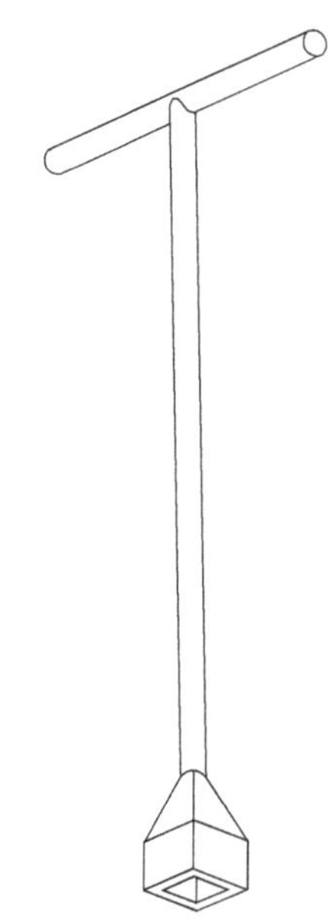

Figure 6-10 Valve key for water main valves

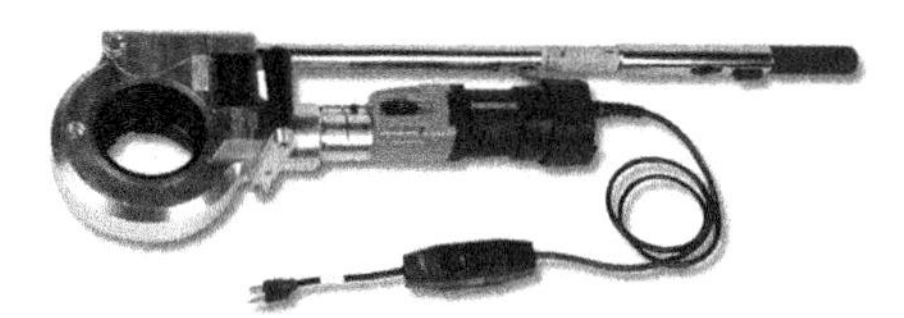

Courtesy of E.H. Wachs.

Figure 6-11 Electric valve operator

corrosion on the gates. It is suggested that the valve be closed about 5 to 10 turns, and then opened two or three rotations, repeating this sequence until the valve is closed.

If a valve does not seat fully on the first try, it is often because some foreign matter has accumulated in the depression at the bottom of the valve body. This can often be cleared away by opening a downstream fire hydrant and then cycling the valve up and down from the closed position so that high-velocity water flowing under the gates will scour the sediment. Using a cheater bar to force a valve that will not seat completely creates a serious potential of breaking the stem. Repairing a broken stem is relatively difficult and time-consuming, and if the stem is broken on a very old valve, there may be a problem with finding the correct replacement parts.

Portable Power Operators

Several types of portable electric- and gasoline-powered tools are available for operating valves (Figure 6-11). They greatly speed the work of exercising

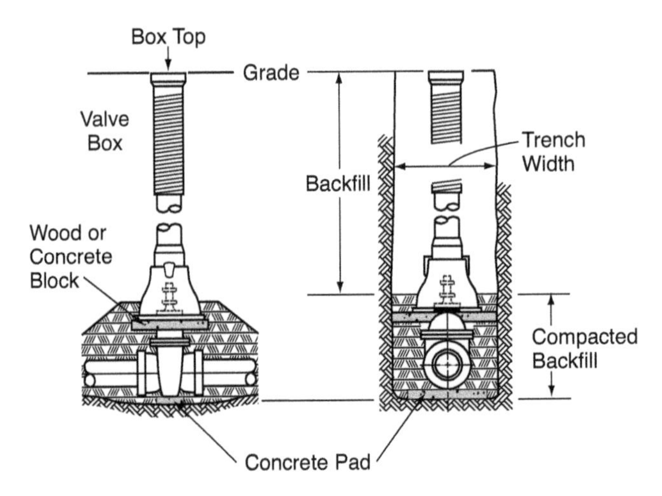

Courtesy of the Ductile Iron Pipe Research Association.

Figure 6-12 Valve-box installation

valves, which will ensure that the valves will be operable when needed. If a water system has large valves in transmission mains, a power operator will save a great amount of time and energy in operating them. The power operator should preferably have a torque-limiting device to ensure that there is not excessive pressure applied to the valve stem. In any event, it is generally best to manually apply the last few turns of travel in both opening and closing the valves.

Valve Boxes and Vaults

Buried valves are made accessible for operation either by installing a valve box over them or by building a valve vault around them. Valve boxes are made in two or more pieces that are adjustable to the particular depth of the valve. Figure 6-12 shows a typical valve-box installation. The bottom of the box is externally supported so that it does not rest on the valve. This is particularly important if the valve box is to be placed under pavement because if the box is supported directly on the valve, the pounding of traffic on top of the box could damage the valve or adjacent piping.

A valve vault is constructed around a valve after it is installed (Figure 6-13). It is important that the bottom of the vault is on firm ground so the vault will not sink and cause the weight of traffic to exert excessive pressure on the water main at the points where the pipe goes through the vault walls.

Valve Inspection and Preventive Maintenance

Distribution system valves are important to the operation of the system and are a critical component to deliver reliable service to customers. Neglect results in inoperable, inaccessible, or lost valves that are needed in emergencies. A practical valve maintenance program can result in cost savings from reduced property or fire losses.

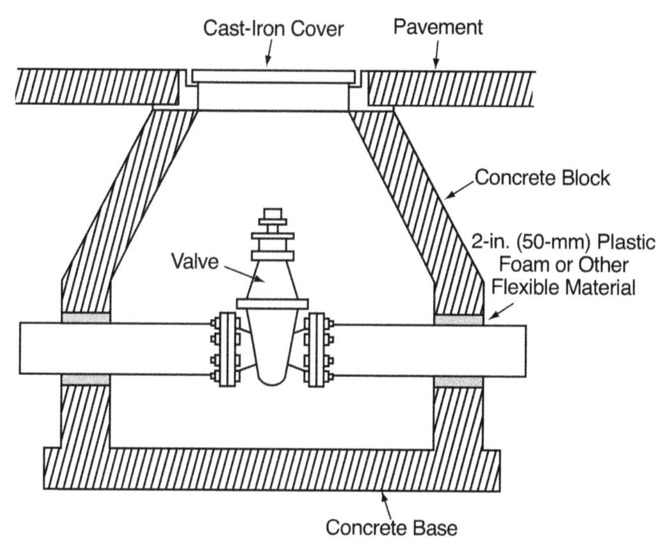

Figure 6-13 Valve vault

The first step in developing an effective valve maintenance program is to establish priorities. Critical customers, high-volume users, or problem areas may need special attention. The numbers of valves that need periodic maintenance can be a barrier when developing a program. However, if a systematic maintenance schedule is followed, the task is manageable. General guidelines for a valve inspection and maintenance program are listed as follows:

- Inspect each valve on a regular schedule (annually if possible) and more frequently for large (16 in. [400 mm] or larger) valves.
- Inspect the valve box or vault, operate the valve several times, and lubricate as required.
- Follow manufacturer's suggestions.
- Cycle all gate valves from full open to full close and back to open at least once every two years. A more detailed procedure is included in AWWA Manual M44, *Distribution Valves: Selection, Installation, Field Testing and Maintenance*. Take caution when exercising large valves in critical single-source transmission mains when closing the valve. Some valves may need less exercise. Follow the manufacturer's guidelines.
- Make repairs promptly. Keep complete records of all operation and maintenance procedures.

BIBLIOGRAPHY

ANSI/AWWA C500. *AWWA Standard for Metal-Seated Gate Valves for Water Supply Service*. Denver, Colo.: American Water Works Association.

ANSI/AWWA C504. *AWWA Standard for Rubber-Seated Butterfly Valves*. Denver, Colo.: American Water Works Association.

ANSI/AWWA C508. *AWWA Standard for Swing-Check Valves for Waterworks Service, 2 In. (50 mm) Through 24 In. (600 mm) NPS*. Denver, Colo.: American Water Works Association.

ANSI/AWWA C509. *AWWA Standard for Resilient-Seated Gate Valves for Water Supply Service*. Denver, Colo.: American Water Works Association.

AWWA. 2010. Principles and Practices of Water Supply Operations—*Water Transmission and Distribution*, 4th ed. Denver, Colo.: American Water Works Association.

AWWA. M44—*Distribution Valves: Selection, Installation, Field Testing and Maintenance*. Denver, Colo.: American Water Works Association.

Chapter 7

Water Main Installation and Rehabilitation

It is important that water mains be properly installed. Careless handling of pipe and fittings and improper pipe installation can cause added maintenance, future repair problems, and possibly even a threat to public health. Rehabilitation of pipelines can restore structurally sound pipelines to new or near-new flow capacity. This practice is usually less costly than pipeline replacement.

PIPE SHIPMENT, UNLOADING, AND STRINGING

Most operations for shipping, unloading, and stringing pipe are basically the same regardless of the pipe material used. The various operations are discussed in general in this chapter. Additional details can be obtained from the pipe manufacturers and the various pipe associations.

Pipe Shipment

Pipe to be used for small installations is generally delivered from a local warehouse. Large-diameter pipe and pipe for a large job may be shipped directly from the factory. Pipe is usually shipped by truck, although in some unusual circumstances, it may be shipped by railroad or by barge. The advantage of truck delivery is that the pipe can be delivered direct to the jobsite. Some purchasers require the ends of pipe be covered to prevent an accumulation of dirt and bugs in the pipe while in transit or storage.

Pipe Unloading

All pipe should be unloaded carefully. Pipe may appear quite durable, but it should not be dropped or allowed to strike other pipe, rocks, or the pavement. Ductile-iron and steel pipe can be physically damaged by a sharp blow. Damage may affect both the lining and exterior coating. A relatively small gouge in polyvinyl chloride (PVC) pipe can cause a stress point that

Courtesy of the Ductile Iron Pipe Research Association.

Figure 7-1 Pipe being lifted with a pipe tong

may result in pipe failure years later. Plastic pipe is particularly susceptible to damage in cold weather.

Before releasing the cables that secure the load to the truck, an inspection should be made of all blocking to make sure the pipes do not roll when the cables are released. Workers should never remain on the load or in front of it after the restraints are released.

If power equipment is not available, small-diameter pipe may be unloaded with skids and snubbing ropes. Use of a derrick or backhoe for lifting can make unloading much faster and safer. No one should ever stand under a lifted load. If a forklift is used, the forks should be padded to prevent damage to the pipe or pipe coating. Plastic pipe should not be handled with an unpadded chain. It is best to use a nylon sling or pipe tong as illustrated in Figure 7-1.

Pipe Stacking

It is often most efficient to unload pipe directly along the proposed installation route so it will not have to be handled again. If the pipe must be stockpiled for later use, it should be carefully stacked in a manner that is well protected. Each manufacturer can provide instructions for stacking a particular type of pipe, but in general, the following rules should apply:

- The stacking area should be an area of flat land that is not subject to flooding.
- The bottom layer of pipe should be laid on timbers thick enough to keep the pipe bells from touching the ground.
- Blocks should be secured at each end of the timbers to keep the pipe from rolling.

- For ductile-iron pipe (DIP) and concrete pipe, boards with blocks at each end should be placed between each layer of pipe.
- Pipe should be stacked with the bells of each layer in opposite directions, with the bells projecting beyond the barrels of the pipe below, taking particular care that they do not touch.
- Plastic pipe should be left in shipping units until needed and should not be stacked higher than 3 ft (0.9 m).
- Plastic pipe should be protected from sunlight by covering it with a canvas or opaque plastic cover, and the cover should be installed in a manner that will allow air to circulate beneath it.
- If there is any potential of vandalism or children playing on the stack, the storage area should be secured with a fence.

Rubber gaskets should be stored in a location out of direct sunlight, out of possible contact with petroleum products, and safe from vandalism. In cold weather, it is desirable to keep the gaskets and gasket lubricant warm until they are ready for use.

Pipe Stringing

In rural areas, pipe is often distributed (strung) along the installation route in advance of excavation. In residential areas, it is generally best to string pipe as the excavation progresses, both to avoid the inconvenience to residents of having the pipe on the street or parkway and because of the danger of vandalism or children injuring themselves playing around the pipe.

If possible, the pipe should be placed close to the ditch and on the side opposite where the excavated dirt will be piled (Figure 7-2). This will make

Courtesy of CertainTeed Corporation.

Figure 7-2 Pipe strung along a work site

pipe handling the easiest during installation. The pipe must also be placed where it will be protected from traffic and the operation of heavy equipment.

The pipe should be laid with the bells in the direction of installation progress so the lengths won't have to be turned as they are lifted into the ditch. Each length must be prevented from rolling by driving substantial stakes into the ground or using blocks. If it is impossible to place pipe where dirt will not get into the ends, then the ends should be covered with a plastic bag or plug to prevent contamination.

EXCAVATION

Excavation for water main installation is both expensive and dangerous. Preparations must be carefully made in advance so the job will run smoothly, efficiently, and safely.

Preparations

Before excavation can begin, a plan must be prepared for the project. The plan should show at least the following details:

- The location and depth of the main that is to be installed
- The location of valves, hydrants, and all fittings that can be anticipated
- The location and depth of all sewer and gas pipes, and electric, telephone, television, and street-light cables in the line of the new construction
- Details of any other obstructions that must be protected or avoided

The plans must generally be submitted in advance to the state drinking water control agency for approval before work is started. If the work at any point will be on state highway or railroad property, the agency's approval will also be required. In some cases, county or city approval may be required. If the pipe installation or the operation of equipment will enter onto private property at any location, property access or usage rights may have to be obtained from the property owners.

Water and Sewer Line Separation

Adequate separation should always be maintained between water mains and sewer lines. The theory is that sewers are likely to leak and saturate the adjoining soil with wastes. If at the same time there is an adjoining leak in a water main, and the main becomes depressurized, the wastes could be drawn into the main.

The general rule is that there should be a horizontal separation of at least 10 ft (3 m) between water and sewer lines, and that the water line should be at least 1 ft (0.3 m) above the sewer. Each state has regulations specifying water and sewer separation, and these should be reviewed before progressing with construction.

Notifying Property Owners and the Public

Owners of property adjoining distribution system work should be notified in advance with details of the work to be done and approximately when work will take place opposite their properties. Some property owners may be inconvenienced by having their driveways unusable for a day or two and their parkway in a mess for a week or two. It is certainly not a time when property owners would want to have parties at their homes, so they should be given ample notice to make personal plans. The property owners should also be given assurances that the work will progress as rapidly as possible and that the work area will be properly restored.

The general public should also be advised in advance of the work that will be done and why it is being done. This is a good opportunity to put a positive twist on an inconvenience. The public can be told how progressive you are in expanding your system and how the community will benefit with improved fire flow and water pressure after this small inconvenience is over. The story can be told in local newspapers and other media in advance, and signs warning drivers and pedestrians of the upcoming work can be posted at the site beginning a week or so before work is to begin.

Plans must also be made in advance to have adequate signs, barricades, flashing lights, and other safety equipment available to protect both workers and the public. State and federal regulations on the methods of placing warning equipment for various classes of streets and highways should be reviewed to be sure that protection of the construction site will be in compliance.

Recording Site Conditions Before Construction

If the work site is in an urban area, it is a good idea to take a series of photographs along the construction route just before the work is started, and include the dates they are recorded. Particular attention should be given to recording the condition of sidewalks, fences, trees, and bushes. If there are later claims that the site was not properly restored, the photos can serve as proof that, for example, the property owner's fence was already broken before the job was started. Or conversely, that there was more damage done than was realized, and it should be restored to the original condition.

It should be kept in mind that site damage can sometimes occur at some distance from where the work is to take place. For example, careless machine operators can break sidewalks or damage trees and bushes on the opposite side of the street while they are maneuvering their equipment.

Identifying Serious Conflicts

In general, it is best to decide in advance how best to avoid difficult conflicts between the proposed main and other utilities. One common problem is an intersection with a crossing sewer line at the same elevation. The grade of a sewer usually cannot be changed, so the grade of the water main must be altered. This will require either special fittings or a gradual change in grade on either side of the crossing. If the exact grade of the sewer cannot be determined by probing or measurements at nearby manholes, it is usually best to make a preliminary excavation to make accurate measurements.

Trenching

By far the most expensive part of pipe installation is the excavation, so it is important that it be done as economically as possible.

Most trenching for water main installation is done with hydraulic backhoes. These machines are easy to operate, quite maneuverable, and provide excellent control. The machine should preferably be sized proportionally to the size of pipe being installed. A machine that is too large will probably do more damage than is necessary, and a machine that is too small will do the job too slowly.

Concrete and asphalt roads and driveways should be scored in advance using an air hammer or diamond-edged power saw. This will minimize damage to surrounding pavement and keep the amount of pavement replacement to a minimum. It is often most economical to employ a firm specializing in pavement cutting to do this work.

As a rule, large pieces of asphalt and concrete should not be used for trench backfill. It is most efficient to have a truck waiting for direct loading when large quantities of pavement are to be removed to eliminate having to rehandle it. If asphalt is removed using a rotomill, the debris may be small enough that it can be used in the upper part of the trench backfill.

Trench Width

The best bucket width to use will be based in great part on pipe size, trench depth, and local experience concerning soil conditions. On the one hand, the trench width should be a minimum, both to aid rapid progress of the job and to minimize site damage and restoration. But the trench must be wide enough to maintain trench wall support and allow workers to work properly.

Trench width should generally be no more than 1 to 2 ft (0.3 to 0.6 m) greater than the pipe diameter. All that is needed is enough room for workers to make up the pipe joints and tamp the backfill under and around the pipe. It is particularly important to maintain trench width as narrow as possible under paved areas because this will minimize the traffic load that will be transmitted down through the backfill and exerted on the pipe. Recommended trench widths for small-diameter pipe are listed in Tables 7-1 and 7-2. If a much wider trench must be used or if unusually heavy surface loads may be exerted on the pipe, the pipe manufacturer or a design engineer should be consulted for special installation recommendations.

When pipe is to be laid on a curve, a limited amount of deflection can be obtained at each pipe joint. The trench width must, of course, be wider around a curve, but particular care should be taken not to over-dig the outside wall of the circle because the pipe must be blocked against a firm surface to prevent outward movement.

Table 7-1 Recommended trench widths for ductile-iron mains

Nominal Pipe Size, in. (mm)	Recommended Trench Widths, in. (mm)
3 (80)	27 (0.69)
4 (100)	28 (0.70)
6 (150)	30 (0.76)
8 (200)	32 (0.81)
10 (250)	34 (0.86)
12 (300)	36 (0.92)
14 (350)	38 (0.96)
16 (400)	40 (1.00)
18 (450)	42 (1.07)
20 (500)	44 (1.12)
24 (600)	48 (1.22)
30 (750)	54 (1.37)
36 (900)	60 (1.52)
42 (1,050)	66 (1.68)
48 (1,200)	72 (1.83)
54 (1,350)	78 (1.98)
60 (1,500)	84 (2.13)

Table 7-2 Recommended trench widths for PVC pipe

	Trench Width	
Pipe Diameter, in. (mm)	Minimum, in. (m)	Minimum, in. (m)
4 (100)	18 (0.46)	29 (0.74)
6 (150)	18 (0.46)	31 (0.79)
8 (150)	21 (0.53)	33 (0.84)
10 (250) and greater	12 (0.31) greater than outside diameter of pipe	24 (0.61) greater than outside diameter of pipe

Trench Depth

In warm climates, the depth at which mains are buried is generally dictated by the depth necessary to sufficiently spread surface loadings and protect the pipe from damage. The minimum cover for mains is typically 2.5 ft (0.8 m), and 18 in. (0.5 m) for water services.

In colder areas, the mains are placed at a depth that is locally considered the maximum depth of frost. Although a water main can withstand having frost around and below it as long as the water is always in motion, the danger of having mains too shallow is that water services will freeze at the point of connection.

Consideration should also be given to the fact that frost penetration is usually the deepest under pavement and driveways that are kept clear of snow. If the weather turns very cold with no snow cover, frost can penetrate quite deep within just a few days.

If adequate burial depth of a main is not possible, the main can be insulated. Closed-cell, expanded polystyrene insulation, 2 in. (50 mm) thick and 2 to 4 ft (0.6 to 1.2 m) wide, placed a short distance above the pipe, has been successfully used to reduce the danger of pipe freezing.

Excavation

When excavating a pipe trench, the excavated material should be piled on the pavement side of the trench. This protects the trench and workers from traffic, and keeps the equipment that will be backfilling the trench on the pavement where its operation can do minimal damage.

The excavated material (spoil) must be placed far enough away from the trench so that it will not fall back into the excavation, and a space should be maintained for workers to walk along the side of the trench. Consideration must also be given to the weight of the excavated material close to the trench and the possibility of its contributing to trench wall failure (cave-in). The general rule is to place the spoil at least 2 ft (0.6 m) back for trenches up to 5 ft (1.5 m) deep, and about 4 ft (1.2 m) back for deeper trenches. Keeping the area adjacent to the trench clear also minimizes the danger of someone inadvertently kicking dirt down on a worker in the trench, or of someone tripping and falling into the trench.

The bottom of the trench should be dug as closely as possible to the exact grade that has been specified. The trench should provide a continuous, even support for the pipe. If there are found to be high points after the pipe has been placed in the trench, the excess dirt will have to be excavated by hand. If there are low points, they should be filled with granular material or other special fill that will provide good support for the pipe.

The excavation depth is often checked using a "story pole." A string is installed on stakes along the side of the ditch slightly ahead of excavation, with its elevation a given number of feet above the ditch bottom as determined by the plans. A marked pole is used to periodically check the excavation depth by resting the pole on the trench bottom and checking to see whether the mark near the top of the pole lines up with the string.

It is usually best not to excavate much ahead of pipe laying. The most important reason for this is that the longer a trench stands open, the more likely it is to cave in. Other reasons are the likelihood of flooding if it should rain and the dangers to curious children and the interested public. An open trench is particularly dangerous after workers have left the site and worse after dark. Most contractors try to have very little open trench left at the end of the day, and then guard it well with barricades, lights, and warning tape. Local regulations often require that the trench be completely filled or protected in a specific way overnight.

Rock Excavation

The excavation for a water main trench may encounter loose boulders, ledge rock (such as hardpan or shale), or solid rock. The possibility that a main cannot be buried to the required depth without rock excavation must be determined before the job is started because it makes quite a difference in how the job progresses. The easiest method is to take periodic test borings or make quick excavations with a backhoe along the route.

If blasting is required, it is usually best done by a firm specializing in the work because it has both the special expertise and insurance required. A detailed record should be kept of the dates of blasting in the event of any claims for damage resulting from the work.

Poor Soil Conditions

Some material considered unacceptable to surround a water main include coal mine debris, cinders, sulfide clays, mine tailings, factory wastes, and garbage. This material should be excavated to well below the grade line of the main and hauled away for disposal in an approved manner. The excavation must then be filled with suitable material that has been trucked in to the site.

Groundwater

If the elevation of groundwater in the area is above the level of the trench bottom, water will enter the ditch. Trying to assemble pipe under water is usually not successful and could result in leaking joints because of silt, leaves, or other material inadvertently caught under the joint gasket. Working in saturated soil is also dangerous because of the greatly increased possibility of cave-ins. In some areas, groundwater levels fluctuate during the year, so selecting the dry time of year for construction may avoid groundwater problems.

If pipe must be installed below the groundwater table, the accepted practice is to dewater the ground in advance of the excavation by installing a system of well points to below the level of trench bottom. The points are connected to a manifold and pumping system, and operated until the ground is safe to excavate. It is important to check with local and state regulatory authorities to see if there are any special requirements on disposal of the pumped water. This type of dewatering can greatly increase the cost of a main installation and must be carefully considered before the job is started.

Causes of Trench-Wall Failure

Soil is generally classified as clay, till, sand, or silt. Firm clay and till with low moisture content can usually be excavated easily and safely. Dry sand requires special care in excavation because it can slip and run easily. Wet till is generally unpredictable and has a high potential for caving in.

Just because soil appears firm and stable does not ensure that the walls of an excavation will not cave in. Some of the primary causes of failure include

- The pressure of water contained in the soil;
- The load of construction equipment working near the edge of the trench;
- The load of excavated soil that has been piled close to the edge of the trench;
- Trench walls that are too steep for the type of soil; and
- Cleavage planes in the soil, usually caused by the previous installation of another utility, such as a gas main or sewer.

Trench wall failures occur most often in winter and early spring when ground moisture is higher. Failures usually give little warning and occur almost instantaneously, but some of the warning signs that workers can look for are

- Cracks in the ground surface parallel to the trench, located about one half to three quarters of the trench depth away from the trench edge;
- Soil crumbling off of the trench wall;
- Settling of the ground surface near the trench; and
- Sudden changes in the color of excavated soil, indicating that a previous excavation was made in the area.

Preventing Cave-in

Lack of shoring or shoring failure is the leading cause of death and injury in underground construction. Appropriate shoring is now required by Occupational Safety and Health Administration requirements and similar laws enacted by each state. Failure to comply with these laws can result in stiff penalties. Regulations generally require trench-wall protection for all trenches deeper than 4 or 5 ft (1.2 or 1.5 m).

The five basic methods of preventing cave-in are sloping, shielding, shoring, sheeting, and use of collapsible metal shoring.

Sloping is the excavation of the trench walls at an angle, based on the cohesive strength of the soil. The angle varies with the type of soil, the amount of moisture in the soil, and surrounding conditions, such as vibration from machinery. The approximate angles of repose for various soil types are shown in Figure 7-3.

Shielding is the use of a steel box in the ditch to protect workers. The box is open at the bottom and top, and is pushed or pulled along the trench as work progresses as a constant shield against caving of the trench walls (Figure 7-4). The box is also often called a *trench shield*, *portable trench box*, *sand box*, or *drag shield*.

The shield does not prevent cave-in. It must fit loosely in the trench to be moved and is designed only to protect workers in case a cave-in should occur. As the shield is moved forward, the last pipe length is usually backfilled immediately to prevent cave-in.

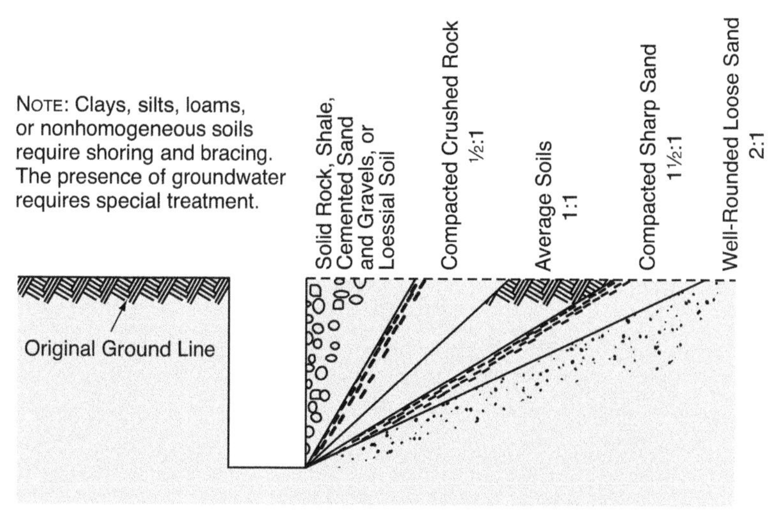

Figure 7-3 Approximate angles of repose for soil types being excavated

Courtesy of American Concrete Pressure Pipe Association.

Figure 7-4 Pipe being installed using a shield

Shoring is basically a framework of wood, metal, or both, installed to prevent caving of trench walls. As illustrated in Figure 7-5, a shoring assembly has three main parts. The uprights are placed directly against the trench walls. Then horizontal stringers are installed to support the uprights. The stringers are often called *whalers* because they are similar to the horizontal members in the hulls of wooden ships. Trench braces or trench jacks are finally placed to keep the stringers separated and tight against the trench walls.

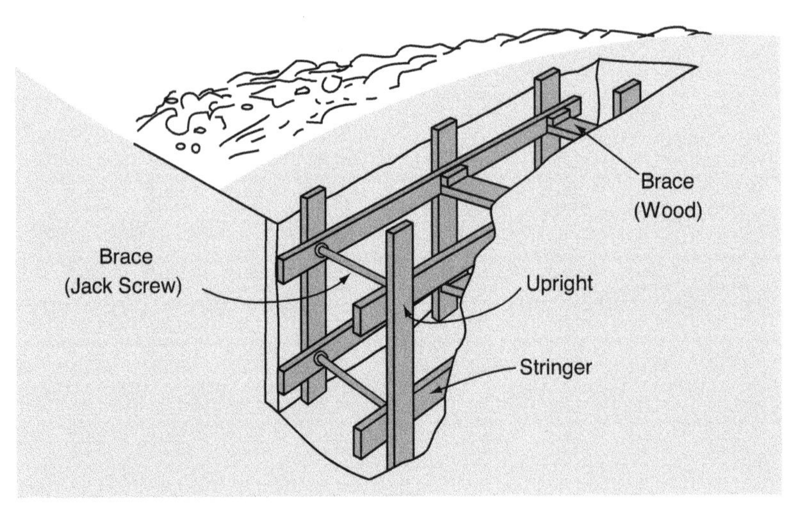

Figure 7-5 Principal parts of a shoring assembly

Shoring should always be installed from the top down, and removed from the bottom up, for maximum protection of workers. It is often necessary to leave the shoring in place as the trench is backfilled; and in some cases, it may be necessary to leave the uprights in place and just cut them off near the ground surface. Ropes are usually fastened to jacks and stringers so they can be removed from the surface and workers will not have to enter the excavation as protection is removed.

Sheeting is the process of installing tightly placed planks against the trench wall when soil conditions are very poor. If the excavation must be very deep in poor soil, steel sheet piling may have to be driven into the ground before excavation can begin. As illustrated in Figure 7-6, the sheeting is supported and later removed in about the same manner as shoring.

The sizes of shoring and sheeting members and the sizes and number of braces depend on trench depth and other local conditions. State regulations should be consulted for further details.

Collapsible shoring assemblies are lightweight panels of steel or aluminum that are available in various panel sizes and with spreaders designed for standard bucket width trenches, such as 24, 30, 42, and 48 in. (0.6, 0.8, 1.1, and 1.2 m). The assemblies can usually be handled by two or three workers (Figure 7-7), and they fold for easy transporting in a small truck. Standard assemblies can be stacked for use in a trench as deep as 12 ft (3.7 m).

PIPE LAYING

As the trench is excavated, the pipe must be inspected and carefully installed on the bottom of the trench.

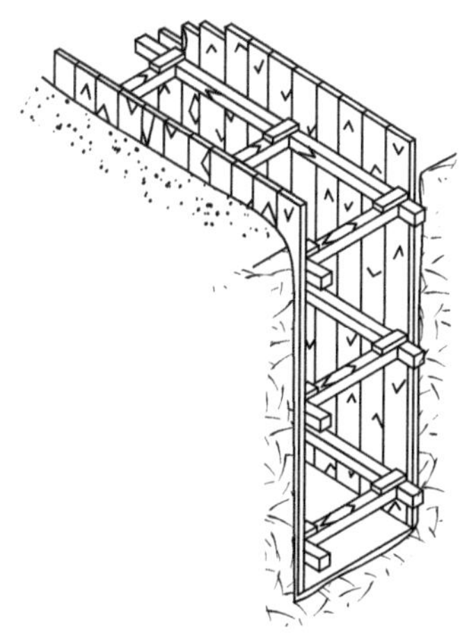

Figure 7-6 A tight sheeting in soft ground

Courtesy of Ultra Shore Products.

Figure 7-7 Collapsible shoring assembly

Pipe Inspection Prior to Installation

After the trench is prepared, but before the pipe is placed in the ditch, the pipe should be inspected. It takes only a few minutes, but it is well worth the time compared to the work required to later remove a defective pipe section. All pipe is inspected as it leaves the factory, but there are several opportunities for damage as the sections are stored and handled before installation. Some of the things to look for are a split in the spigot end, damage to the exterior, and damage to the lining. Another way of checking metal pipe is to tap it gently with a hammer. If the pipe does not ring, it may be cracked and should be rejected and inspected further before being used.

The pipe should also be relatively clean as it is laid. It should be inspected for oil, dirt, grease, animals, and other foreign matter. If mud has accumulated in the pipe while it was strung out, the pipe may have to be washed out with a hose. If there is any appreciable amount of foreign matter in the pipe, it should be swabbed with a strong hypochlorite solution before it is installed.

Pipe Placement

Pipe lengths should never be tossed or rolled into the trench from the top. They should be lowered by hand with ropes or with mechanical equipment. If the pipe is lowered using a sling, it is usually necessary to do some hand excavation to remove the sling. If a pipe tong is used for handling the pipe, removal is greatly simplified.

Pipe having bell ends is usually laid with the bells facing in the direction that work is progressing, but this is not mandatory. Contractors often find it easier to work with the bells facing uphill if the slope is steeper than 6 percent.

When work is not in progress, the open end of the pipe should be plugged to prevent dirt, animals, and water from entering. Although a small piece of plywood is often used for this purpose, a pipe plug made for the type of pipe joint is much better.

Pipe Bedding

The trench bottom must be leveled so that the barrel of the pipe will have firm support along its full length. A leveling board is usually used to check for high spots or voids. As illustrated in Figure 7-8, it is important that undue weight is not placed on the pipe bells. Normal practice is to hand excavate "bell holes" at each bell location both for this purpose and to provide a free area around the bell for joint assembly.

If there are many small rocks in the soil or the soil is unstable, it is often specified that special bedding material be placed on the trench bottom. In this case, the trench is overexcavated, and a few inches of granular material is spread on the bottom of the trench before the pipe is placed.

After the pipe has been placed, the load-bearing capacity of the pipe should be increased by compacting the backfill beneath the pipe curvature (haunching). As illustrated in Figure 7-9, a pipe that is supported only on the center of the bottom has a great amount of pressure exerted from above the pipe, which is tending to deform the pipe. If the bedding is compacted up to the springline (center) of the pipe, the support is greatly increased; and if compacted bedding is placed completely around the pipe, supporting strength is increased by 150 percent.

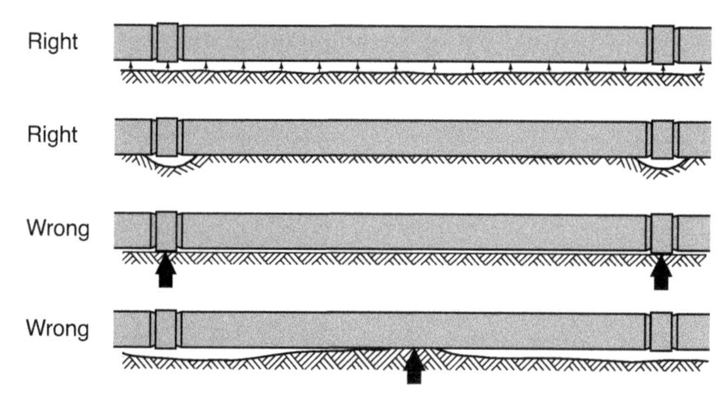

Courtesy of J-M Manufacturing Company, Inc.

Figure 7-8 Correct and incorrect pipe bedding

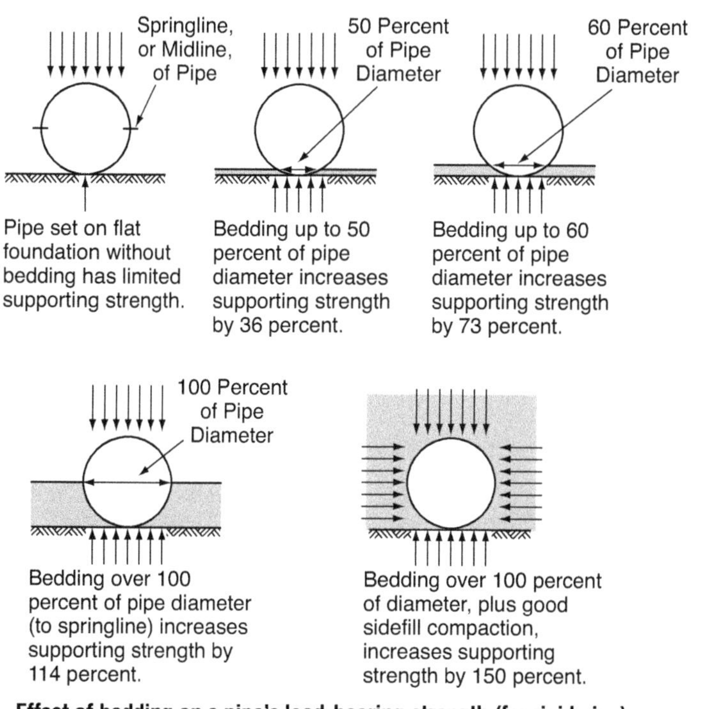

Figure 7-9 Effect of bedding on a pipe's load-bearing strength (for rigid pipe)

Pipe Jointing

One of the most important procedures in making up all types of joints is that the pipe bell, spigot end, and gasket must be kept clean. All sand, dust, tar, and other foreign matter must be carefully wiped from the gasket recesses of the bell, and the spigot must be smooth and free of rough edges.

A small piece of a leaf caught under the gasket can cause a leak that will prevent the new pipeline from passing the pressure test.

Push-on Joints

The four general steps required for making up push-on joints are illustrated in Figure 7-10. One particularly important point is that the spigot end must be pushed completely "home" into the bell. All full lengths of pipe have a painted line near the spigot end. This line should be all the way to the face of the bell when the joint is complete.

The spigot end of all push-on joint pipe has a beveled end that allows it to easily slip into the gasket. If a pipe is cut to make a short length, the cut end must be beveled so that it will not cut or jam in the gasket. The bevel on

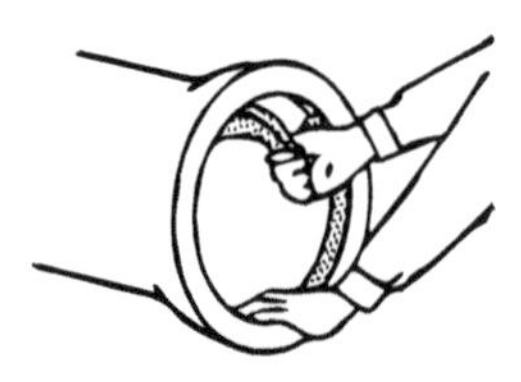

1. Thoroughly clean the groove and the bell socket of the pipe or fitting; also clean the plain end of the mating pipe. Using a gasket of the proper design for the joint to be assembled, make a small loop in the gasket and insert it in the socket, making sure the gasket faces the correct direction and that it is properly seated. NOTE: In cold weather, it is necessary to warm the gasket to facilitate insertion.

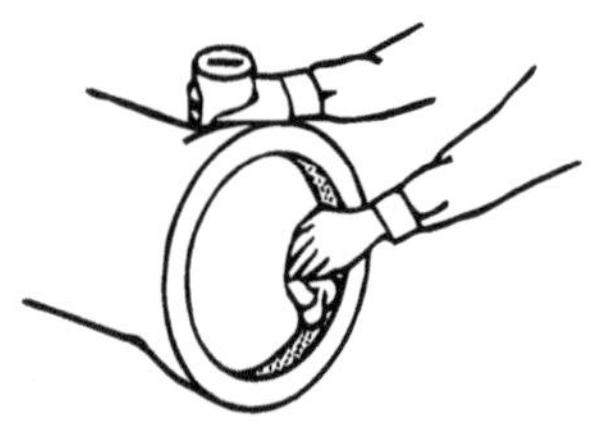

2. Apply lubricant to the gasket and plain end of the pipe in accordance with the pipe manufacturer's recommendations. Lubricant is furnished in sterile containers, and every effort should be made to protect against contamination of the container's contents.

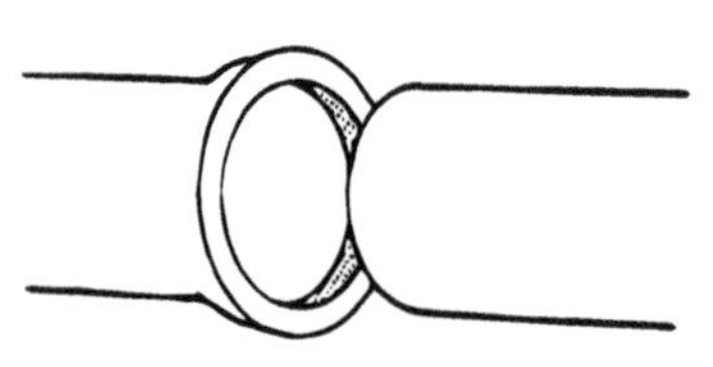

3. Be sure that the plain end is beveled; square or sharp edges may damage or dislodge the gasket and cause a leak. When pipe is cut in the field, bevel the plain end with a heavy file or grinder to remove all sharp edges. Push the plain end into the bell of the pipe. Keep the joint straight while pushing. Make deflection after the joint is assembled.

4. Small pipe can be pushed into the bell socket with a long bar. Large pipe requires additional power, such as a jack, level puller, or backhoe. The supplier may provide a jack or level puller on a rental basis. A timber header should be used between the pipe and jack or backhoe bucket to avoid damage to the pipe.

Figure 7-10 Assembling a push-on joint

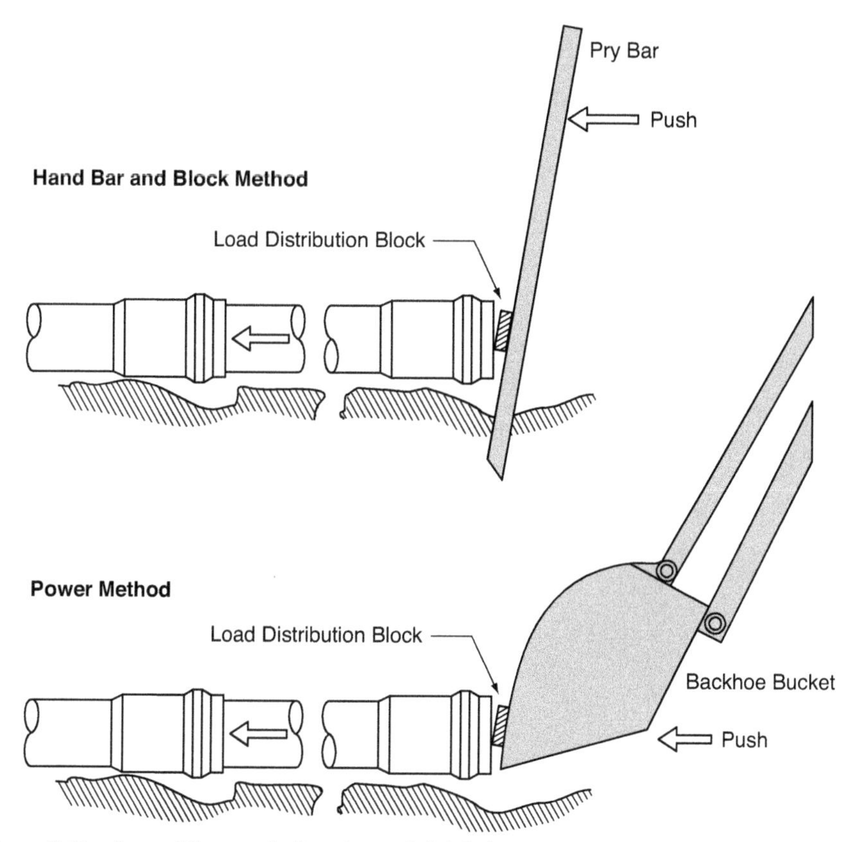

Figure 7-11 Assembling small-diameter push-joint pipe

DIP can be made with a portable electric grinder or a special grinder made for the purpose. The bevel can be made on PVC pipe with a rasp file or with a special beveling tool. After the end of a cut piece is beveled, a measurement should be made of the "home" mark on a full pipe length, and a similar mark placed on the cut end using a felt-tipped marker. This will ensure that the pipe is inserted the full distance when it is installed.

If a push-on joint is to be deflected, it must first be inserted with the two pipes in line, and then the deflection can be made. If an attempt is made to push the pipe in at an angle, it may jam or not make up a tight joint.

Small-diameter pipe can be pushed home using a pry bar and block of wood, as illustrated in Figure 7-11. Larger pipe joints are usually made up using a come-along or by pushing with the bucket of a backhoe.

The assembly of concrete pipe joints is similar, except that the round gasket is mounted in a groove on the spigot end before insertion. Concrete pipe joints must be made up completely tight all around the pipe. If there is to be any deflection, it must be achieved by the use of special fittings.

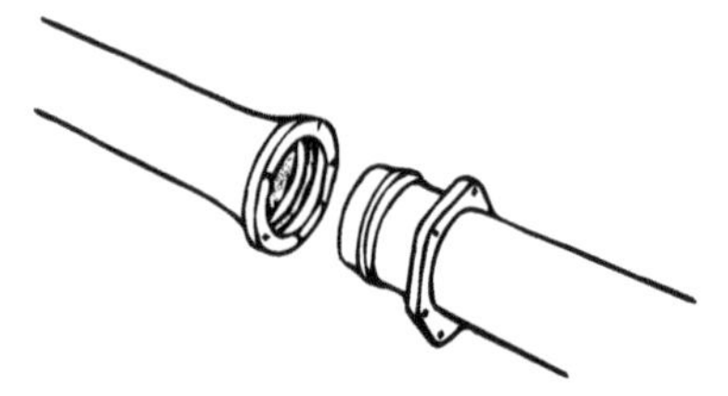

1. Clean the socket and the plain end. Lubrication and additional cleaning should be provided by brushing both the gasket and the plain end with soapy water or an approved pipe lubricant meeting the requirements of ANSI/AWWA C111/A21.11 just prior to slipping the gasket onto the plain end for joint assembly. Place the gland on the plain end with the lip extension toward the plain end, followed by the gasket with the narrow edge of the gasket toward the plain end.

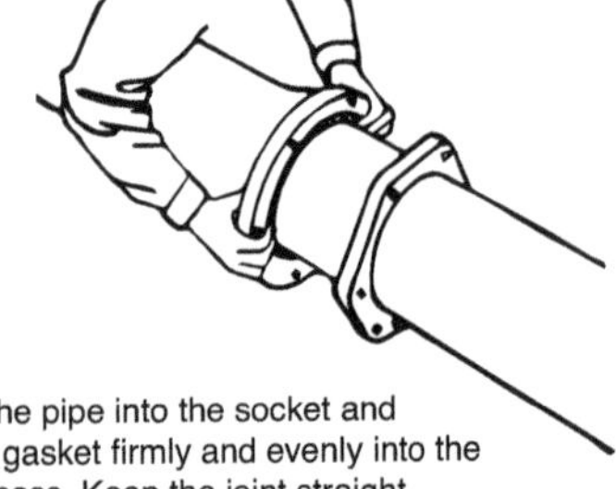

2. Insert the pipe into the socket and press the gasket firmly and evenly into the gasket recess. Keep the joint straight during assembly.

3. Push the gland toward the socket and center it around the pipe with the gland lip against the gasket. Insert bolts and hand-tighten nuts. Make deflection after joint assembly but before tightening bolts.

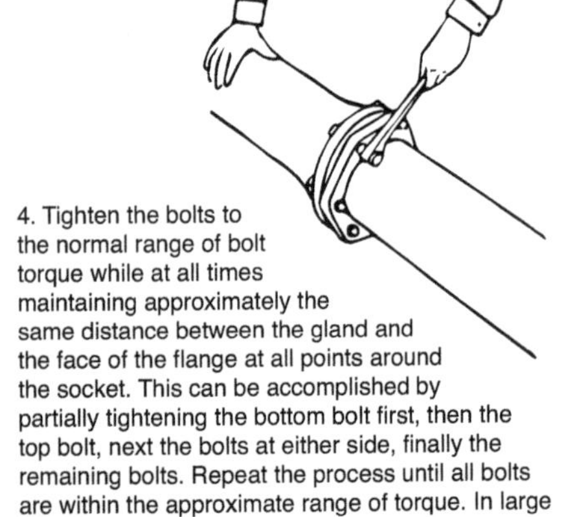

4. Tighten the bolts to the normal range of bolt torque while at all times maintaining approximately the same distance between the gland and the face of the flange at all points around the socket. This can be accomplished by partially tightening the bottom bolt first, then the top bolt, next the bolts at either side, finally the remaining bolts. Repeat the process until all bolts are within the approximate range of torque. In large sizes (30–48 in. [762–1,219 mm]), five or more repetitions may be required. The use of a torque-indicating wrench will facilitate this

Figure 7-12 Assembling a mechanical joint

Mechanical Joints

Mechanical joint fittings are made up as illustrated in Figure 7-12. Although these joints are more expensive and take longer to assemble than push-on joints, they make a very secure seal and allow quite a bit of deflection.

CONNECTING TO EXISTING MAINS

A new main can be connected to an existing main by either inserting a tee or by making a pressure tap.

Tee Connections

One method of connecting a new main to an existing main is by inserting a tee in the old main. Doing this means that the existing main will be out of service for a period of several hours if all goes well—longer if all does not go as planned. If this is acceptable, the first thing to do is to determine whether the valves on either end of the section hold securely. If they do not, they may have to be repaired before proceeding.

A method of inserting a standard tee using two rubber-joint sleeves is illustrated in Figure 7-13. The outside diameter of the old main should be carefully measured and sleeves purchased having the correct gaskets for the pipe size. If the old main is cast iron, the pipe will usually have to be cut in three places in order to break it out. A valve should also be ready to install on the outlet of the new tee so the main can be flushed and put back in service as quickly as possible.

Pressure Taps

Most connections to old mains are currently made with a pressure tap that avoids having to take the old main out of service. As illustrated in Figure 7-14, the first step in the process is to clean the old main and install a split tapping tee. A special tapping valve having a flange on one side and a mechanical joint fitting on the other is then bolted to the tee, and the drilling machine is fastened to the valve outlet.

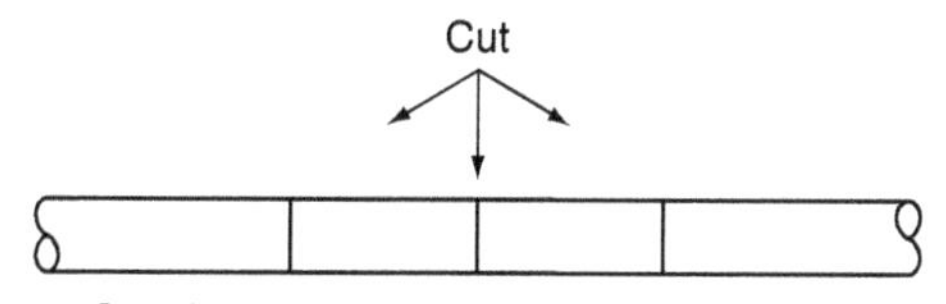

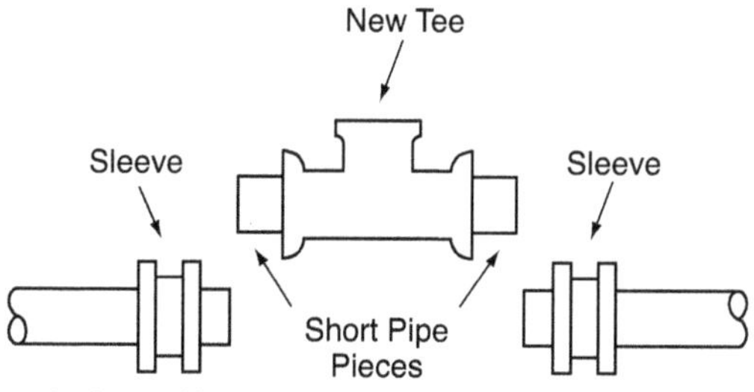

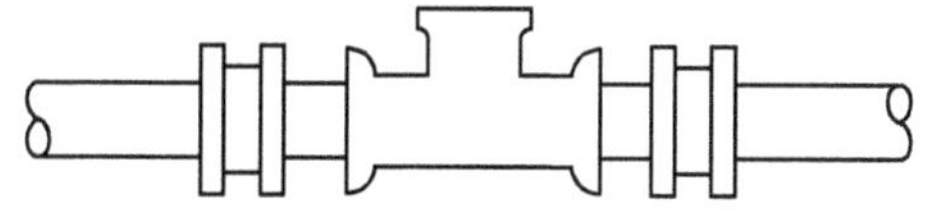

Figure 7-13 Steps in using short pieces of new pipe and rubber-joint sleeves to install a tee

Courtesy of Mueller Company, Decatur, Ill.

Figure 7-14 Preparing to tap a large connection

Courtesy of T.D. Williamson, Inc.

Figure 7-15 Coupon cut from main

The drilling machine extends through the open valve to cut a circle of material from the main using a shell cutter (Figure 7-15). The drill is then retracted, the valve closed, the drilling machine removed, and the connection is completed.

THRUST RESTRAINT

Water in motion can exert tremendous pressure if it is suddenly stopped or there is a change in the direction of flow. The thrust against tees, valves, bends, reducers, and fire hydrants can cause a leak or completely push fittings apart if they are not suitably restrained or blocked. The pressure can be several times normal if there is a surge or water hammer in the system caused by such things as quickly closing a hydrant.

Plastic pipe is particularly smooth, so it is very vulnerable to slipping out of a push joint if not adequately restrained. Polyethylene bags placed around DIP for corrosion control provide almost no soil friction, so the pipe can slide within the bag if not well restrained.

Typical horizontal blocking is illustrated in Figure 7-16. The block can be made of poured concrete or other material that will not disintegrate over time. Wood should not be used for blocking. The block should be centered on the thrust force and should cradle the fitting but not cover the joint fittings. It is important that the ditch not be over-dug at changes in pipe direction, otherwise there will be no firm, undisturbed soil to block against.

Similar blocking should also be placed when pipe is laid around a curve. Even though the curve is made by deflecting the pipe joints and appears

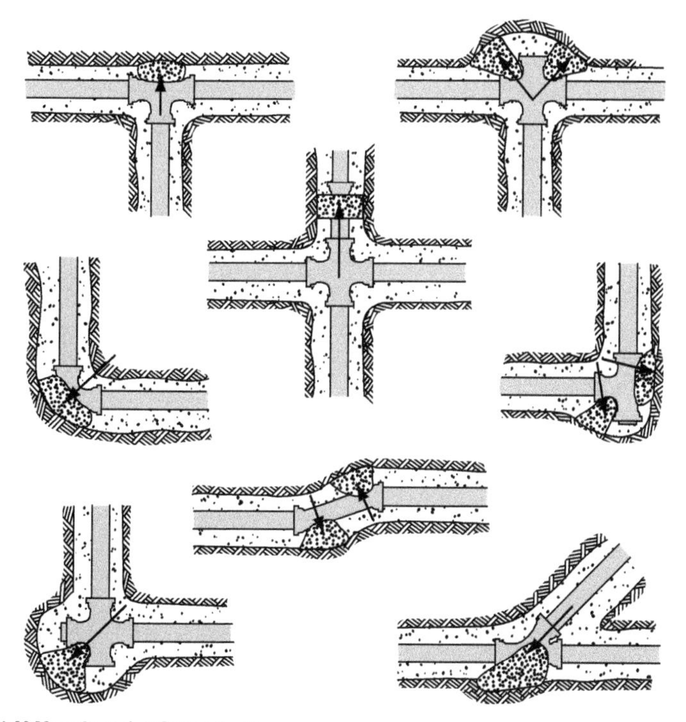

Courtesy of J-M Manufacturing Company, Inc.

Figure 7-16 Common concrete thrust blocks

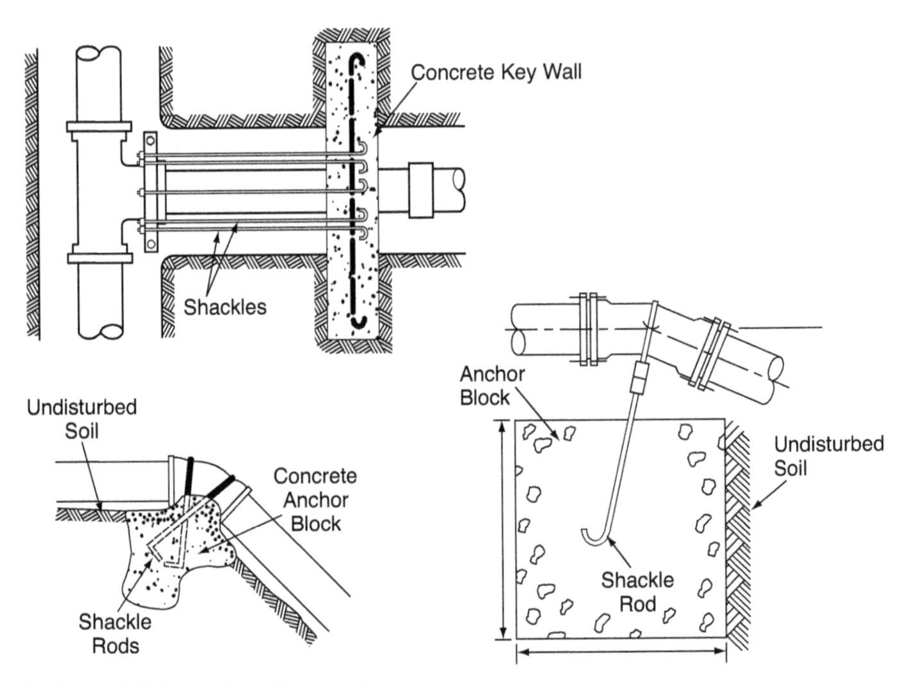

Courtesy of J-M Manufacturing Company, Inc.

Figure 7-17 Various types of thrust anchors

quite gradual, the fresh backfill around it may not be adequate to prevent the line from pushing outward if there is water hammer. PVC pipe is particularly vulnerable to movement if not firmly blocked against the outer wall of the trench.

When there is no undisturbed earth behind a horizontal bend or tee to place a block against, and also for vertical bends, tie rods or thrust blocks made of large masses of concrete may have to be used (Figure 7-17). Several companies now offer restraining fittings that have set screws or clamps that can be used in place of blocking.

AIR RELIEF

When a water main is laid over uneven ground, a bubble of air will be trapped at each high point as the main is filled. If the bubbles are allowed to remain in the pipe, the flow of water will be restricted much as though there is a partly closed valve at these points. For small-diameter mains, opening one or more hydrants will generally create enough velocity to blow the bubbles out of the main. For large mains, however, there is no practical way of flushing the bubbles out.

One method of releasing air from large mains is to locate a fire hydrant at each high point. To be completely effective, the hydrant should be

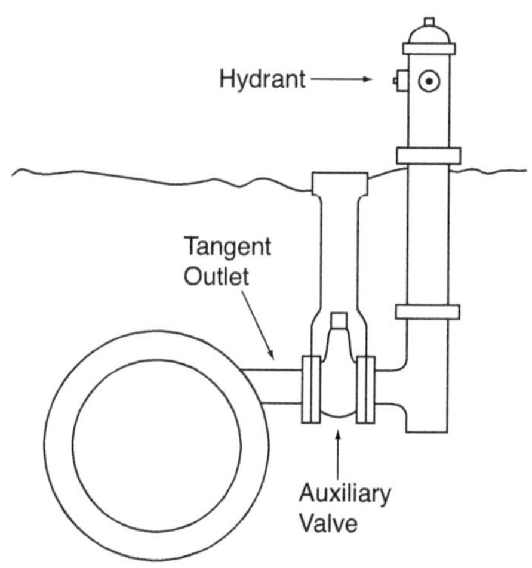

Figure 7-18 Fire hydrant installed on a tangent outlet located on a high point of a large-diameter water main

connected to a tangent tee that draws from the top of the main (Figure 7-18). The only problem with this method is, if there is entrained air in the water, the bubbles will gradually re-form over a period of time, so the air will have to be manually released periodically from each hydrant.

A method for releasing air automatically is to install an air relief valve at each high point. The valve is often located in a pit to facilitate maintenance, and the air is vented through a screened pipe that extends above ground level (Figure 7-19).

BACKFILLING AND MAIN TESTING

After water main pipe and fittings have been installed in the trench, the excavation must be backfilled with suitable material and the pipe tested for leaks, then flushed, disinfected, and tested for bacteriological quality before the public may be allowed to use water from the new section of the system.

Backfilling the Trench

Backfill material placed around and over the newly installed pipe serves several purposes besides just filling up the excavation. The most important is that it distributes the surface loads. For example, the downward pressure from a heavy truck passing over a pipe in a shallow trench with uncompacted fill could damage the pipe because the load is transmitted directly downward. If the backfill is properly compacted, the load is transmitted outward to the trench walls and the pipe will not be damaged.

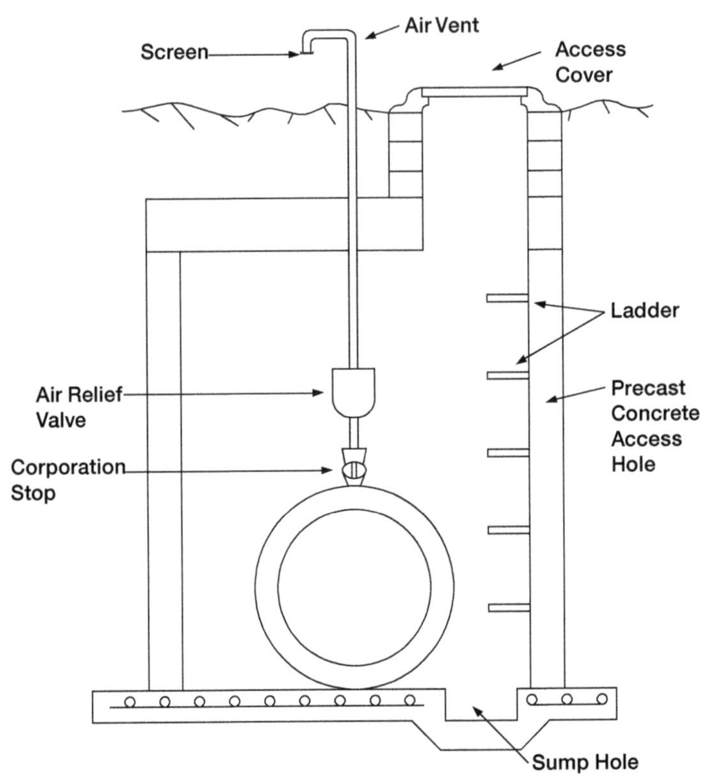

Figure 7-19 Installation of an air relief valve at a high point on a large-diameter water main

Another important function of properly placed backfill is to provide lateral restraint of the pipe to keep it from moving in the event of water hammer in the system.

Placing Backfill

Only clean sand or selected soil should be used for the first layer of backfill. It should contain no peat, stones, roots, construction debris, or frozen material and should have enough moisture to permit thorough compaction. If the excavated soil is not suitable for the first layer that will surround the pipe, other backfill material should be brought in.

Initial Backfill Layers

The first layer of backfill should be placed equally on both sides of the pipe, up to about the center of the pipe. This material should then be compacted, a process often called *haunching*.

A hand tamper is often used for haunching, but the work goes much faster if a pneumatic tamper is available. Compacting the soil under and on each side of the pipe up to the center of the pipe (springline) provides valuable support so the pipe will withstand the weight that will later be placed on it from above.

Backfill practices above the springline vary depending on the type of pipe used, local soil conditions, and regulatory requirements. In general, another layer of good-quality material should be placed over the pipe and again compacted to protect and secure the pipe. This layer should be at least 6 in. (150 mm) over the top of the pipe for pipe up to 8 in. (200 mm) in diameter, and thicker for larger pipe as recommended by the pipe manufacturer.

Backfilling the Remainder of the Trench

The backfill material used to fill the remainder of the trench does not have to be selected, placed, or compacted as carefully as the first two layers. In general, the excavated material may be used unless the job specifications or regulatory agency direct otherwise. If soil backfill is not compacted in some manner, the trench will continue to settle for years. There are four methods of dealing with settling.

If the construction is in a rural area where the ground surface does not have to be restored immediately, the easiest method of backfilling is to carefully place the soil in the trench to prevent voids, and then finish with a neat mound of excess fill over the top of the trench. The ditch will then be settled by rain and gravity over a period of time. If the ditch settles to below grade, someone could be injured from stepping into it, so someone must accept responsibility for patrolling the ditch periodically for several years to provide more fill as necessary.

The next easiest method of settling the backfill is by using water. If the backfill material is granular and drains well, just flooding the ditch may be sufficient. If the backfill is clay, however, settling must be done by jetting. This is usually done by using a 1½-in. (38-mm) galvanized pipe connected to a fire hose, with the pipe repeatedly pushed all of the way to the level of the pipe in the trench. If done thoroughly, this method will provide compaction of the backfill to within 5 percent of maximum density. On the negative side, this method requires a very large quantity of water, particularly in a wide trench, so if there is a scarcity of water, or the water must be purchased at a rather high rate, it may not be practical.

If compaction of the fill is necessary but it is not practical to do it with water, the fill is usually placed in the trench in layers and tamped with power equipment. Pneumatic or gasoline-engine tampers may be used by a crew working up and down the trench line, but this should be used only where there is no danger of trench wall cave-in caused by the vibration. A boom-mounted tamper is also available that is operated from a crane so workers do not have to be in the ditch. This method provides a fair degree of compaction depending on the type of soil, but there will usually be some settling of the earth at the surface over a period of years.

In areas such as street crossings where the surface must be repaved as soon as possible, it is usually quickest and most economical to backfill the entire trench in that area with granular backfill, such as sand or fine stone. Most engineers and local authorities now specify this because it virtually eliminates future settling of the pavement.

Pressure Testing the Main

After the trench has been at least partially backfilled, the new main must be tested to determine whether there are any leaks. The test may be performed on one section at a time between valves, or the installer may wait and test the entire job at one time.

Many years ago, when only lead joints were used on cast-iron pipe, it was always assumed that there would be leaks. The only thing the installer could do was hope that it would not be too much. Now, an installation of all mechanical, push-on, or other rubber-gasket joints will have virtually no leakage unless something is defective.

Testing Procedure

The following general procedures apply to testing of all types of pipe.

- If poured concrete blocking is used, allow at least 5 days before testing. If high early strength concrete was used, this time can be shortened.
- Make sure the valves at all connections with existing mains are holding completely tight.
- Fill the new main with water and be sure that air has been released at all high points. If chlorine tablets have been installed in the pipe as work progressed, be sure to fill the main very slowly so the tablets will not be dislodged.
- Close all fire hydrant auxiliary valves.
- Connect a pressure pump to a corporation stop in the main. The pump must have a pressure gauge and connection to a small tank of makeup water.
- Apply partial pressure and again check that all air has been removed from the system. Allow the pipe to stand with pressure on it for at least 24 hours to stabilize.
- Pump up the pressure as specified in the applicable AWWA standard. The minimum is usually 1.5 times the operating pressure, or 150 psi (1,030 kPa) for a period of 30 minutes.
- Examine the piping and fittings for visible leaks or air that was not previously released.
- Again pump up the pressure and wait for at least 2 hours.
- During the waiting period, pump up the pressure as required to maintain the minimum test pressure. Pump from a calibrated container and record the quantity of makeup water used.
- Compare the amount of leakage (the quantity of water required to bring the pressure back up) with the suggested maximum allowable leakage in the appropriate AWWA standard.

Table 7-3 Flow rates and number of hydrants required for flushing mains

Flow Rate*				
in.	(mm)	gpm	(L/sec)	Number of Hydrants Open†
6	(150)	200	(13)	1
10	(250)	600	(38)	1
16	(400)	1,600	(100)	2

*Based on 2.5 ft/sec (0.76 m/sec) at 40 psi (280 kPa) pressure.
†Based on hydrant with one 2½-in. (63-mm) outlet.

Rapid loss of pressure will usually be caused by an open valve, a cracked or broken pipe, or a joint that slipped out after it was made up. These leaks are usually relatively easy to locate by continuing to apply pressure until the water comes to the surface.

One possible cause of a slow drop of pressure is that all of the air was not completely removed before testing began. In this case, the amount of apparent leakage will usually be less on each subsequent repeat test.

A slow leak is often difficult to locate. If the initial test was on a long section of main, it will probably be necessary to locate a small leak by performing tests on small sections between valves to narrow down the location. Leak detection equipment can then be used to locate the leak. Another alternative is to continuously subject the main to the highest pressure that can be safely applied, and wait for the water to come to the surface.

Flushing, Disinfection, and Testing

All new sections of water main must be thoroughly flushed, disinfected, and tested for bacteriological quality before the water can be used by customers.

Flushing

Flushing is primarily required to remove any mud and debris left in the pipe from the installation. If the workers were not careful in plugging the end of the main when it was unattended, the flushing might also produce dead animals and other foreign matter.

One or more fire hydrants should be used for flushing so that a velocity of at least 2.5 ft/sec (0.76 m/sec), and preferably 3.5 ft/sec (1.1 m/sec), is obtained in the pipe. This velocity should be maintained long enough to allow two or three complete changes of water, and for the water to run visibly clean.

Table 7-3 lists the number of hydrants required to adequately flush smaller mains. If a hydrant is not to be installed at the end of a new main, a temporary blowoff connection should be provided. For very large mains where there are insufficient hydrants or plant capacity to supply the water needed for flushing, the new pipe should be cleaned with poly pigs.

Disinfection and Testing

Disinfection of mains and storage facilities is discussed in chapter 8. This includes disinfection procedures, bacteriological testing, and the process for returning the mains and facilities to service.

Site Restoration and Inspection

The construction site should be promptly restored to its original condition. Disturbed grass should be replaced with sod or seeded and watered if necessary to get good initial growth, street pavement and curbs should be replaced in kind, and ditches and culverts should be checked for proper drainage.

All construction debris must be removed from the site, and private driveways, walks, fences, lawns, and bushes must be returned to their original conditions.

Damaged roots can cause some trees to die, sometimes several years later. If there is a question regarding damage to trees, a qualified tree expert should be consulted. In some cases, pruning and feeding may help to save damaged trees.

As soon as all restoration is considered complete, a responsible person should make a detailed final inspection of the entire jobsite, and make note of anything that was missed. This is also a good time to complete as-built drawings of the construction, showing any changes from the original plans and noting exact measurements to fire hydrants and valves. It is also a good idea to make a final check of all valves to make sure they are in the full-open position, and to flow test all new hydrants and record the results in the system records.

Pipeline Rehabilitation

Identifying pipeline segments that are in need of renovation can involve several separate but related practices. Scheduled main pressure and flow testing can be used to determine if mains need to be rehabilitated or replaced. Main breaks and leaks in an area may be an indication that mains need to be considered for renovation. And customer complaints upon investigation may alert system operators to problem pipeline segments.

Pipeline rehabilitation methods are available for a variety of situations. Metal pipelines can be cleaned and lined if the pipe walls are still structurally sound. Lining materials can include cement mortar, epoxy, plastics, membranes, or other cured resin products. Trenchless methods are available to replace pipelines, in some cases, in place.

Pressure and Flow Tests

Pressure can be checked at points on the distribution system by attaching a pressure gauge to a fire hydrant (Figure 7-20). After the gauge is attached, the hydrant should be opened several turns to be sure the drain valve is closed, and air must be bled through the petcock before a reading is taken.

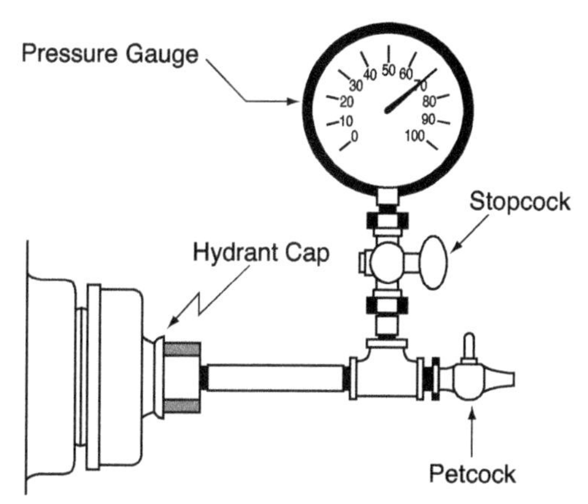

Figure 7-20 Hydrant pressure gauge

It is usually advisable to take readings during periods of high and low water use to determine the difference. Maximum water use on most water systems is usually around supper time on a summer evening.

A more satisfactory way of recording pressure fluctuations is to install a chart recorder on a hydrant for at least 24 hours. To prevent vandalism of the recorder, a barrel is placed over the hydrant. The fire department must be informed that the hydrant is out of service.

If there are no hydrants at locations that require pressure testing, it is possible to attach a pressure gauge to a sill cock on a customer's building. To do this, obtain customer permission first, and be sure no water will be used in the building during the test.

A wide variation in pressure between readings taken at night and at maximum use times of the day is a sure indicator that the system is being stressed from normal usage and will probably be considerably worse when fire flow is required.

It is also advisable to periodically check the capacity of key locations on the system by conducting fire flow tests. The testing will indicate whether the system should be reinforced with new transmission mains or would benefit from cleaning the mains. It will also indicate other problems, such as valves inadvertently left closed after a main repair.

A permanent record of pressure and flow test information should always be kept at each location so that comparisons can be made with future tests to identify any changes in the system operation.

Flow-testing procedures are detailed in AWWA Manual M17, *Installation, Field Testing, and Maintenance of Fire Hydrants*.

Cleaning Water Mains

Most water distribution systems have some accumulation of sediment that has settled in the bottom of the mains. How much sediment exists is a function of the quality of the water source, treatment, and the type of piping.

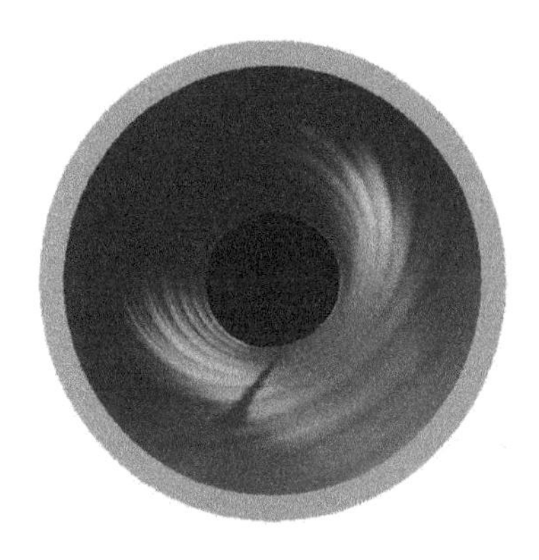

Smooth Pipe

Courtesy of J-M Manufacturing Company, Inc.

Tuberculated Pipe

Courtesy of Girard Industries.

Figure 7-21 Effect of tuberculation

In most cases, the sediment can be adequately removed by periodically flushing the main. If old piping is badly corroded, it may be necessary to clean and line the pipe to maintain good water quality and to provide good flow.

Old, unlined, cast-iron water pipe often develops tuberculation that gradually constricts flow and causes water discoloration due to the corrosion products (Figure 7-21). It is possible to clean these pipes using one or more of the following cleaning techniques, but in most cases, the cleaning process is not a permanent solution. Unless the cleaned pipe is lined or the corrosiveness of the water is reduced, experience has shown that regrowth of the incrustation is rather rapid. It is also occasionally necessary to clean pipe other than old cast iron if there are accumulations of slime or deposits from iron bacteria.

Cleaning Preparations

A pipe-cleaning operation requires very careful preparation. The mains to be cleaned must be carefully mapped, and all hydrants and valves tested for proper operation. Customers must be notified of the times that their water will be turned off, and police and fire departments notified of the work to be done.

A way of dealing with possible water hammer must also be considered. If, for instance, a valve is operated too rapidly, or a cleaning pig is suddenly caught and stops moving, a very serious pressure surge can result. The best method of relieving excessive pressure is to install a pressure relief valve on a fire hydrant located near the point where pressure is being applied to the cleaning instrument.

Air Purging

Small mains up to 4 in. (100 mm) in diameter can sometimes be successfully cleaned by air purging. All services must be shut off and a blowoff valve opened at the end of the section. Spurts of air from a large compressor are then forced in with the water at the upstream end, and the air–water mixture will usually remove all but the toughest scale.

Swabbing

Polyurethane foam plugs, called *swabs*, may be forced through a pipe by water pressure to remove slime, soft scale, and loose sediment. Swabs are made of soft foam for use in mains with severe tuberculation or where there may be changes in the cross section of the pipe. They will not significantly remove hardened tuberculation, and they wear out quickly. Hard foam swabs are used in newer mains and where only minor reductions in diameter are expected.

An experienced crew can swab several thousand feet of main a day if the job is properly planned. Swabs should be operated at the speed recommended by the manufacturer. If they travel too fast, they remove less material and wear out faster. A pitot tube should be used to measure the flow on the end of the pipe section and the water entering the section throttled to obtain the optimum speed of the swab.

Preparations include turning off all water services on the pipe section, closing the valves on any connecting mains, and making provisions for inserting and retrieving the swab. The time should be noted when the swab is launched so the approximate exit time can be anticipated. If the travel time becomes too long, the swab is probably stuck, in which case the flow should be reversed to push it back to the starting point. Enough swab runs should be made so that the water runs clear within 1 minute following the run.

Pigging

Pipe-cleaning pigs are stiff, bullet-shaped foam plugs that are forced through a main using water pressure. They are harder, less flexible, and more durable than swabs. Although they are able to remove tougher encrustation, their limited flexibility reduces their ability to change direction and they are more likely to be caught at a constriction in the piping.

Pigs of different grades are classified as bare pigs, cleaning pigs, or scraping pigs. *Bare pigs* are made of high-density foam and are usually sent through a tuberculated main first to determine whether there are any obstructions. *Cleaning pigs* have tough polyurethane ribs applied in a crisscross pattern, and they will remove most types of encrustation and growths from the main. A bare pig is sometimes sent behind an undersized cleaning pig to maintain the water seal. *Scraping pigs* have spirals of silicon carbide or hardened steel wire brushes that will remove hard encrustation and tuberculation. Cleaning and scraping pigs that are somewhat smaller than

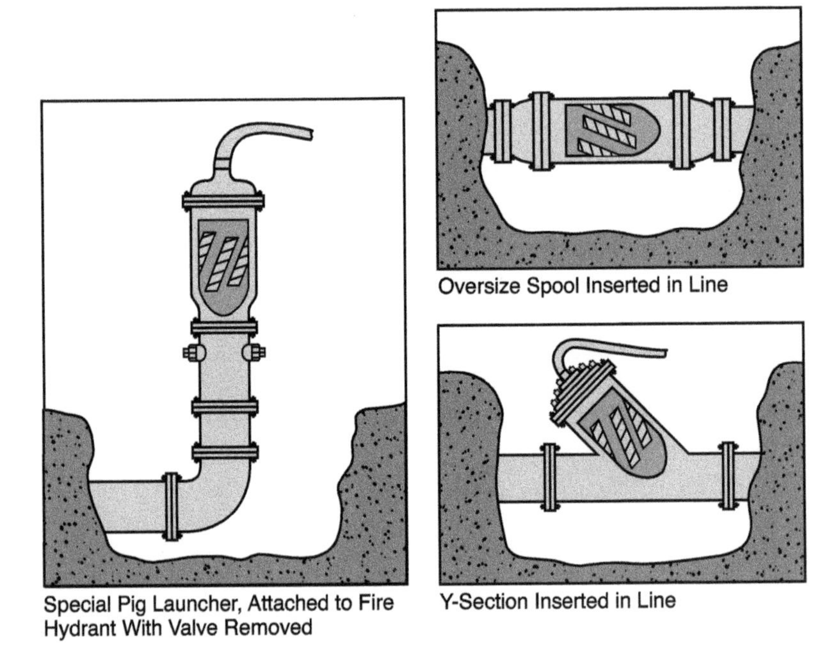

Courtesy of Girard Industries.

Figure 7-22 Launching methods for pipe-cleaning pigs

the pipe diameter are usually sent through first to gradually remove layers of encrustation.

Launching Pigs

There are no easy methods of launching pigs. Three methods of launching pigs are illustrated in Figure 7-22. One method requires that the top of a hydrant be removed and a special adapter connected to the hydrant base. A separate source of water is required to force the pig into the main, and then as it enters the main, it is pushed by system pressure. The other two methods require that a section of main be removed to install the launcher. In some cases, it is also possible to remove the gates from a gate valve and insert the pig into the main through the body of the valve.

If a pig should become lodged in a main, the main must be exposed and opened to remove the pig. To facilitate locating a pig in this situation, a small radio transmitter called a *sonde* can be installed in the pig and the signal can be picked up by a pipe locator. There are numerous pitfalls in the operation of pigging, so it is recommended that a first-time water utility crew work with someone experienced in using the equipment and performing the operation.

Other Cleaning Methods

Badly encrusted mains may also be cleaned using metal scrapers consisting of a series of units having high-carbon spring-steel blades. The sections are free to rotate, and the unit is pushed through the main by water pressure acting against pusher cones.

If deposits in a main are particularly thick or dense, mechanical cleaning with a power drive is usually required. The work is generally done by a special contracting firm using a unit similar to a sewer rodding machine to pull a cutter through the main. Another method for cleaning pipelines is "ice pigging." Ice is manufactured onsite using a low temperature created by dry ice or carbon dioxide refrigerant. The ice slurry is inserted into the pipeline, and water flow moves it through the pipeline. The ice scours the pipe walls and cleans the pipe. Good results have been shown where this method has been used.

Following any of the pipe-cleaning procedures, the main must be thoroughly flushed and then disinfected before it is returned to service.

Lining Water Mains

Cleaning a tuberculated water main can usually restore the interior to close to that of a new pipe, but if the quality of the water remains the same, the tuberculation will recur at a very rapid rate. In some cases, it may be possible to slow the corrosion rate by altering the water quality or adding corrosion inhibitors. In most cases, it is best to coat the pipe interior to prevent further corrosion. Although this process is quite expensive, the cost must be compared to the cost of totally replacing the pipe. In a rural area, it may be less expensive to simply replace the pipe. But in an urban area, with all of the problems of working around trees, driveways, other utilities, and obstructions, it will usually be less expensive to line the existing pipe. One exception would be if the existing pipe is undersized and should be replaced with a larger main.

Cement–Mortar Lining

In the cement–mortar lining process, after the main has been cleaned, a thin layer of cement mortar is applied to the pipe walls. As illustrated in Figure 7-23, small-diameter pipe is lined by remote-controlled equipment pushed through the pipeline. Larger-diameter pipe can be lined with a machine that allows an operator to enter the pipe on a carriage to control the operation.

If there are customer services on the main, the buildings must be served with water from a temporary piping system during the lining operation. This is usually accomplished by installing a temporary 2-in. (50-mm) pipe on the ground surface and making connections to each service line. After the lining operation has been completed, it is necessary to dig up each

Figure 7-23 A cement–mortar lining machine for use in small-diameter pipe

service connection to remove the mortar from the corporation stops and reconnect the water services to the main.

Epoxy Lining

A thin coating of epoxy resin lining is applied to a cleaned iron or steel pipeline to complete this process. Materials approved for this process are certified by ANSI/NSF Standard 61.

Prior to lining, the pipe must be cleaned to remove any material so that the epoxy will adhere to the pipe surface. The pipe surface is then dried by swabbing or other means. Pipe temperature must be above the minimum recommended by the manufacture (usually 38°F [3°C]).

The epoxy resin is typically a two-component system. The resin is applied using special machines that have heated reservoirs where pumps control the resin and hardener as it is applied. The lining machines often are computer controlled with low-thickness alarms. The epoxy is sprayed by a spinning applicator head. The head is pulled to the far end of the pipe segment and then winched back through the pipe at a regulated speed. A typical application results in a coating 1 mm thick.

After resin application, the ends of the pipe are closed and the resin cures for at least 16 hours (manufacturer recommendations may differ). The pipe is then flushed with clean water, disinfected, and returned to service if all tests are satisfactory.

Slip-Lining

In this process, several lengths (40 ft [12 m] each) of flexible thermoplastic pipe are fused end to end and inserted into a metal pipeline. The plastic pipe is usually high-density polyethylene (HDPE) but can be other approved plastics. The outside pipe diameter is typically at least 10 percent smaller than the inside diameter of the metal pipe. This provides sufficient space for an easy insertion.

Slip-lining can be used when existing pipelines are not structurally sound. Also, this lining method is useful when pipes are covered by buildings, railroads, bridges, rivers, or other obstacles.

One problem is that the size of the slip-line pipe is smaller than the original pipe. Flow capacity, therefore, may be somewhat reduced. This is offset by the reduced friction of the plastic pipe compared to the old, unlined pipe. Additionally, the plastic pipe is not susceptible to corrosion. If there are elbows in the pipeline, slip-lining may not be possible.

Close-fit slip-lining techniques are modifications to the standard insertion method. The advantage to these methods are that the inside pipe diameter is not significantly reduced, thus the new pipe retains most of the original flow capacity. Also, it is possible using these methods to select different liner wall thicknesses so that the pipe can be fully structural if needed.

The close-fit techniques deform the pipe for insertion and then use various methods to restore the pipe to its circular form against the original pipe wall. The insertion pipe can be folded, compressed, or formed. The pipe is then restored to the circular form using the material memory, pressure, or heat.

Cured-In-Place Lining

This lining method involves inserting a polymer fiber tube or hose that contains a thermoset resin system into a pipeline to be renovated. The resin is cured creating a pipe within a pipe or a semi-rigid liner adhering to the original pipe wall. Curing can occur under ambient conditions or heat can be applied by steam or water. Several materials can compose the tube including felt, woven polyester, or elastomeric membrane.

Internal Joint Seals

Large pipelines (16 in. [400 mm] and greater) often develop leaks in joints. Internal joint seals eliminate these troublesome leaks in all types of pipes and working pressures of 300 psi (2,068 kPa) or higher. The seals have several lip seals on the outer edges that can completely cover the pipe on either side of the joint. Its low profile and edge design allows water to flow without creating turbulence. Most internal joint seals are manufactured of ethylene propylene diene monomer (EPDM) and are certified to ANSI/NSF Standard 61 for use in potable water. The seals are commercially available in two widths (standard and extra wide).

Installation of the seals involves several steps:

1. Pipeline preparation,
2. Joint filling,
3. Surface preparation of the joint area,
4. Surface lubrication,
5. Positioning,
6. Placing retaining bands,
7. Expanding the seal,
8. Pressure testing, and
9. Completion report.

Trenchless Main Replacement

Earlier in this chapter, pipeline installation methods were discussed in some detail. These methods require an excavation trench to lay the pipe. A trenchless method has been developed where an existing pipeline is being replaced. This method is called pipe bursting. The method breaks and moves the existing pipe so that new pipe can be inserted in the void. The new pipe is most often the same or larger diameter.

The installation system uses a pneumatic, hydraulic, or static bursting unit. This unit forces itself through the existing pipe by busting and fragmenting the pipe and expanding into the surrounding soil. The unit head pulls the new replacement pipe along behind. Polyethylene pipe is often chosen as the replacement pipe material. After the replacement pipe has been completely installed, disinfected, and tested, the service lines can be reconnected and the pipeline returned to full service.

BIBLIOGRAPHY

ANSI/AWWA C600. *AWWA Standard for Installation of Ductile-Iron Water Mains and Their Appurtenances*. Denver, Colo.: American Water Works Association.

ANSI/AWWA C603. *AWWA Standard for Installation of Asbestos–Cement Pressure Pipe*. Denver, Colo.: American Water Works Association.

ANSI/AWWA C605. *AWWA Standard for Underground Installation of Polyvinyl Chloride (PVC) Pressure Pipe and Fittings for Water*. Denver, Colo.: American Water Works Association.

ANSI/NSF Standard 61. *Drinking Water System Components*. Ann Arbor, Mich.: NSF International.

AWWA. 2010. Principles and Practices of Water Supply Operations—*Water Transmission and Distribution*, 4th ed. Denver, Colo.: American Water Works Association.

AWWA. M11—*Steel Pipe—A Guide for Design and Installation*. Denver, Colo.: American Water Works Association.

AWWA. M17—*Installation, Field Testing, and Maintenance of Fire Hydrants*. Denver, Colo.: American Water Works Association.

AWWA. M23—*PVC Pipe—Design and Installation*. Denver, Colo.: American Water Works Association.

AWWA. M28—*Rehabilitation of Water Mains*. Denver, Colo.: American Water Works Association.

Fullan, E. 2012. Handling Asbestos-Cement Pipe: A New Approach. *Opflow*, July.

Mays, L.W. 2000. *Water Distribution Systems Handbook*. Denver Colo.: American Water Works Association; New York: McGraw-Hill.

Chapter 8

Disinfection of Water Mains and Storage Facilities*

Disinfecting water mains and storage facilities is a recommended practice for most water systems. ANSI/AWWA Standards C651, C652, and C655 describe requirements for these procedures. Most water systems and many regulatory agencies have adopted these procedures and have included them in the regulations and policies that apply to water utility operation.

Consult the ANSI/AWWA standards for disinfection water facilities (C651, C652, and C655) for the specific requirements. Also, detailed step-by-step procedures (including many handy reference tables and calculation aids) to accomplish disinfection of pipelines and storage facilities can be found in the AWWA publication *Disinfection of Pipelines and Storage Facilities Field Guide* (Lauer and Sanchez 2006). Information on advancements in field disinfection methods including the use of ozone is also available through AWWA publications and technical conferences.

DISINFECTION OF NEW PIPELINES

Operators should follow these important steps to provide the maximum public health protection when installing new or replacement drinking water mains and other pipelines intended for potable water service. Each of these steps is discussed in some detail in the following paragraphs.

1. Inspection
2. Sanitary Construction Methods
3. Flushing and Cleaning
4. Preventing Backflow During Installation
5. Providing Temporary Service
6. Chlorination (chemical disinfection)

* This chapter is taken largely from *Disinfection of Pipelines and Storage Facilities Field Guide*, by William C. Lauer and Fred J. Sanchez, American Water Works Association, Denver, Colo., 2006.

7. Final Flushing
8. Bacteriological Testing
9. Connection to Distribution System
10. Documentation

Inspection

Examine pipes and appurtenances for any damage, paying close attention to joints. Any damage to the pipe ends or gasket areas may result in leakage. Reject any damaged sections. The pipe sections need to be clean and free of blemishes on the pipe interior. Make sure that coatings and linings are fully cured before completing installation. Before placing pipe in a trench, thoroughly inspect it for contamination and any defects.

Sanitary Construction Methods

Installation methods for new water mains should include the adoption of sanitary procedures that can improve disinfection. These procedures are critical to achieve satisfactory bacteriological quality. Although it may seem time-consuming or even unnecessary to perform each sanitary practice, experience has shown that they actually save time because they reduce the failure rate of bacteriological tests. Failed tests can necessitate re-chlorination, or sometimes require re-installation. These sanitary practices are described in the following paragraphs.

Keep Pipe Clean and Dry

Protect the interior of pipes and pipeline fittings from contamination during installation. Seal any openings with watertight plugs or other suitable plugs at any time the construction is halted and personnel are not present to ensure that animals or debris do not enter the pipe.

As pipe is delivered to the site, and if it is strung along the trench, take precautions to prevent material from entering the pipe. Any delay in the installation of delivered pipe can lead to possible contamination. Consider covering the ends of the pipe if it is not installed on the delivery day (Figure 8-1). Pipe may be purchased and delivered with covered ends; this practice prevents foreign material from entering the pipe.

Joints

Connect joints in the trench before stopping work. Plugs should remain in place to prevent water or mud from entering the pipe.

Packing Materials

Use packing made of molded or tubular rubber rings, rope or treated paper, or other approved (ANSI/NSF Standard 61) materials. Jute or hemp materials are not approved for this purpose.

Courtesy of Denver Water.

Figure 8-1 Protected pipeline along trench

Sealing Materials

Only use sealing materials, gaskets, and lubricants certified according to ANSI/NSF Standard 61. Keep lubricants in clean, closed containers and apply with clean applicator brushes (tools).

Wet-Trench Construction

Sometimes it is not possible to prevent water from entering the pipe as it is installed. Water entering the pipe should contain approximately 25 mg/L chlorine. This can be accomplished by adding chlorine granules or tablets to each length of pipe as it is lowered into the wet trench or by treating the trench water itself.

Flooding by Storm or Accident During Construction

If the new main is flooded during construction, drain the pipe and flush (if possible) with potable water until the flood water is removed. Chlorinate the flooded section so that a 25-mg/L residual remains for at least 24 hours. Drain the chlorination water from the main (follow acceptable disposal procedures for high chlorine water). Upon construction completion, disinfect the main using the continuous-feed or slug method described later in this chapter.

Flushing and Cleaning

This may be the single most important step to ensure that the pipe passes the bacteriological tests needed before the main can be used for potable water service. Extra time and care taken with this process will result in less re-work, re-testing, and a quicker return to service.

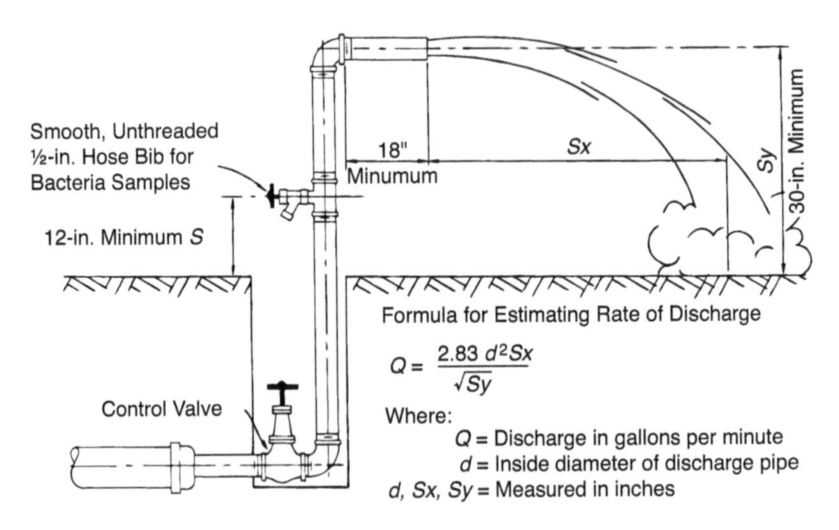

Note: This figure applies to pipes up to and including 8-in. (200-mm) diameter.

Figure 8-2 Trajectory discharge method of flow estimation and illustration of combination blowoff and sampling tap

Fire hydrants are usually used for flushing. In some cases, a blowoff connection is used if one has been installed. A velocity of at least 2.5 ft/sec (0.76 m/sec) is needed to remove dirt or other debris from the pipe. A low-velocity flush (1 ft/sec [0.3 m/sec]) is adequate to "rinse" a clean pipe or to refresh the water within a main. Regardless of the velocity, flush long enough to result in two or three complete changes of the water within the pipeline. The flow may be measured by a flowmeter or estimated using the trajectory discharge method (Figure 8-2).

Large-diameter pipelines may require excessive water to flush adequately. In these cases, it may be necessary to employ poly pigs or swabs (power washers and hydraulic jet sprayers are sometimes used) to clean the pipe without using great volumes of water (Figure 8-3).

If dirt enters the pipe, remove it by flushing with potable water, then swab the interior surface of the pipe with 1 to 5 percent chlorine solution (sodium or calcium hypochlorite). If there is material adhering to the interior of the pipe that cannot be flushed out, then a more aggressive cleaning may be necessary. Hydraulically propelled swabs, poly pigs, or other suitable device including mechanical scrapers may be used to remove the attached material. Follow this cleaning with an application of chlorine solution as described previously.

A more aggressive method of cleaning larger-diameter pipelines is to use power washers or hydraulic spray jets. The equipment, used commonly for cleaning sewer collection system piping, is suitable for this purpose (Figure 8-4). These devices may be self-propelled and can do a superior job

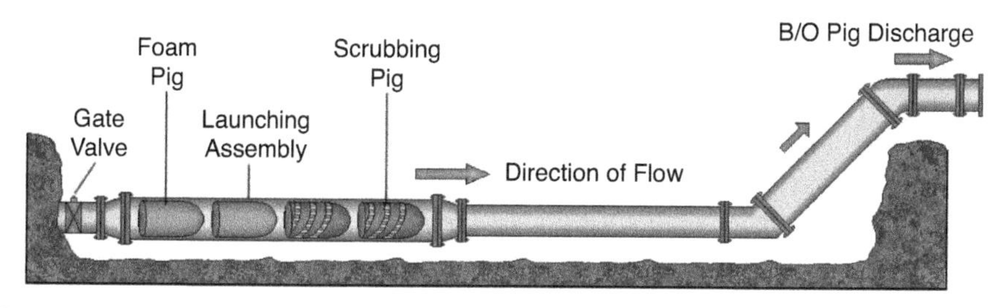

Courtesy of Nicole Peschel.

Figure 8-3 Pipeline pigging illustration

Courtesy of Denver Water.

Figure 8-4 Hydraulic jet cleaning

cleaning the pipe interior. Some high-pressure power washers can clean an entire pipe section from one end.

Preventing Backflow During Installation

High chlorine levels used for disinfection of new pipelines or other nonpotable conditions may contaminate the potable distribution system. Therefore, prevent backflow from newly installed (but untested) pipelines to the active distribution system. This can be accomplished with a physical separation (Figure 8-5) between the systems or by the use of an approved backflow prevention device. Hydrostatic pressure testing of new pipelines is another period when backflow is a possibility.

Providing Temporary Service

When new pipelines are installed as replacements, interruptions in water service may occur. It is common practice in the United States to install temporary piping to provide continuous service when the supply is disrupted for more than several hours. Temporary supply lines are typically installed aboveground, and residences and businesses are connected by flexible hose to outside spigots or in meter pits.

The use of temporary water lines can significantly increase the project's time and cost. One of the main reasons for this is that the temporary network must be disinfected, and the water must pass bacteriological testing

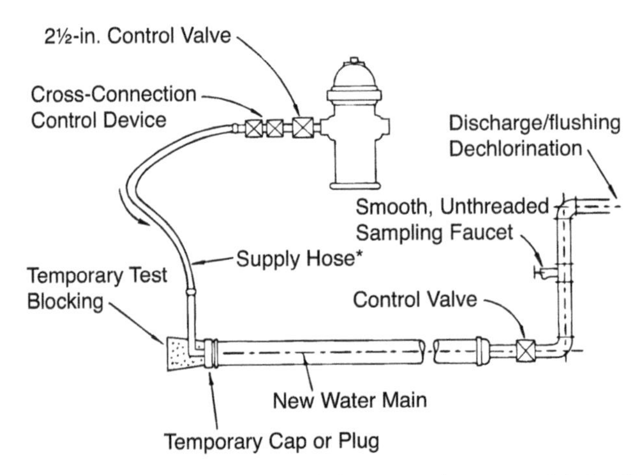

*Clean potable-water hose only. This hose must be removed during the hydrostatic pressure test.

Note: Applies to pipes with diameters 4 in. (100 mm) through 12-in. (300 mm). Larger sizes must be handled on a case-by-case basis.

Figure 8-5 Suggested temporary flushing/testing connection

before customers can be connected. Additionally, the public can view the temporary lines as an unsightly nuisance that may impede vehicular traffic and cause hazards for pedestrians.

Many other industrialized countries do not provide temporary service lines. Instead they select alternatives such as supplying bottled water or locating water trucks in the area. Some suppliers consider lengthy interruptions in water service as acceptable. A decision to install temporary service lines or other alternative water sources requires careful consideration of several factors including site conditions, fire flow requirements, expected duration of the service interruption, cold temperatures, backflow prevention devices, internal plumbing, number of customers affected, and the type of customer (e.g., hospitals, schools, or individual residences).

Temporary pipelines are made of a variety of materials including steel, polyvinyl chloride (PVC), and high-density polyethylene (HDPE). Avoid using fire hoses or rubber hoses. Components must be certified to NSF Standard 61 and approved for potable water service. Generally, installation follows the same procedures as for permanent pipelines.

Disinfection procedures are the same as for permanent pipelines (ANSI/AWWA Standard C651, *Disinfecting Water Mains*). Temporary pipelines must be flushed (usually at 2.5 ft/sec [0.76 m/sec]) to remove any dirt or debris. Chlorination is conducted according to ANSI/AWWA Standard C651 and as described in the section that follows. Bacteriological testing is then conducted before the temporary service is placed into use.

Table 8-1 Pipeline disinfection methods summary

Method	Chlorine Dosage	Holding Time	Minimum Chlorine Residual	Comments
Tablet or granules	25 mg/L	24 h, 48 h if temperature <5°C	Detectable	Add enough to fill entire pipeline segment
Continuous feed	25 mg/L	24 h	10 mg/L	Add enough to fill entire pipeline segment
Slug	100 mg/L	3 h	50 mg/L at any point	Fill % of pipeline with slug and move the slug slowly to give 3-h contact

Chlorination

Three methods of chlorination for pipeline disinfection are tablet or granules, continuous feed, and slug. Factors to consider when selecting one of these methods are the length and diameter of the main, the type of joints, availability of materials, equipment required, training of personnel, and safety concerns.

A summary of the pipeline disinfection methods is included in Table 8-1. A more detailed description of the procedures for each method is discussed later in this chapter.

Follow all safety precautions when performing these procedures (see Chlorine Institute references in the bibliography at the end of this chapter). Regulatory requirements may apply to ensure that these procedures are conducted in a safe manner. Consult appropriate regulatory agencies and requirements before beginning any chlorination procedure.

Final Flushing

Flush heavily chlorinated water from the main following the required holding period (usually at least 24 hours). Make sure that all main fittings, valves, and branches are flushed. Measure chlorine residual of the flush water exiting the chlorinated main until the residual is no greater than the feedwater.

Flush-water containing high chlorine concentrations may require treatment to avoid adverse environmental impact. Consult appropriate regulatory agencies regarding requirements for the disposal of heavily chlorinated water. It may be necessary to dechlorinate the water prior to disposal. Some of these procedures may not be approved by every regulatory agency. Follow regulatory agency directives regarding the approved treatment procedures for your area (dechlorination is discussed later in this chapter).

Bacteriological Testing

Perform bacteriological tests on water in the new pipeline after flushing but before it is connected to the distribution system. The tests must satisfy all applicable regulatory requirements. A certified laboratory must perform the analyses. The results must be negative (no coliform present) before connecting the pipeline and releasing it for use by customers. The testing protocol is listed in ANSI/AWWA Standard C651, *Disinfecting Water Mains*.

Connection to Distribution System

When all bacteriological tests are satisfactory, the new main may be connected to the distribution system and placed in service. If the final connection is one pipe length (18 ft [5.5 m]) or less, the new pipe, fittings, and valve(s) required for the final connection may be spray-disinfected or swabbed with a 1–5 percent chlorine solution just prior to being installed. If the final connection is greater than one pipe length (18 ft [5.5 m]), disinfect the connection pipe and have bacteriological samples tested as previously described. Connect the pipe to the distribution system after obtaining satisfactory test results. Cover the ends of the pre-disinfected connection pipe during the testing and before the connection is complete.

Documentation

Keep accurate records to document the installation conditions and test results (including water quality testing). The records should identify persons that conducted tests and that released the pipeline for service. Complete the as-built plans that identify the types and locations of valve boxes, hydrants, and other appurtenances. Record the valve opening direction and any flow test results for hydrants.

PIPELINE DISINFECTION METHODS

Tablet (or Granules) Method

Calcium hypochlorite tablets (or granules) are placed in the water main as it is being installed. The main is then filled with potable water. This procedure cannot be used on solvent-welded plastic or screwed-joint steel pipe. Fire or explosion may result from the reaction of calcium hypochlorite and joint compounds. This method also cannot be used if the main must be flushed before it is disinfected, since this process would remove the calcium hypochlorite from the main.

If this method is used, great care must be taken to ensure that the pipeline is clean, dry, and free of dirt, debris, and other contamination. The pipeline must be filled slowly to ensure that the granules are not moved to one end of the pipeline and to allow time for the tablets to dissolve.

Chlorination Steps and Precautions

1. Place calcium hypochlorite granules or tablets at the upstream end of the first section of pipe, at the end of each branch main, and at 500-ft (152-m) intervals.

2. Use an ANSI/NSF Standard 61 certified adhesive to attach tablets inside and at the top of each section of the main.
3. Place an equal number of tablets at each end of a given pipe length. Place one tablet in each hydrant, hydrant branch, and other appurtenance.
4. Gently fill the main with water. Ensure that the filling rate does not result in a velocity greater than 1 ft/sec (0.3 m/sec). Eliminate air pockets.
5. Contain the water in the pipe for at least 24 hours. Hold the water for at least 48 hours if the water temperature is less that 41°F (5°C). Optionally, water may be supplied from a temporary connection with an appropriate backflow prevention device. Bleeding a small amount of water from the pipe occasionally during the holding period will ensure that chlorinated water is well distributed throughout the pipe.
6. A detectable chlorine residual must be present at each sampling point (see "Bacteriological Testing" section earlier in this chapter) after the holding period.

Continuous Feed Method

In the continuous feed method, water from the distribution system flows slowly into the new pipe section while a concentrated chlorine solution is added. Chlorine solution is injected using a solution-feed chlorinator or pumped from a concentrated hypochlorite solution (Figure 8-6). This method is acceptable for general application.

Chlorination Steps and Precautions

1. Flush the new main at a velocity of at least 2.5 ft/sec (0.8 m/sec) unless site-specific conditions prevent this practice. Large mains (greater than 24 in. [600 mm]) may be broom-swept.
2. Hydrostatic testing must be completed prior to disinfection.
3. Place (optional) calcium hypochlorite granules or tablets in the pipe sections as previously described in the tablet method. This practice provides a strong chlorine concentration during the initial filling of the pipeline and is used to disinfect annular spaces at pipe joints.
4. Supply a constant flow rate from a temporary, backflow-protected source or other approved supply source. Determine the flow using a Pitot gauge in the discharge (or other type of meter), measuring the time to fill a precalibrated container, or by using the trajectory discharge method (shown in Figure 8-2).
5. Inject the concentrated chlorine solution no more than 10 ft (3 m) from the beginning of the new main. The dosage should be at least 25 mg/L free chlorine. Prepare a 1 percent chlorine solution by mixing 1 lb (454 g) of calcium hypochlorite or 0.8 gal (3 L) of 10 percent strength sodium hypochlorite solution in 8 gal (30.3 L) of water. Adjust the chlorination continuous feed rate to match the pipeline fill rate to achieve 25 mg/L chlorine dosage.

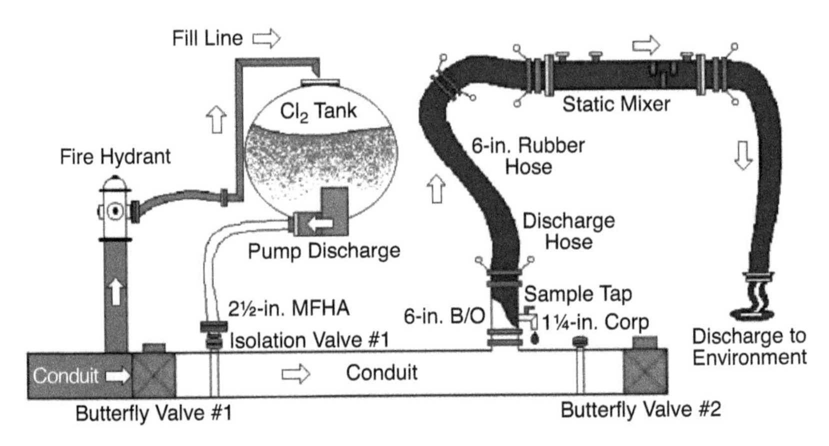

Courtesy of Nicole Peschel.

Figure 8-6 Chlorination by continuous-feed method

6. Measure the chlorine residual regularly to verify the concentration (high-range chlorine test kit or applicable method in current edition of *Standard Methods for the Examination of Water and Wastewater* or AWWA Manual M12, *Simplified Procedures for Water Examination*). Take samples for chlorine testing near the point of injection and at the end of the pipeline.
7. Retain the chlorinated water in the main for at least 24 hours. Operate valves and hydrants connected to the treated section to ensure that they are also disinfected.
8. At the end of the 24-hour holding period, the water within the treated main must have a chlorine residual of at least 10 mg/L free chlorine.
9. A solution-feed vacuum-operated chlorinator and booster pump may be used for the chlorine application. Gasoline or electrically powered chemical-feed pumps may be used to inject hypochlorite solutions.
10. Make sure that chlorine solution feed lines are compatible with these chemicals and that the feed system can withstand the feed pressure. The feed system must be free of leaks.
11. Take precautions to control any water discharges (dechlorination may be necessary).

Slug Method

The slug method requires that a section of new main be highly chlorinated, then valves are operated and water is slowly removed from the end of the main, causing the "slug" of highly chlorinated water to move along the newly installed pipeline (Figure 8-7). As this slug passes tees, crosses, and hydrants, the adjacent valves are operated to ensure the disinfection of all appurtenances and branches.

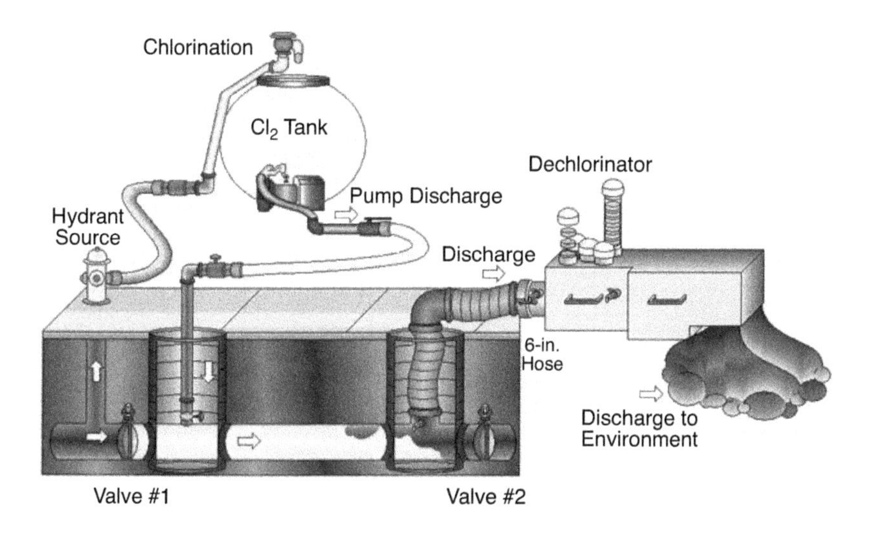

Courtesy of Nicole Peschel.

Figure 8-7 Slug method pipeline disinfection example

Chlorination Steps and Precautions

1. Follow steps 1–4 in the "Continuous Feed Method" section.
2. Inject the concentrated chlorine solution no more than 10 ft (3 m) from the beginning of the new main. The dosage should be at least 100 mg/L free chlorine.
3. Measure the chlorine residual (near the injection point) regularly to verify the concentration (high-range chlorine test kit or applicable method in current edition of *Standard Methods for the Examination of Water and Wastewater* or AWWA Manual M12, *Simplified Procedures for Water Examination*).
4. Apply chlorine continuously to develop a solid column, or slug, of chlorinated water that will expose all interior surfaces to a concentration of approximately 100 mg/L for at least 3 hours. Measure the chlorine residual at the end of the slug to verify the concentration.
5. Measure the free chlorine residual as the slug moves through the new pipeline. If the residual drops below 50 mg/L at any point, stop the procedure and add chlorine to the slug area to restore the free chlorine concentration to at least 100 mg/L. Choose sample locations that will verify the minimum contact time of 3 hours.
6. Operate valves and hydrants as the slug passes them to disinfect appurtenances and pipe branches.
7. High calcium hypochlorite concentrations (greater than 100 mg/L) may raise the pH of the water to unacceptable levels. Measure the pH of the chlorinated water when using this method to ensure that

optimum pH is maintained. Remember that if the pH is greater than 10, virtually all of the chlorine residual is present as hypochlorite ion, which is a less effective disinfectant than hypochlorous acid.

BASIC PIPELINE DISINFECTION CALCULATIONS

Flushing Rate

What is the flush flow rate to achieve at 2.5 ft/sec (0.76 m/sec) velocity in a pipeline?

$$\text{flow rate (gpm)} = 2.5 \text{ ft/sec} \times 60 \text{ sec/min} \times \text{volume gal}$$

Example 1: What is the flushing flow rate in gpm needed to achieve 2.5 ft/sec velocity in a-6 in. pipeline that is 500 ft long?

The volume of pipeline per foot,

$$\text{volume (gal/ft)} = 3/12 \times 3/12 \times 3.14 \times 1 \text{ ft} \times 7.48 \text{ gal/ft}^3 = 1.47 \text{ gal/ft}$$

So, to move 2.5 ft/sec,

$$\text{gpm} = 2.5 \times 60 \times 1.47 = 220 \text{ gpm}$$

How long do you need to flush at this rate to change the water out at least three times?

$$\text{time (min)} = 3 \times \text{volume of pipeline (gal)/flush rate (gpm)}$$

$$\text{min} = 3 \times 1.47 \text{ gal/ft} \times 500 \text{ ft/220 gpm} = 10 \text{ min}$$

Volume of Pipeline

$$\text{volume (gal)} = \pi r^2 \times L \times 7.48 \text{ gal/ft}^3$$

R (pipeline radius must be in ft)

Example 2: What is the volume (gal) of an 8-in. pipeline that is 200 ft long?

Convert r to ft, r = 8 in./2 = 4 in. or 4/12 ft = 0.33 ft

$$\text{volume (gal)} = 3.14 \times 0.33 \times 0.33 \times 200 \times 7.48 = 517 \text{ gal}$$

Tablet Method

Table 8-2 shows the number of 5-g calcium hypochlorite tablets needed to produce a chlorine residual (if there is no demand) of 25 mg/L in a 20-ft (6-m) pipe length.

Example 3: How many calcium hypochlorite 5-g tables are needed for 100 ft of 8-in. pipeline to achieve a 25-mg/L dosage?

Table 8-2 Tablets for 25 mg/L chlorine dosage for each 20-ft (6-m) pipe length

Pipe Diameter		
inches	**millimeters**	**Number of Tablets**
4	100	1
6	150	1
8	200	2
10	250	3
12	300	4
16	400	7

From Table 8-2, it can be seen that two tablets are needed for each 20-ft pipe length. There are 100/20 pipe lengths = 5. So, 5 × 2 = 10 tablets.

Chlorination Feed Rate—Continuous Feed Method or to Form a Slug

$$\text{chlorine solution feed rate (gal/min)} \times \text{solution strength (mg/L)} = \text{water fill rate (gpm)} \times \text{chlorine dosage in pipeline (mg/L)}$$

Example 4: A 1 percent chlorine solution is used to disinfect 1,000 ft of 6-in. pipe. Water is being pumped into the pipeline at 20 gpm. A dosage of 25 mg/L is needed. What is the chlorine solution feed rate (gal/h) needed for this procedure? How many gallons of 1 percent chlorine solution are needed to complete the job?

$$1\% \text{ chlorine solution} = 10{,}000 \text{ mg/L chlorine}$$

$$\text{chlorine solution feed rate (gpm)} = 25 \text{ mg/L} \times 20 \text{ gpm} \div 10{,}000 \text{ mg/L} = 0.0025 \times 20 \text{ gpm} = 0.05 \text{ gpm}$$

$$\text{this is } 0.05 \text{ gpm} \times 60 \text{ min/h} = 3 \text{ gal/h}$$

$$\text{volume of pipeline} = \pi \times 3/12 \times 3/12 \times 1{,}000 \times 7.48 \text{ gal/ft}^3 = 1{,}468 \text{ gal}$$

$$\text{time to fill} = 1{,}468 \text{ gal}/20 \text{ gpm} = 73.4 \text{ min}$$

$$\text{gallons of chlorine } 1\% \text{ needed} = \text{solution feed rate} \times \text{time} = 0.05 \text{ gpm} \times 73.4 \text{ min} = 3.67 \text{ gal}$$

DISINFECTION FOR PIPELINE REPAIRS

Broken pipelines, severe service line leaks, and other damaged distribution system components require emergency repairs. An action plan is needed to deal efficiently with these situations. The plan should include coordination with police, fire, and street or highway department personnel and describe personnel training requirements, discuss the need for accurate records, and ensure the availability of critical repair parts. Careful planning can lead to the quick restoration of water service.

When making emergency repairs, follow these steps to reduce the chance of microbial contamination and to safely return the pipeline to service:

1. Ensure that repair parts are available and in good sanitary condition before they arrive at the site.
2. Employ sanitary repair practices and check the trenches for potential contamination.
3. Use prudent chlorination procedures.
4. Take samples for bacteriological testing.
5. Document the event and the procedures used to repair and return the pipeline to service.

Condition and Storage of Repair Parts

Store a carefully selected inventory of spare (repair) parts so that they can be delivered quickly to a site of an emergency main break. Either a parts supplier or the utility property is a suitable site for locating a parts inventory as long as they are accessible when they are needed.

Inspect all pipes, fittings, valves, clamps, and other appurtenances upon delivery. Provide security to protect these materials while they are in storage. Store pipe and other parts in a manner to avoid casual water and dirt from entering. Remove any visible foreign matter from the interior of stored repair parts. Use and maintain watertight end caps or other protective coverings to prevent possible contamination.

Sanitary Repair Practices

Take care to prevent water, dirt, and other material from entering the damaged pipeline. Divert surface water from the construction site, using barriers if necessary. Dewater the excavation so that water is below the pipe invert. Before beginning a repair, clean the interior of the pipe where it has contacted the soil or backfill material. Remove soil beyond reach by low-velocity flushing in both directions. If possible, maintain flow or positive pressure in the leaking pipeline to prevent backflow of contaminated water. This is particularly important until the leak is exposed and secured.

Repair crews need to be aware that their actions could contaminate the water supply during a pipeline repair. Crews must ensure the cleanliness of all cables, pipes, and hoses drawn through the inside of the pipelines or appurtenances. Disinfection (using chlorine dip or swab solutions) of hand tools (such as saws) used in the repair is a prudent step. Equipment operators should also use caution to avoid introducing soil or other material into the repaired pipeline.

Chlorination Procedures

When an existing main is opened, due to a break or leak, the excavation is usually wet and may be contaminated. ANSI/AWWA Standard C651, *Disinfecting Water Mains*, requires the application of liberal amounts of hypochlorite to the open trench to reduce potential pollution. Calcium hypochlorite tablets are used for this application because they dissolve slowly to continuously release chlorine as the repair is completed.

Apply chlorine solution (1 percent hypochlorite solution is common) to the interior of pipe and fittings used for pipeline repair. The solution can be swabbed with an applicator or sprayed on the interior surfaces of the pipeline and repair parts.

Disinfect the section of repaired main whenever possible. The slug method is usually used for this purpose since it takes less time to accomplish. The repaired section is isolated, all service connections shut off, and then the main is flushed before chlorinating. The disinfection chlorine dose may be as much as 300 mg/L and the contact time as short as 15 minutes.

Calcium hypochlorite granules or tablets require some time to dissolve. Therefore, this chemical may not be the best choice if the main must be returned to service quickly.

The repaired main is then flushed (dechlorinate if necessary) until the chlorine residual is no higher than the distribution system feedwater (distribution systems that use chloramines require testing for free and combined chlorine).

Thoroughly flush the repaired main regardless of whether it is chlorinated or not. Flushing is the best way to remove any foreign material that may have entered the main during the repair. Flush until the water is clear and the chlorine residual is equivalent to the distribution system feedwater.

Tapping sleeves are used to install a tap in the main without the need to shut down the main. When using these devices, clean the exterior of the main in the tap location. Also, apply calcium hypochlorite to the interior of the sleeve. After the tap is made, it is impossible to disinfect the annulus without a pipeline shutdown. The space between the sleeve and the pipe is only about ½ inch (13 mm), so a very small amount of calcium hypochlorite applied to the interior of the sleeve is all that is necessary to result in a high chlorine dosage.

Bacteriological Testing

After the main is placed back into service (some regulatory agencies may require satisfactory bacteriological test results *before* the main is placed back into service), take samples from the repaired main and test for total coliform bacteria. In most cases, samples should be taken from both upstream and downstream of the repair. If a positive total coliform test is obtained, take corrective action. Actions may include additional flushing or chlorination of the repaired section. ANSI/AWWA Standard C651,

Disinfecting Water Mains, requires continued sampling daily until two consecutive negative results are obtained.

Documentation

Keep accurate records for future reference. Record the date, location, type of break, materials used for the repair, personnel involved in the repair, procedures used, and test results. These records are useful if problems develop or to determine a pattern of pipeline breaks in a specific area.

DISINFECTION OF STORAGE FACILITIES

Disinfection of newly constructed potable water storage facilities (e.g., tanks, standpipes, underground basins, covered reservoirs) is required. Additionally, all storage facilities taken out of service for inspection or maintenance must be disinfected before they are returned to service. The procedures discussed in this section follow the requirements of ANSI/AWWA Standard C652, *Disinfection of Water-Storage Facilities*. Consult the standard for specific language and other requirements. Contact applicable regulatory agencies to determine if there are local requirements that apply to the disinfection of potable water storage facilities.

Follow these steps to properly disinfect potable water storage facilities.

1. Use sanitary construction methods.
2. Employ approved chlorination (chemical disinfection) procedures.
3. Remove and properly dispose of highly chlorinated water (full storage facility chlorination method).
4. Perform bacteriological testing.
5. Document the release to potable service.

Sanitary Construction or Maintenance Practices

Remove all materials not part of the structure or operating equipment. This may include mold on walls or ceilings, and algae attached to surfaces. Clean the interior walls, floors, and other items that are part of the structure with high-pressure water, sweeping, scrubbing, or other effective methods. Material removed by cleaning procedures must then be completely removed. Check all penetration screens to ensure that openings are protected. Check the ceiling of the storage facility for cracks (look for daylight). Finally, any materials that are placed in the facility following the cleaning must be clean and sanitary.

Chlorination

Three techniques are used for the chlorination of water storage facilities: (1) the full storage facility chlorination method, (2) the surface application method, and (3) the chlorinate-and-fill method. These techniques are discussed in detail later in this chapter.

Follow all safety precautions when performing these procedures (see Chlorine Institute references in the bibliography at the end of this chapter). Working on or in a storage facility can be hazardous. Regulatory requirements may apply to ensure that these procedures are conducted in a safe manner. Consult appropriate regulatory agencies and requirements before beginning any chlorination procedure. Some precautions include the following:

- Personal protective equipment such as goggles, gloves, and breathing apparatus
- Special fans or other ventilation equipment for use within storage facilities
- Adequate lighting
- Electrical equipment that is specifically designed for safe use in a wet environment
- Protective clothing
- Safety cages or cables to prevent accidents on ladders or entry steps
- Confined-space entry regulatory requirements

Removal of Highly Chlorinated Water

Discharge regulations (applicable locally) may dictate the methods of disposal for highly chlorinated water. This is particularly applicable to the full storage facility chlorination method. When using this technique, the entire volume of the storage facility contains water with a chlorine residual greater than 10 mg/L. Environmental protection regulations may require special provisions or permits prior to disposal of this water. Dechlorination may be required (methods are described in this chapter). Contact proper authorities prior to disposal to determine all applicable regulations and requirements. Notify the applicable wastewater treatment facility when discharging to a sanitary sewer.

Bacteriological Testing

Sample water from the storage facility following completion of the chlorination method and test for total coliform bacteria according to the latest edition of *Standard Methods for the Examination of Water and Wastewater* (chlorine residual, free and combined, is tested to verify conformance with the chlorination requirements and to restore the water to potable quality). If the test is negative for total coliform, the facility can be released and placed into service. If the test is positive for total coliform, take repeat samples until two consecutive samples (taken 24 hours apart) are negative. The facility may be re-chlorinated according to the procedures above and then tested again for total coliform as previously described.

Take samples from a sample tap on the outlet piping or from one that is directly connected to the facility. Ensure that the water drawn from the tap represents the water within the facility. Additional samples may be advisable to ensure uniform disinfection results. Storage facility hatches are suitable for this purpose. It is recommended to sample the facility fill-water to make sure it does not contain total coliform bacteria.

Additional Test for Odor

Test the water in the facility for odor. Offensive odors may indicate the presence of trace substances that may cause customer complaints. If an offensive odor is detected, take steps to eliminate it before the facility is placed into service. This may include further cleaning or allowing additional time for odors to dissipate before returning the tank to service.

Release to Potable Service

Document the disinfection procedure and submit required records to applicable regulatory agencies. Connect the facility to the distribution system, notifying system operators that the facility is now active. If disinfection is conducted by contractors, require full documentation of all aspects of the disinfection process and certified test results.

FULL STORAGE FACILITY CHLORINATION METHOD

This method generally involves chlorinating the entire volume of the storage facility so that the water within the facility has a residual chlorine concentration of at least 10 mg/L at the end of the required retention period. Liquid chlorine, sodium hypochlorite solution, or calcium hypochlorite tablets (or granules) are suitable for this procedure.

The chlorination process may involve a continuous fill of the facility with chlorinated water with a concentration of at least 10 mg/L. Alternatively, the chlorination can be performed in two steps, adding the entire amount of chlorine to disinfect the facility to a small amount of water covering the floor and then filling (after a holding period) the facility with potable water to achieve 10 mg/L throughout.

Liquid Chlorine

When using a liquid chlorine solution, feed liquid chlorine into the water, filling the facility so that the concentration is constant. Portable equipment that should be available includes a liquid chlorine cylinder, a gas-flow chlorinator, a chlorine ejector, safety equipment, and the appropriate connections to inject the chlorine solution into the fill stream. Trained operators are needed for this operation. Safety regulations may require personnel certification for emergency response and first aid training.

Sodium Hypochlorite

When using a sodium hypochlorite solution, add it to the feedwater with a chemical feed pump or by pouring it into the fill-stream. In each case, add the solution to ensure adequate mixing and to result in a constant chlorine concentration. Convenient locations to pour sodium hypochlorite are the cleanout or inspection manhole in the lower level of the facility, the riser pipe of an elevated tank, or a roof manhole. Pour in the hypochlorite solution when the water is 1 to 3 ft (0.3 to 0.9 m) deep if possible.

Calcium Hypochlorite

When using calcium hypochlorite, granules or tablets broken into small pieces (¼ in. [6 mm] or smaller), can be added to the facility as described for the sodium hypochlorite solution in the previous paragraph. Take care to place the granules or tablets on a dry surface unless safety precautions are taken to ensure adequate ventilation or provide personal protective equipment.

Chlorination Steps and Precautions

1. Retain the chlorinated water in the facility for at least 6 hours when continuous feed is used from the gas-chlorinator or chemical feed pump. If a high chlorine concentration is added by pouring sodium hypochlorite or by placing calcium hypochlorite granules or tablets and filling the storage facility, then the water should be retained for at least 24 hours. If a two-step process is used, the tablets or granules are allowed to dissolve in a small amount of water (a few feet) before the facility is filled to capacity. The chemical may take several hours to dissolve.
2. Measure the chlorine residual of the water following the required retention period. The free chlorine residual must be at least 10 mg/L. If the chlorine residual is below 10 mg/L, add more chlorine and retain for another period as described in step 1. Continue this procedure until a free chlorine residual of at least 10 mg/L is obtained following the required retention period.
3. Reduce the chlorine residual of the water within the storage facility by draining (dechlorinate if necessary) and refilling with potable water. Another method of achieving this result is to hold the water until the chlorine residual is reduced and blend the remaining water with potable water.

SURFACE APPLICATION METHOD

In this method, a 200-mg/L chlorine solution is applied to all surfaces within the storage facility that will be in contact with the water when the facility is completely full.

Chlorination Steps and Precautions

1. Apply the chlorine solution to all exposed surfaces with suitable brushes or spray equipment. Portable spray equipment is adequate for small tanks.
2. Add chlorine to drains or drain piping so that the chlorine residual of the pipe when filled with water shall be at least 10 mg/L. Overflow piping does not need to be chlorinated. Make sure the drain line value is closed to avoid the need for dechlorination.
3. Allow at least 30 minutes for contact of the strong chlorine solution with the treated surfaces.
4. Introduce potable water and remove the water from treated drains and drain piping.
5. Fill the facility to overflow level with potable water and return to service.

CHLORINATE-AND-FILL METHOD

With this method, chlorine and water are added so that about 5 percent of the storage volume of the facility is filled and this water has a chlorine residual of approximately 50 mg/L. After a retention period of at least 6 hours, the facility is filled to overflow with potable water. The resultant chlorine residual is suitable for potable use.

Chlorination Steps and Precautions

1. Add chlorine and water to fill about 5 percent of the storage volume and result in a chlorine residual of approximately 50 mg/L. Liquid chlorine and sodium hypochlorite are usually used for this purpose. Calcium hypochlorite granules or tablets can also be used, provided that adequate mixing is available to ensure that the chemical fully dissolves and is distributed evenly throughout the facility. Follow procedures described for the full facility chlorination method.
2. Retain this water within the facility for at least 6 hours.
3. Fill the remainder of the storage facility with potable water and hold for at least 24 hours.
4. Remove highly chlorinated water from drains and drain piping.
5. Test the free chlorine residual of the water in the full facility to ensure it is at least 2 mg/L. Add more chlorine if the residual is below this value and re-test after another 24-hour retention period. Repeat this process until satisfactory results are obtained.

BASIC STORAGE FACILITY DISINFECTION CALCULATIONS

Storage Facility Volume

$$\text{volume of a cylinder (gal)} = \pi r^2 H \times 7.48 \text{ gal/ft}^3$$

$$\text{volume of a rectangle or square (gal)} = LWH \times 7.48 \text{ gal/ft}^3$$

Example 5: What is the maximum volume of water that can be stored in a cylindrical tank that is 10 ft in diameter and 20 ft high with an overflow at 18 ft from the base?

$$\text{volume (gal)} = 3.14 \times 5 \times 5 \times 18 \times 7.48 = 10{,}569 \text{ gal}$$

Storage Facility Walls Surface Area

$$\text{wall surface area for a cylindrical tank (ft}^2) = \pi dH$$

Example 6: What is the surface area for a cylindrical tank 10 ft in diameter and 20 ft high?

$$\text{area (ft}^2) = 3.14 \times 10 \times 20 = 648 \text{ ft}^2$$

Chlorination Spray Solution

To make a 200-mg/L chlorine spray solution, dilute bleach with a measured volume of water. Usually the strength of liquid sodium hypochlorite bleach is given in percent.

$$1\% = 10{,}000 \text{ mg/L}$$

Example 7: How many liters of 5 percent bleach are needed to make 150 gallons of 200 mg/L chlorine spray solution?

$$5\% \text{ chlorine} = 50{,}000 \text{ mg/L, so dilute by } 200/50{,}000 = 0.004$$

Therefore, you need

$$150 \text{ gal} \times 0.004 = 0.6 \text{ gal or } 0.6 \times 3.785\text{L/gal} = 2.27 \text{ L}$$

Chlorine Amount for Full Facility Method

This method requires a 10-mg/L chlorine dosage for full volume of the facility. The following equations are used to calculate the amount of chlorination chemical (pure liquid chlorine, liquid sodium hypochlorite bleach, or calcium hypochlorite) needed for this procedure:

$$\text{pure chlorine needed in pounds (lb)} = \text{volume (gal)} \times 10 \text{ ppm} \times 8.34 \text{ lb}/1{,}000{,}000 \text{ gal}$$

$$\text{pure chlorine needed in gallons (gal)} = \text{volume (gal)} \times 10 \text{ ppm} \times 1/1{,}000{,}000 \text{ gal}$$

$$\text{amount of chlorination chemical in lb} = \text{lb pure chlorine} \div \%\ \text{purity}/100$$

$$\text{amount of chlorination chemical in gal} = \text{gal pure chlorine} \div \text{purity}/100$$

Example 8: How many gallons of 10 percent sodium hypochlorite bleach are needed to disinfect (full facility method) a 20-ft diameter cylindrical tank that has a maximum water depth of 30 ft?

$$\text{volume (gal)} = \pi r^2 H \times 7.48 \text{ gal/ft}^3 = 3.14 \times 10 \times 10 \times 30 \times 7.48 = 70{,}461.6 \text{ gal}$$

$$\text{amount of 10\% sodium hypochlorite (gal)} = V \times 10 \text{ ppm} \times 1/1{,}000{,}000 \div 10/100 = 7.1 \text{ gal}$$

Chlorine Amount for Chlorinate-and-Fill Method

This method requires the operator to chlorinate 5 percent of tank volume to 50 mg/L, then fill and hold for at least 24 hours. The following equations can be used to calculate the amount of chlorination chemical needed for this procedure.

$$\text{5\% volume (gal)} = \text{tank volume (gal)} \times 0.05$$

$$\text{pure chlorine (gal) for 50 mg/L dosage} = \text{5\% volume} \times 50/1{,}000{,}000 \text{ ppm}$$

$$\text{pure chlorine (lb) for 50 mg/L dosage} = \text{5\% volume} \times 50/1{,}000{,}000 \text{ ppm} \times 8.34 \text{ lb/gal}$$

$$\text{amount of chlorine chemical needed (gal)} = \text{pure chlorine (gal)} \div \%\ \text{strength}/100$$

Example 9: How many gallons of 10 percent sodium hypochlorite bleach is needed for the chlorinate-and-fill method for a cylindrical tank 20 ft in diameter and 30 ft of water depth maximum?

$$\text{volume of tank (gal)} = \pi r^2 H \times 7.48 \text{ gal/ft}^3 = 3.14 \times 10 \times 10 \times 30 \times 7.48 = 70{,}461.6 \text{ gal}$$

$$\text{5\% volume} = 70{,}461.6 \times 0.05 = 3{,}523 \text{ gal}$$

$$\text{pure chlorine (gal) for 50 mg/L dosage} = 3{,}523 \text{ gal} \times 50/1{,}000{,}000 = 0.18 \text{ gal}$$

$$\text{gal of 10\% bleach} = 0.18 \text{ gal}/0.1 = 1.8 \text{ gal of 10\% sodium hypochlorite bleach}$$

Figure 8-8 Diver disinfection

UNDERWATER INSPECTION OF STORAGE FACILITIES—DISINFECTION PROCEDURES

Storage facilities can be inspected by divers (Figure 8-8) or remote operated vehicles while the facility is filled with potable water. This procedure must be conducted by trained and certified personnel. Great care must be exercised to ensure that the potable water contained in the facility is not contaminated.

The procedure described here and in ANSI/AWWA Standard C652, *Disinfection of Water-Storage Facilities*, includes the minimum requirements. Contractors and owners may employ additional procedures to increase the safety and effectiveness of these procedures. Underwater inspection of a potable water storage facility may be prohibited by regulatory agencies in some locations. The applicable regulatory agencies should be contacted to determine their requirements before beginning this procedure.

In most cases, underwater inspection is performed by contractors to the utility. The following steps are recommended to ensure that the inspection is conducted safely and successfully.

1. Prejob meeting—Discuss safety and procedures.
2. Storage-facility isolation—Clarify operating conditions.
3. Storage-facility access—Establish entry and exit procedures.
4. Initial water quality—Sample water in facility for testing.
5. Equipment and personnel requirements—Use dedicated equipment and qualified personnel.
6. Equipment disinfection—Follow requirements of ANSI/AWWA Standard 652.
7. Post-inspection chlorine residual—Test residual and add disinfectant if necessary.
8. Bacteriological testing—Conduct bacteria testing.
9. Affidavit of compliance—Contractors should provide this documentation.

DECHLORINATION OF HIGHLY CHLORINATED WATER

Dechlorination is the practice of partially or totally removing the chlorine residual. There may be a number of reasons for this practice. Highly chlorinated water used for the disinfection of facilities, pipelines, and appurtenances is commonly dechlorinated upon discharge. In some areas, even low chlorine residual drinking water must be dechlorinated before it can be drained to storm sewers or directed to waterways. Water discharged during pipeline break emergencies must occasionally be dechlorinated.

Dechlorination practices generally have not been optimized to the degree of chlorination operations. Local conditions and regulations have driven the need to dechlorinate and the development of methods that address these situations. Some of the most important factors that need to be considered are the flow (or volume) to be dechlorinated, the chlorine residual level (initial and final), the time (or distance) that dechlorination must be achieved, the location, the chemistry of the water to be dechlorinated, the impact on the receiving waterway or facility, and any regulatory requirements. System operators should evaluate all of the applicable factors and select the dechlorination method that best suits their situation.

The information provided here reflects the array of practices currently used by water utilities to dechlorinate water in the field. These are not the only methods but are the most common ones. In some cases, the performance values presented are from empirical field observations and may not predict the performance of a practice used in another situation. Operators should test any dechlorination procedure to determine the exact conditions for that application.

Dechlorination Practices Used in the Field

There are a number of nonchemical and chemical dechlorination methods. The selection of a method depends on site-specific factors such as cost, availability of chemicals, containment logistics, availability of specialized equipment, and regulatory approval.

Nonchemical techniques include retention in holding ponds, land application, groundwater recharge, discharge through hay bales and other natural obstructions, and discharge into sanitary sewers. These methods have the advantages of simplicity because they avoid the issues connected with the storage, handling, and safety concerns related to dechlorination chemicals.

Chemicals commonly used for dechorination in the field include sodium bisulfite, sodium metabisulfite, sodium sulfite, sodium thiosulfate, calcium thiosulfate, ascorbic acid, and sodium ascorbate. Sulfur dioxide is a common dechlorination chemical, but it is not commonly used in field applications and its use is not included in this discussion. Chemicals have advantages over nonchemical methods because they usually require less time to affect dechlorination.

Some of the dechlorination chemicals pose potential health concerns if not handled properly and may cause adverse environmental impacts. For example, sodium bisulfite and sodium metabisulfite are skin, eye, or respiratory tract irritants. Sulfite-based chemicals can cause water quality concerns by depleting dissolved oxygen in receiving streams. Some dechlorination chemicals produce hydrochloric acid and, therefore, decrease water pH. When selecting a chemical for dechlorination, it is important to consider the by-products of the reaction and to receive approval from the appropriate regulatory agency.

Treat and Test Field Method

The amount of chemical needed for dechlorination (and to some degree, the effectiveness of nonchemical methods) can be calculated as indicated in this chapter. As a practical matter, though, most operators use the "treat and test" method of dechlorination in the field. The operator uses whatever method they choose for dechlorination (nonchemical or chemical) and then measures the result. If the desired result is not obtained, then the method is adjusted until it is. This is a trial-and-error method.

In some cases, a combination of dechlorination methods is used to provide backup while the system is adjusted. Frequent, periodic samples are taken and tested to ensure that the dechlorination is successful. If a chemical method is used, the feed amount is adjusted so that the desired chlorine residual is just attained. This is necessary to avoid an overfeed situation. Overfeeding some dechlorination chemicals can have adverse effects on receiving waters.

Calculating the amount of dechlorination chemical necessary for the volume and concentration present is recommended so that enough chemical is on hand to complete the job. The tables and equations in this chapter are provided to perform these calculations. It is also recommended that some extra chemical (at least 20 percent) is taken to the site, given that some of the calculations are approximate and local conditions may affect the exact amount needed.

Frequent chlorine residual testing is needed when employing this method. Reliable, certified test methods and field test kits must be used for this purpose. Personnel must be trained in the proper use of this test equipment. Regulatory agencies may require documentation of the test results and other applicable information, such as the initial volume, the chlorine residual, dechlorination method used, and the amount of dechlorination chemical.

Common reducing agents used in the field include sodium bisulfite ($NaHSO_3$) and sodium sulfite (Na_2SO_3), sodium metabisulfite ($Na_2S_2O_5$), sodium thiosulfate ($Na_2S_2O_3$), calcium thiosulfate (CaS_2O_3), ascorbic acid (vitamin C), and sodium ascorbate ($NaC_6H_7O_6$). A comparison of these chemical dechlorination agents is provided in Table 8-3.

Chemical Feed Techniques

Several methods are commonly used for feeding dechlorination chemicals in the field. Some of these may require the availability of electricity. Portable generators can provide electricity in areas where it may not be readily available.

Gravity Feed Method

The gravity feed method typically involves adding dechlorinating solution from a container equipped with a spigot that is placed above the water flow path. The discharge spigot on the container can be adjusted to provide a minimum dechlorinating solution feed rate into the water flow, based on calculations involving the concentration of the dechlorinating solution, water flow rate, and residual chlorine concentration in the flow stream.

Wherever possible, dry dechlorinating agent should be mixed directly within the container prior to use, rather than using pre-mixed dechlorinating solutions provided by suppliers. This minimizes the volumes of dechlorinating solution needed on field vehicles. As part of the dechlorination procedure, samples can be collected downstream of the feed point; analyzed for pH, dissolved oxygen, and residual chlorine; and, if necessary, the chemical feed rate can be adjusted to ensure a nondetectable residual chlorine concentration prior to discharge.

Gravity feed systems are simple to operate, have minimal equipment requirements, have been used effectively by various utilities, and are inexpensive (low-density polyethylene carboys equipped with spigots). Unless adjusted, chemical feed rates can be expected to decrease slightly over time,

Table 8-3 Comparison of common dechlorination agents

Agent	mg/L Dose at pH 7.0 (mg/L to dechlorinate 1 mg/L chlorine)*	pH of 1% Solution	Decomposition/ Off-Gassing	Oxygen Scavenger
Sodium bisulfite	1.5	4.3	Sulfur dioxide	Strong
Sodium metabisulfite	1.4	4.3	Sulfur dioxide	Strong
Sodium sulfite	1.8	8.5–10.0	Sulfur dioxide	Moderate
Sodium thiosulfate	1.9	7.0	Sulfur dioxide	Weak
Calcium thiosulfate	1.0	6.5–7.0	Sulfur dioxide	Weak
Ascorbic acid	2.5	6.0*	None	None reported
Sodium ascorbate	2.8	7.0	None	None reported

*All dechlorination dosage estimates are from ANSI/AWWA Standard C655, *Field Dechlorination* (dosage amount may differ based on variable conditions). Dechlorination capacity must be checked for each application using site-specific conditions.

however, as the available head within the carboy decreases during use. A disadvantage of using this technique is that it involves field testing and calculations for flow rates and water quality parameters to adjust chemical feed rate. Field maintenance crews sometimes prefer a method that does not involve field calculations.

Chemical Metering Pumps

This method for injecting a dechlorinating solution is similar to the gravity feed method, except that a chemical metering pump is used to inject the dechlorinating solution from a container into the water flow. Chemical metering pumps are capable of delivering chemical solutions over a wide range of flows. In addition, the flow rates are adjustable and the pumps provide a constant chemical feed rate.

Relative to a gravity feed system, this type of system requires more equipment (e.g., storage container, pump, energy source, and tubing) and costs are significantly higher. Although chemical metering pumps may provide field personnel with a more reproducible method, this feature involves much higher cost and greater operator ability and attention during dechlorination.

Venturi Injector Systems

Venturi injectors are differential pressure injection devices that allow for the injection of liquids (e.g., dechlorinating solutions from hydrants) into a pressurized water stream. This method is well suited for pressurized water releases such as those through hydrants.

Water main discharges are constricted by a regulator valve or gate valve and routed through a venturi injector system. Dechlorinating solution is drawn into the injector from a plastic container, with the chemical feed rate controlled by a metering valve. Venturi injector units should include a flowmeter installed near the metering valve to measure the chemical feed rate (e.g., a rotameter), a threaded fitting at the upstream end of the pipe to attach adapters (e.g., reducing sections), and a fitting at the downstream end of the pipe for attachment of flexible discharge piping (e.g., a hose).

Spray Feed Systems

The dechlorinating solution can be sprayed into the flow (on pipe walls, surfaces, or pipeline appurtenances) via a backpack sprayer similar to the type used for pesticide and herbicide application or chemical fire extinguisher. The advantages of this technique are that the chemical feed rate is fairly constant (given a steady pressure within the solution chamber). Hence, dosages can be approximated fairly accurately, and a piped or channelized flow is not required to effectively feed the chemical. This method is typically more effective in adding dechlorinating chemicals to sheet flows than the other alternatives previously described.

A significant disadvantage of a spray feed system is that it requires the equipment to be set up at a stationary point and monitored for adequate chemical and pressure. If a stationary point is not available, a person will have to don the sprayer and continuously apply the dechlorinating agent to the flow.

Flow-Through Systems

Flow-through systems (Figure 8-9) include any method where the solid chemical is held stationary and the flow allowed to run over, around, and/or through it. Examples include pumping chlorinated water through a container filled with dechlorinating agent, or laying permeable bags of the chemical in the flow path. For this application, the dechlorinating agent is in tablet or powder form.

The advantages of flow-through systems are that they are simple and can be used for sheet flow applications as well as for channelized or pumped flow. The disadvantages are that there is limited control over the dosage, overdosing or underdosing to significant levels could occur, and it may be difficult, in some cases, to tell when the chemical has been used up and must be replaced.

Also, due to the contact time required for dissolution of the powder or tablet, this method is more suitable for low- and medium-velocity discharges. This method may, however, be suitable for dechlorinating releases from unidirectional flushing where the velocity of the flow is approximately 5.0 ft/sec (1.5 m/sec), provided that equipment is specifically designed for this flow.

A variation in the application of a flow-through system is the automatic tablet dispenser. This device is similar to the tablet feeders currently used for disinfection in many water/wastewater treatment plants. The feeder typically consists of a nonmoving housing and automatic feed tubes. The tubes are inserted down through a removable top cover of the feeder into the stream of water. The lower end of each tube is slotted to permit free flow of water through the tubes to ensure good contact between the water and dechlorination tablets. As the stream of water flows past the feed tubes containing dechlorination tablets, the dechlorination agent is released into the water by dissolution. Commercial tablet dispensers are available for a wide variety of flow rates and chlorine concentrations.

Flow Control Measures

During planned and unplanned water releases, it may be necessary to construct flow control measures to prevent the water from entering directly into a water body and to provide an opportunity for better mixing of the dechlorinating agent. Construction of berms, swales, ditches, or redirection pipes are common methods used to control the flow of released water.

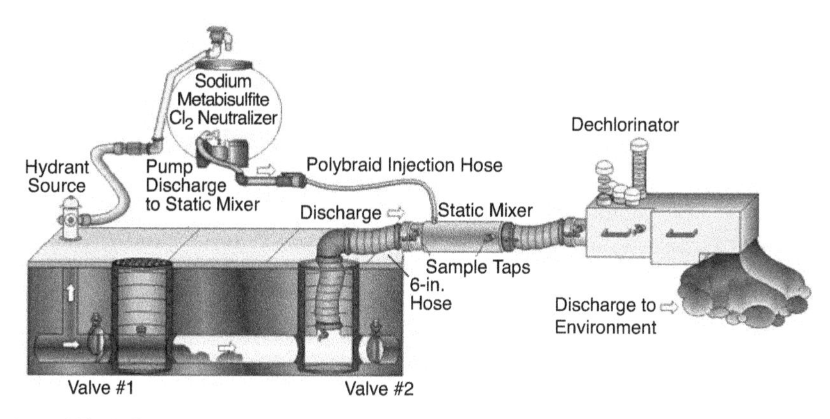

Courtesy of Nicole Peschel.

Figure 8-9 Pipeline dechlorination system

BASIC DECHLORINATION CALCULATIONS

Dechlorination Chemical Amount for a Fixed Volume and Chlorine Residual

dechlorination chemical amount (lb) = dose from Table 8.3
× chlorine residual (mg/L)
× volume of chlorinated water (gal) × 3.785 L/gal ÷ (1,000 mg/g × 454 g/lb)

Example 10: What is the amount (lb) of sodium sulfite needed to dechlorinate the contents of a cylindrical tank 20 ft in diameter at a water depth of 35 ft with a chlorine residual of 10 mg/L?

sodium sulfite (lb) = 1.96 × 10 mg/L × 3.14 × 10 × 10 × 35 × 7.48
× 3.785 L/gal ÷ (1,000 mg/g × 454 g/lb) = 13.43 lb

Dechlorination Chemical Feed Rate to Match Discharge Rate

chemical feed rate (gpm) = withdrawal rate (gpm)
× chlorine residual (mg/L) × chemical dose from Table 8.3
× % solution strength ÷ 10,000 ppm/%

amount of chemical solution needed (gal) = volume of chlorinated water (gal) ÷ withdrawal rate (gpm) × dechlorination chemical feed rate (gpm)

Example 11: A 10-in.-diameter pipeline contains 200 ft of 100 mg/L chlorinated water. This slug is being withdrawn slowly (4.5 gpm) to disinfect the pipeline. What is the feed rate needed of a 1 percent sodium hypochlorite solution to dechlorinate this water as it is withdrawn from the pipe, and how much of this chemical will be needed to complete the entire job?

chemical feed rate (gpm) = withdrawal rate (gpm) × dose from Table 8.3 × chlorine residual (mg/L) × % strength/10,000/% = 0.088 gpm or 5.3 gal/h

chemical solution needed (gal) = $\pi r^2 L$ × 7.48 gal/ft^3 ÷ withdrawal rate (gpm) × dechlorination feed rate (gpm) = 3.14 × 5/12 × 5/12 × 200 × 7.48 ÷ 4.5 × 0.088 = 15.9 gal

BIBLIOGRAPHY

ANSI/AWWA C651. *AWWA Standard for Disinfecting Water Mains.* Denver, Colo.: American Water Works Association.

ANSI/AWWA C652. *AWWA Standard for Disinfection of Water-Storage Facilities.* Denver, Colo.: American Water Works Association.

ANSI/AWWA C655. *AWWA Standard for Field Dechlorination.* Denver, Colo.: American Water Works Association.

ANSI/NSF Standard 61. *Drinking Water System Components.* Ann Arbor, Mich.: NSF International.

APHA, AWWA, and WEF (American Public Health Association, American Water Works Association, and Water Environment Federation). 2012. *Standard Methods for the Examination of Water and Wastewater*, 22nd ed. Edited by E.W. Rice, R.B. Baird, A.D. Eaton, and L.S. Clesceri. Washington, D.C.: APHA.

AWWA. 2001. *Good Practices for Preventing Microbiological Contamination of Water Mains.* Denver, Colo.: American Water Works Association.

AWWA. M12—*Simplified Procedures for Water Examination.* Denver, Colo.: American Water Works Association.

Chlorine Institute. 2000. *Chlorine Basics*, 7th ed. Pamphlet 1. Washington, D.C.: Chlorine Institute.

Chlorine Institute. 2001. *Personal Protective Equipment for Chlor-Alkali Chemicals*, 5th ed. Pamphlet 65. Washington, D.C.: Chlorine Institute.

Chlorine Institute. 2008. *Water and Wastewater Operators' Chlorine Handbook*, 2nd ed. Pamphlet 155. Washington, D.C.: Chlorine Institute.

Chlorine Institute. 2011. *Sodium Hypochlorite Manual.* Pamphlet 96. Washington D.C.: The Chlorine Institute.

Chlorine Institute. Bookstore Website, http://bookstore.chlorineinstitute.org/index.php/ci-catalog.

Lauer, W.C., and F.J. Sanchez. 2006. *Disinfection of Pipelines and Storage Facilities Field Guide.* Denver, Colo.: American Water Works Association.

Chapter 9

Fire Hydrants

Fire hydrants are so seldom used for fighting fires in small communities that it is easy to forget how important they are. Hydrants that operate properly and provide adequate flow can make the difference between losing or saving valuable property and even lives in the event of a serious fire.

HYDRANT USES

Fire hydrants have several uses in a water distribution system. The most obvious is for public protection fighting fires. But there are several other uses that should be addressed in the process of operating and maintaining a water distribution system.

Firefighting

It is important to a community to have fire hydrants that are properly spaced and fed by an adequate distribution system. Adequate fire hydrants can make the difference between losing and saving buildings that catch fire, and having a good system in place can result in substantial savings in fire insurance for residents.

In most communities, the water utility has the responsibility for installing and maintaining hydrants. In some cases, the fire department will assume responsibility for maintaining and flow-testing hydrants. Water distribution system operators have both moral and legal responsibilities to keep hydrants in good working order.

Other Hydrant Uses

Other frequent uses for fire hydrants include

- Flushing water mains to clear the pipes of rust and sediment;
- Furnishing large volumes of water for flushing sewers to clear them of accumulated solids;
- Filling tank trucks for rural firefighting, street washing, tree spraying, and other uses; and

- Construction uses, such as supplying water for settling dust during demolition and mixing mortar for construction before a permanent water service is installed.

In general, uses other than for fighting fires should be discouraged because water is wasted in the process and use of hydrants by inexperienced persons could cause damage to the hydrants. This is particularly true in very cold weather when a hydrant may freeze during or after use.

Unauthorized use of fire hydrants should be absolutely prohibited, and all police and public works employees should be instructed to immediately report any use that they notice which may not be authorized. If hydrant use by anyone other than experienced system workers is necessary, it should be allowed only under specific conditions. In general, it is best to issue a hydrant use permit both for record purposes and to be sure that all specific conditions are understood. The conditions specified by many water utilities include the following:

- Only a standard hydrant wrench may be used for operating hydrants—repeated use of a pipe wrench on the five-sided nut will round off the nut so a hydrant wrench will no longer work.
- Very old hydrants should not be used by inexperienced persons—hydrants often will not seat well when closed and may be inadvertently left running. Also, an inexperienced person might break the stem while forcing it in an attempt to get the hydrant to stop leaking.
- Hydrants on busy streets should not be used unless absolutely necessary—their use could unnecessarily cause traffic tie-ups or accidents.
- Unless there is some specific reason why it is necessary, hydrants on the end of dead-end mains should not be used—there is usually sediment accumulation in dead-end mains that will be needlessly stirred up.
- Hoses should not be left attached overnight or when the user is not present—if firefighters should have to use the hydrant, they will be delayed by having to remove the hose.
- If use from a hose will require that the water be intermittently turned on and off (e.g., for mixing batches of mortar), an auxiliary hose valve should be used. The hydrant should be opened at least halfway and left in that position until the use is completed—repeated opening and closing of the hydrant valve may damage the hydrant and may flood the underdrain system so the hydrant will not drain properly.

Some water utilities have eliminated the problems caused by tank trucks being filled from random points on the system by establishing one watering point and insisting that all tankers be filled at that location. The watering point is usually located at the city yard where the operation can be closely observed. It is set up to make the operation simple and eliminate any

Courtesy of Neptune Technology Group Inc.

Figure 9-1 Fire hydrant meter

possibility of a cross-connection. It is also convenient to meter the water at this point and charge the users for the amount of water taken.

Charging for Water Used From Hydrants

From the standpoint of water use accountability, it is a good idea to meter water taken from fire hydrants. Many water utilities have one or more special hydrant meters (Figure 9-1) available for contractors to use when they require water for construction. The contractor must be informed that the hydrant should be flushed before the meter is installed the first time so that accumulated sediment will not lodge in the meter. The meter must also be protected from vandalism and freezing and needs to be removed at night. Some water system operators also assemble a metering system for small hoses using a regular home-type water meter.

Besides recording the exact amount of water used by contractors, using a meter also discourages them from wasting water. When the water is not metered, there is a strong tendency by workers to let the water run rather than shut it off between uses.

The problem with metering hydrant flow is that the meters receive considerable abuse when they are thrown in the back of a truck with the picks and shovels at the end of each day. For this reason, many utilities find it easier to just estimate the water use and charge a lump sum for a hydrant use permit to cover both the cost of water and the water system's labor in monitoring the use by the contractor.

Problems Caused by Hydrant Operation

The most common problem caused by hydrant operation is that it stirs up sediment that has accumulated in the mains. This is sure to cause "rusty water" complaints by customers unless the section of main is thoroughly flushed until the water is clear. This problem can be somewhat reduced by having a regular hydrant flushing program to minimize the amount of sediment buildup.

Once an area of the system has been stirred up from a major hydrant use, such as for fire fighting, it may be possible to bleed off some of the discolored water by opening some hydrants to partial flow. The velocity in the mains, however, must be kept low enough to not stir up more sediment. When the water is discolored, customers should be informed that the water is safe to drink, but they should not wash clothes until the discoloration has cleared. If any customers have washed clothes, they will probably complain that their white clothes came out yellow or brown. Products are available, packaged in small bottles, that will usually remove the discoloration when the clothes are rewashed with the chemical. Many water systems keep a supply of these bottles at the city hall or other location to give to customers when they have this problem.

Another problem that can be caused by hydrant use is water hammer. If a hydrant on a large main is closed very quickly, the pressure buildup caused by abruptly stopping the large quantity of water traveling through the main can be several times normal pressure. This could result in moving the hydrant backward if it is not firmly blocked, or other system damage, including blowing out adjoining-home water services. All hydrant users should accordingly be warned to open and close hydrants slowly.

TYPES OF FIRE HYDRANTS

The four general types of fire hydrants available include dry barrel, wet barrel, warm climate, and flush.

Dry-Barrel Hydrants

Dry-barrel hydrants are used in locations with freezing weather and are equipped with a main valve and a small drain valve in the base that allows the barrel to drain when the main valve is closed. An additional advantage of dry-barrel hydrants is that there is no flow of water from a broken hydrant because the main valve is underground.

Dry-barrel hydrants are further classified as wet top or dry top. Wet-top hydrants (Figure 9-2A) are constructed so that the threaded end of the operating mechanism is not sealed from water when the hydrant has water in it. Dry-top hydrants (Figure 9-2B) have this mechanism sealed to reduce the possibility of the threads becoming fouled from sediment or corrosion.

The three general types of main valves used in dry-barrel hydrants are illustrated in Figure 9-3. In a standard compression hydrant, the valve closes with water pressure against the seat, which assists in providing a good seal.

Older-style dry-barrel hydrants were made with a single-piece barrel that extended from the foot piece to the bonnet. A disadvantage to this

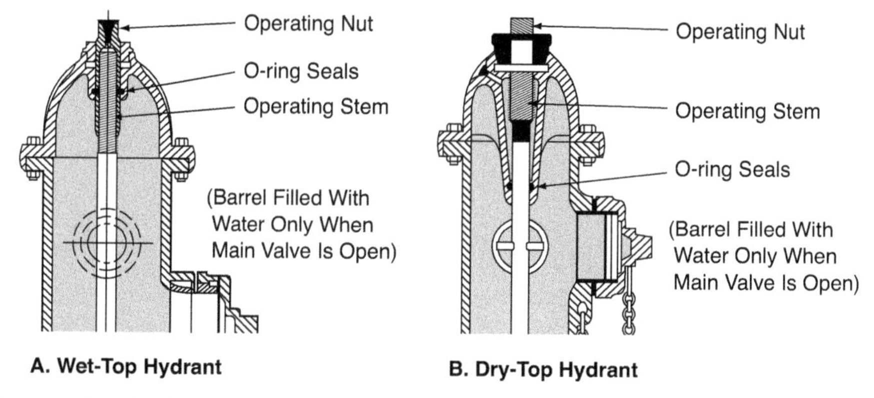

Courtesy of Mueller Company, Decatur, Ill.

Figure 9-2 Wet- and dry-top hydrants

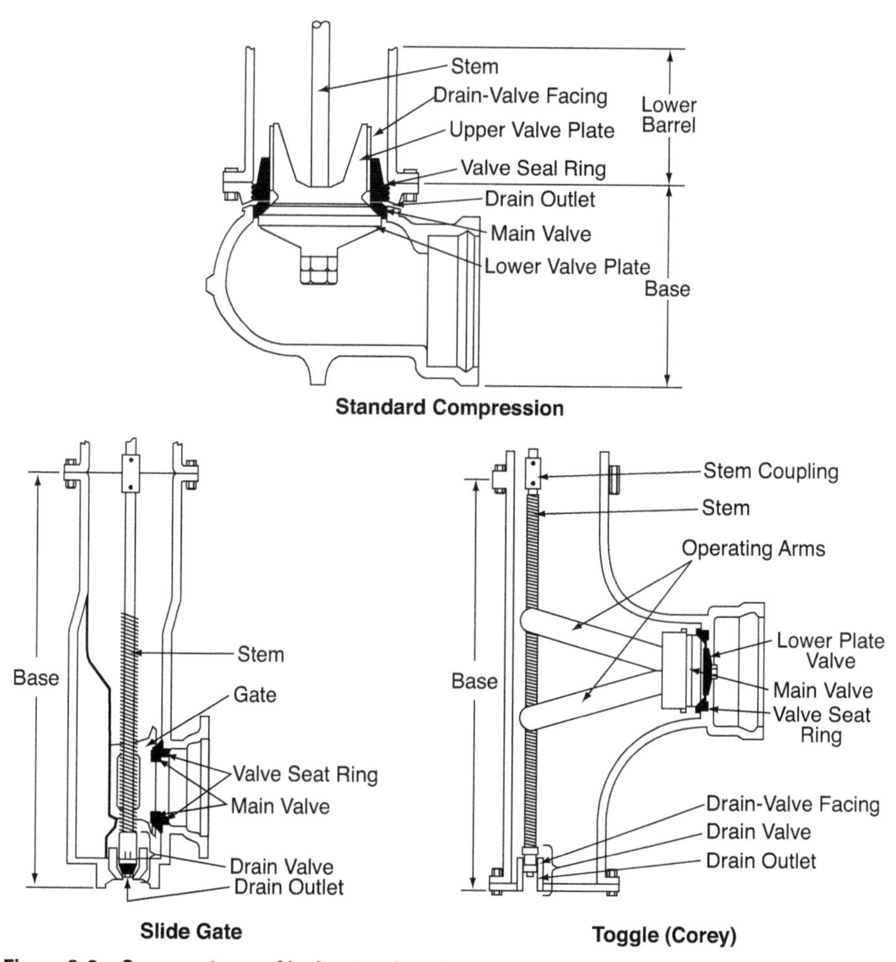

Figure 9-3 Common types of hydrant main valves

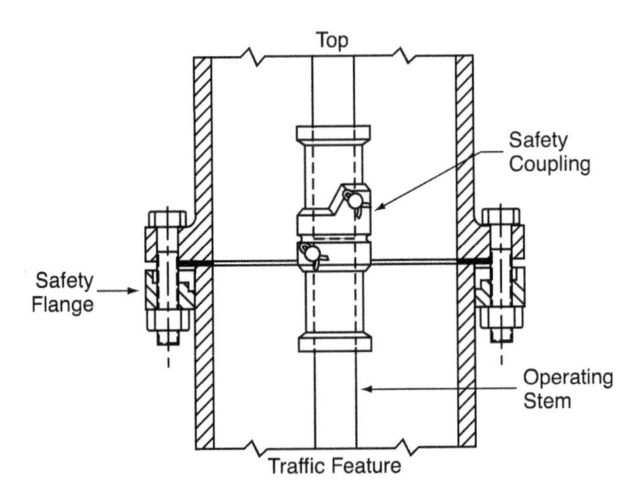

Courtesy of American Cast Iron Pipe Company.

Figure 9-4 Detail of one type of breakaway flange and stem coupling

design is that if the hydrant is broken (usually by being struck by a vehicle), the hydrant must be dug up to the foot piece to replace the barrel. In such an accident, the operating stem is also usually bent beyond repair and must be replaced. If the hydrant is broken when there is frost in the ground, it requires considerable effort to make the repair; sometimes the hydrant must be taken out of service until the ground thaws.

In recent years, all manufacturers have been using the breakaway, or traffic hydrant design, which uses a flanged coupling just above the ground line as illustrated in Figure 9-4. In most cases, all that is broken when a hydrant is struck is the cast-iron safety flange and the safety coupling on the operating stem. The hydrant can usually be easily restored to service using a new flange and coupling. The design of dry-barrel hydrants is covered in AWWA Standard C502, *Dry-Barrel Fire Hydrants.*

Wet-Barrel Hydrants

Wet-barrel hydrants are filled with water all the time, so they are not suitable for locations with freezing weather. The hydrant itself has no main valve, but each nozzle has a separate valve. The design of wet-barrel hydrants is covered in AWWA Standard C503, *Wet-Barrel Fire Hydrants.*

Warm-Climate Hydrants

A warm-climate hydrant has the main valve located at the ground line so that the lower barrel is always full of water and under pressure. The main valve controls flow from all outlet nozzles, and there is no drain mechanism.

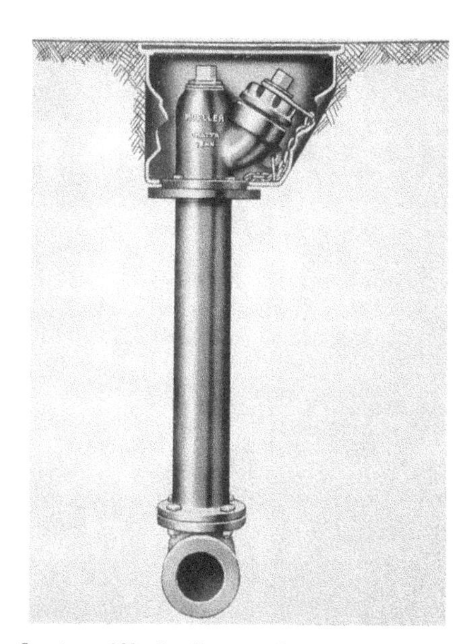

Courtesy of Mueller Company, Decatur, Ill.

Figure 9-5 Flush hydrant

Courtesy of Kupferle Foundry Company.

Figure 9-6 Flushing hydrant

Flush Hydrants

Flush hydrants are used where a regular hydrant would be objectionable, such as on airport taxiways or in pedestrian malls. The entire standpipe and head are below ground and are accessible through a cover that is flush with the surface (see Figure 9-5).

Flushing Hydrants

Small-diameter hydrants are available for installation on the end of a small pipeline, solely for the purpose of flushing the line (Figure 9-6). Flushing hydrants are not intended for fire use. They are generally available with a 2-in. (51-mm) pipe thread inlet and a standard 2½-in. (63.5-mm) nozzle, so a fire hose may be used to divert discharge flow, if necessary.

Auxiliary Valves

Every fire hydrant should have an auxiliary valve installed between the hydrant and the main, so each hydrant may be individually turned off for repair. The type of auxiliary valve most often used is directly connected to the hydrant by a flange. One of the advantages to this arrangement is that the valve cannot separate from the hydrant. Another advantage is that it places the valve at the same location in relation to each hydrant, thus they are easy to find.

HYDRANT INSTALLATION

Fire hydrants must be handled and installed carefully. Carelessness can result in future maintenance problems or even failure of a hydrant when it is needed.

New Hydrant Inspection

Before installing a hydrant, make sure that it meets local standards. Most hydrants installed in the United States open counterclockwise and have operating nuts that measure 1¼ in. (32 mm) from flat to point. But some water systems got started with other standards and have chosen to stay that way for uniformity. Many communities also have a special thread (other than National Standard) for the 2½-in. (63.5-mm) ports. It is much easier to check that these are all correct before installation than to find out after the hydrant is in place.

The depth of burial should also be carefully computed and checked to ensure that the hydrant to be installed will stand the correct distance above the ground. The break-flange should be located about 2 in. (50 mm) above the ground surface. Hydrants are generally purchased to coincide with the water main depth of burial required by the utility. But occasionally, a main may be a little shallower, or may be considerably deeper, due to the need to go over or under an obstruction, such as a sewer.

If the main is considerably deeper than normal at the hydrant location, it is usually best to angle the hydrant lead up to the normal depth using angle fittings. If the main is much shallower than normal, it may be necessary to order a special hydrant with a shorter barrel. If the main is only 1 ft (0.3 m) or so deeper than normal, the regular hydrant can be installed and then the upper section can be raised later using extension kits that are available in 6-in. (150-mm) increments.

Finally, the new hydrant should be opened and closed to ensure that no damage occurred during shipping or storage. If hydrants must be stored outside in freezing weather, they should be placed with the inlet facing down to prevent rain and snow from entering and freezing.

Installation Location

Fire hydrants are usually located at street intersections so that hose can be strung in any direction for fighting a fire. Hydrant spacing typically ranges from 300 to 600 ft (90 to 180 m). Mid-block hydrants are usually best

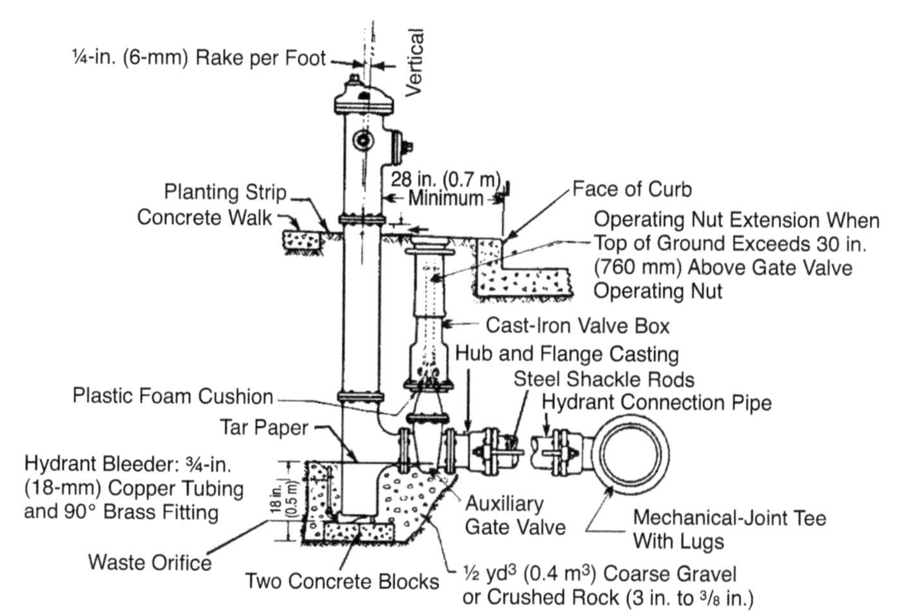

Figure 9-7 Hydrant auxiliary valve installation

located opposite a lot line, as they are least likely to conflict with a driveway and will be least objectionable to the property owners.

Hydrants should be located far enough back from a roadway to minimize the danger of being struck by vehicles. The minimum setback from the face of a curb should be about 2 ft (0.6 m), which is generally beyond large-vehicle overhang and far enough back to place the auxiliary valve behind the curb. On the other hand, they should be close enough to the pavement so a hard suction can be connected between a pumper truck and the hydrant without danger of the truck getting stuck in mud or snow.

In areas with heavy snowfall, consideration should also be given to locations where the hydrant will be least likely to have snow piled on it during snow removal.

Hydrants should not be located close to buildings. Firefighters usually will not position their truck in a location where a building wall will fall on the truck if the building collapses.

Installation Procedures

Fire hydrants must be set on a firm footing that will not rot or settle. A flat stone or concrete slab is ideal (Figure 9-7). As the hydrant is set, a carpenter's level should be used to ensure that it is plumb in all directions.

The hydrant must also be securely blocked or restrained from movement because the force against it will be tremendous if the valve is closed quickly. If possible, the excavation for the hydrant should be made carefully so there will be undisturbed earth directly behind the foot piece. Solid concrete blocks can be tightly wedged between the hydrant and earth, or a small quantity of concrete can be mixed on the job and poured into the void. If poured concrete is used, care must be taken not to block the drain hole.

If the ground behind the hydrant is soft or otherwise unsuitable for blocking, the hydrant must be restrained in some other fashion. One method of restraint that works well is to install at least two lengths of threaded rod between the auxiliary valve and the tee on the main. If mechanical joint fittings are being used, the rod can be substituted for two of the bolts and double nuts used to tighten the joints. If push-on joints are being used, the rod can be fastened through bell clamps.

To facilitate quick removal of the water from the hydrant barrel when the main valve is shut, a pocket of coarse gravel or crushed rock should be placed in the excavation before the hydrant is set. The top of the gravel should be slightly above the drain opening and should be covered with heavy polyethylene or tar paper to prevent the gravel from being clogged with dirt.

If the hydrant barrel will not fully drain, because of a high water table, common practice is to plug the drain hole. If there is any danger of ground frost at the location, special note must be made of these hydrants so that they will be pumped out by hand after every use.

Testing New Hydrants

New hydrants installed in conjunction with installation of a new water main should not be tested during main testing. All hydrant auxiliary valves should be closed during the pressure test. After main testing is completed, the following hydrant test procedures should be followed:

1. Open the auxiliary valve, remove a cap, and fill the hydrant with water.
2. Replace the cap, but leave it loose to let air escape, then tighten it.
3. Apply pressure up to a maximum of 150 psi (1,000 kPa) by using a pressure pump connected to one of the nozzles. If it is not practical to apply higher pressure, system pressure will suffice.
4. Check for leakage at flanges, outlet nozzles, and at the operating stem.
5. Tighten any loose bolts and correct any leakage.
6. Flush the hydrant to remove any foreign material.
7. For dry-barrel hydrants, shut the main valve, immediately remove one cap, and place the palm of a hand over the nozzle. If the barrel is draining properly, a noticeable vacuum can usually be felt. If there is still a question of proper drainage, a string with a small weight attached can be dropped down the barrel and inspected to determine whether it comes out wet.

HYDRANT MAINTENANCE

Maintenance of fire hydrants must be a continuing program. The lives and property of residents depend on hydrants working properly when needed. Water system operators should also keep in mind that fire hydrants are often the only visible signs of the water utility seen by most customers, so hydrant appearance can go a long way in promoting the utility's public image. Shabby hydrants immediately bring to mind a poorly run utility.

Hydrant Inspection and Repair

All hydrants should be inspected regularly, at least once a year, to ensure that they operate satisfactorily. Scheduled hydrant maintenance contributes to distribution system reliability by ensuring operability (chapter 3). Hydrant inspection programs can be combined with main flow testing and pipeline inspection programs. With preplanning, hydrant exercising can be part of an organized preventive flushing program that can be used to clear mains of sediment and improve customer satisfaction.

Many water systems in northern regions do their hydrant inspection in the fall to make sure that the barrels are drained, the hydrants are cleared of vegetation, and snow flags are installed if necessary. In freezing climates, dry-barrel hydrants may require two inspections per year. It is common to inspect hydrants in the spring and then again in the fall. Winter inspections are especially important for dry-barrel hydrants that are installed in areas with high groundwater levels. Dry-barrel hydrants with permanently plugged drains must be pumped out after each use and then inspected. It is advisable to inspect all types of hydrants after each use. This is particularly important in freezing weather to make sure the barrel has drained completely.

Some water utilities have inspection crews equipped to make repairs immediately if any problems are found. Other systems have crews that perform only inspection and, if they find that repairs are necessary, submit repair requests to followup repair crews. If, during inspection, a hydrant is found to be inoperable, an out-of-service sign should be hung on it or a plastic bag placed over it and the fire department notified that the hydrant is not operable.

Inspection procedures for dry- and wet-barrel hydrants and maintenance tips are included in AWWA Manual M17, *Installation, Field Testing, and Maintenance of Fire Hydrants*. The AWWA publication *Hydrant Maintenance Field Guide* is also an inexpensive, pocket-sized reference that is handy for use by field crews.

Hydrant Operation

Fire hydrants are designed to be operated by one person using a 15-in. (380-mm) long wrench. A length of pipe (cheater) should not be used on the wrench. If a hydrant cannot be operated with a standard wrench, it should

Table 9-1 Standard hydrant color scheme to indicate flow capacity

	Usual Flow Capacity at 20 psig (140 kPa [gauge])*		
Hydrant Class	**gpm**	**(L/min)**	**Color†**
AA	1,500	(5,680)	Light Blue
A	1,000 to 1,499	(3,785 to 5,675)	Green
B	500 to 999	(1,900 to 3,780)	Orange
C	Less than 500	(Less than 1,900)	Red

*Capacities are to be rated by flow measurements of individual hydrants at a period of ordinary demand. See *AWWA Standard for Dry-Barrel Fire Hydrants*, C502, for additional details.

†As designed in Federal Standard 595C, General Services Administration, Specification Section, Washington, D.C.

be repaired or replaced. Some very old hydrants may not seat completely on the first try, but will usually respond to reopening and closing the valve several times. A person operating a hydrant must always make sure that the valve is tightly shut and the barrel has drained before leaving the site.

Hydrant Painting

Fire hydrants should be painted colors that are easily visible to the fire department. Red, orange, and yellow are generally the most visible, although not always the favorites of property owners who would prefer less visible hydrants. If darker colors are used, do not allow bushes or other vegetation to obscure the hydrants.

Many water systems now color code their hydrants to indicate the capacity and assist the fire department in deciding which hydrants are the best to connect to. Most systems just paint the hydrant top or caps the code color. The color code scheme suggested by AWWA is shown in Table 9-1.

Hydrant Records

A record card, sheet, or computer record should be maintained for every hydrant in the water system. Basic information that should be recorded at the time of installation includes the make, model, depth of burial, location measurements to various markers, information on the auxiliary valve, and flow performance at the time of installation.

An entry should then be made on the record each time the hydrant is inspected, and notes made of any repairs that are made. It is important that the records be accurately maintained because they could serve as proof of a good hydrant maintenance program in the event that the utility is charged with negligence if a hydrant should fail to operate properly in an emergency.

BIBLIOGRAPHY

ANSI/AWWA C502. *AWWA Standard for Dry-Barrel Fire Hydrants*. Denver, Colo.: American Water Works Association.

ANSI/AWWA C503. *AWWA Standard for Wet-Barrel Fire Hydrants*. Denver, Colo.: American Water Works Association.

AWWA. 2010. Principles and Practices of Water Supply Operations—*Water Transmission and Distribution*, 4th ed. 2010. Denver, Colo.: American Water Works Association.

AWWA. M17—*Installation, Field Testing, and Maintenance of Fire Hydrants*. Denver, Colo.: American Water Works Association.

AWWA. M19—*Emergency Planning for Water Utilities*. Denver, Colo.: American Water Works Association.

AWWA. M27—*External Corrosion—Introduction to Chemistry and Control*. Denver, Colo.: American Water Works Association.

AWWA. M31—*Distribution System Requirements for Fire Protection*. Denver, Colo.: American Water Works Association.

AWWA. M36—*Water Audits and Loss Control Programs*. Denver, Colo.: American Water Works Association.

Deb, A.K., K.A. Momberger, Y.J. Hasit, and F.M. Grablutz. 2000. *Guidance for Management of Distribution System Operation and Maintenance*. Denver, Colo.: Awwa Research Foundation and American Water Works Association.

Federal Standard 595C. *Colors Used in Federal Procurement*. Washington, D.C.: General Services Administration.

Kirmeyer, G.J., M. Friedman, and J. Clement. 2000. *Guidance Manual for Maintaining Distribution System Water Quality*. Denver, Colo.: Awwa Research Foundation and American Water Works Association.

Kirmeyer, G.J., M. Friedman, K. Martel, and A. Sandvig. 2002. *Guidance Manual for Monitoring Distribution System Water Quality*. Denver, Colo.: Awwa Research Foundation and American Water Works Association.

NFPA. 1963, 2009. *Standard for Fire Hose Connections*. Quincy, Mass.: National Fire Protection Association.

Chapter 10

Water Storage

PURPOSES OF WATER STORAGE

To the casual observer, the only purpose of water storage tanks may be to supply large quantities of water during a fire or in the event of failure of the water source. In actuality, there are many other functions of a properly designed storage system. Properly managing storage facilities can improve distribution system reliability to provide an uninterrupted water supply (chapter 3). Utilizing storage to minimize pumping can also contribute to the efficient use of energy and lower operating costs. Water quality can also be enhanced by reducing water age in storage.

Equalizing Supply and Demand

Domestic water use usually changes throughout the day and night, as illustrated in Figure 10-1. There is usually relatively heavy use in the morning as customers prepare breakfast and begin other household duties, then use slacks off at midday. Near supper time there is greatly increased use because almost everyone is at home and customers water their gardens, wash cars, and perform other duties that require water. Then about 10:00 p.m., when most people go to bed, use falls off very sharply. The size of the peaks vary with the season, and water systems with large industrial customers may have unusual use patterns as shifts change or process water is used.

In Figure 10-1, the peak-hour demand is about 175 percent of the average demand for the day. So if this water system had no storage at all, it would need a plant capacity almost double the average requirement to meet the peak-hour use requirement. It would also mean that the system would need several pumps of various sizes that could be switched on and off to match the demand. If the water system has a treatment plant, operation would become increasingly difficult because filters and other equipment would have to be operated at highly variable rates.

With adequate storage, water can be treated and pumped at a more uniform rate and does not have to directly follow customer use. During the

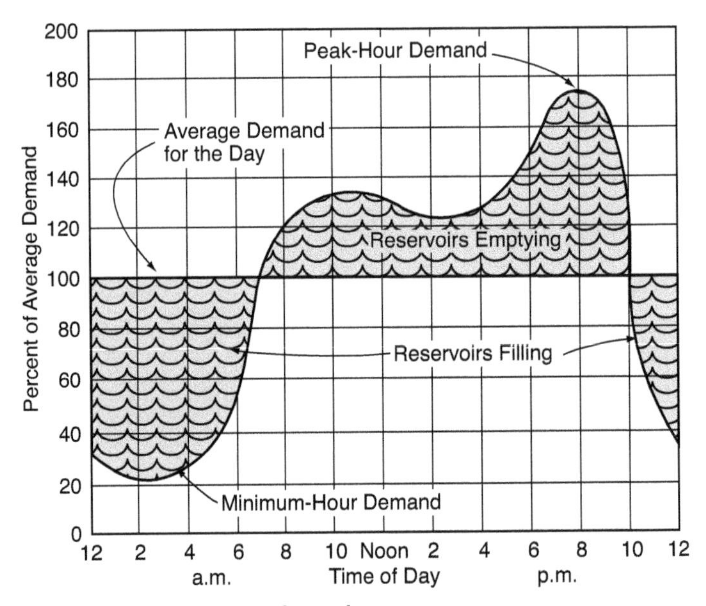

Figure 10-1 Daily variation of system demand

day, water is taken from storage, and during the night, the storage reservoirs are refilled.

A water system that purchases water from another water system has a somewhat different situation. It is usually offered the following three choices:

1. It can draw directly from the selling water utility and have no storage of its own. In this case it will have to purchase the water at a very high rate because the selling utility must furnish either additional storage or pumping capacity to meet the maximum demand.
2. It can furnish enough storage to meet its entire daytime use, and draw water to fill the reservoirs only at night when the selling utility is in its low-use mode. The rate for this water is often called a dump rate and may be quite low because the cost to the selling utility is minimal.
3. A third option is somewhere in between. One arrangement is that the purchasing system can draw water 24 hours a day, but only at a set rate through a throttling valve. It must then have enough storage to cover the additional water required during the high-use periods.

Increased Operating Convenience

Some small water systems with their own treatment plants have found it most convenient to operate the plants only one or two shifts during the day, and then let the systems "coast" at night, using water from storage. The cost of the additional storage required must be balanced against reduced personnel costs. There must also be an arrangement to quickly start up the plant in the event of unusual use, such as a main break or fire.

Balancing Pumping Requirements

Water use is continually changing, depending on the time of day, day of the week, weather conditions, and even such factors as which shows are on television. If it is necessary to frequently turn pumps on and off to meet the demand, it adds to the wear on the pumps and significantly increases electrical costs.

By having some elevated storage in the distribution system, the cycling of pumps can be minimized. For example, if the pump in use is larger than required to meet demand, water in the elevated tank will rise. So, just before the tank would overflow, the pumps are switched to use a smaller unit. The tank level will then probably slowly fall, and at some preset point, such as when the tank is down to half full, the pumps are switched again. In this way, it is often possible to operate on one pump size or combination of pumps for several hours without changing.

Decreasing Power Costs

If the local electric utility has special power rates for off-peak use (usually at night), it may be possible to make a significant savings in power cost by providing additional storage and operating the larger pumps only during the off-peak period to fill the tanks. This plan would be particularly applicable to a water system that can construct a large ground-level storage tank on a high point of ground so it will gravity-feed the system during the daytime.

Emergency and Fire Requirements

One of the principal reasons for distribution system storage is to meet the very high demands for water during a fire. Other unusual demands that must also be considered are unusually high use during large main breaks, temporary loss of power, or problems that might disrupt the water source. In all of these cases, the objective is to try to supply as much water as is required without loss of pressure.

If a water utility has a sizable quantity of elevated storage, the quantity of water available may be sufficient to meet the emergency demands through gravity feed. If most of the system storage is in ground-level reservoirs, pumps must be activated quickly to maintain system pressure. Fire demand can account for as much as 50 percent of the total required distribution system storage.

It is not advisable to rely on storage to make up any shortfall between the quantity of water available from the water source and quantity needed to meet maximum day demand. Maximum day demand can, on occasion, last for several days.

Surge Relief

As large pumps are stopped and started and when valves or hydrants are closed quickly, extremely high pressure surge or water hammer can damage the system piping and customer services. Elevated tanks that are directly connected to the system help to absorb any surges by allowing the shock wave in the water to travel up the riser and into the upper tank section.

Increasing Chlorine Detention Time

Requirements under the federal Surface Water Treatment Rule specify the length of time that chlorine must be in contact with water before the water reaches the first customer. In cases where there is not sufficient detention time in the water treatment process, additional detention can be provided by directing the water into a storage reservoir before it enters the distribution system. The reservoir is then serving two functions. The only special provision for this type of reservoir is that it must have baffles to channel the water in a long path so that the required contact time between the chlorine and water is maintained.

Blending Water From Different Sources

Occasionally, a water system will draw water from two or more sources having different qualities, such as hardness and temperature. A relatively common example is a system that primarily uses groundwater, but because the supply is inadequate, it is supplemented with water from a surface water source.

Residential customers generally dislike having the quality vary and will complain if the changes are significant. Industrial customers are also often disturbed by changes in water quality. If water is used for cooling, they will be bothered by quick changes in water temperature. If water is incorporated into their products, such as soft drinks, consistent water quality is essential.

Another reason for wanting to blend water from different sources would be that a utility has one source with a fluoride concentration exceeding the maximum contaminant level and another source with low fluoride. When water from sources with dissimilar quality is properly blended in a reservoir before distribution, the quality furnished to customers can be maintained relatively uniform.

TYPES OF STORAGE FACILITIES

Facilities for storing water in the distribution system may be either elevated or at ground level. Storage facilities fall into the general classifications of elevated tanks, standpipes, reservoirs, and hydropneumatic systems.

Courtesy of CB&J.

Figure 10-2 A 1,000,000-gallon single-pedestal spheroidal elevated tank painted to look like a hot-air balloon

Elevated Tanks

An elevated tank consists of a tank supported by a steel or concrete tower (Figure 10-2). Most elevated tanks are designed to float on the system—in other words, they are directly connected to a system main, and the overflow point of the tank is the maximum system pressure. Occasionally, for some reason, a tank is not high enough and the maximum system pressure would cause it to overflow. In this case, the tank must be connected to the main through an altitude valve as described in chapter 6.

Most elevated tanks are constructed of steel. Early steel tanks were assembled with rivets, but tanks are now welded to provide a smooth appearance and to make them easier to maintain. Welding also allows tanks to be designed in new, more pleasing designs. The spherical and spheroidal tanks are very popular for small- to medium-sized tanks, both because they are visually more pleasing and because they are easier to maintain than tanks with individual legs. The single-pedestal tank is applicable to large tanks and offers the added advantage of providing space for water system storage in the pedestal.

Standpipes

A tank that has a height greater than its diameter is referred to as a *standpipe*. Standpipes have the advantage of providing storage for a great quantity of water at a cost proportionally not much greater than the cost of a pedestal tank. The disadvantages are that their great bulk may make them visually objectionable to residents, and the tremendous weight of the stored water means that they can only be constructed where foundation conditions are exactly right.

In most cases, only a portion of the water in the top of the standpipe will provide usable system pressure. When the water level falls to less than 70 ft (21 m) from the ground surface, there will be less than 30 psi (207 kPa) of pressure, which is generally the minimum that should be maintained in the distribution system. For this reason, most standpipes are constructed with an adjacent pumping station that can be used to boost the pressure of water from the lower section of the standpipe if it should be needed.

Standpipes may be constructed of welded steel or concrete. Standards for the construction of welded steel tanks are covered in AWWA Standard D100, *Welded Steel Tanks for Water Storage.*

Standpipes are also available that are field-assembled by bolting together uniformly sized steel panels. The panels are factory coated by hot-dipped galvanizing, coating with fused glass, or epoxy coated to provide long-term corrosion protection. A watertight seal is achieved by using a gasket or sealant between the panels. Guidelines for the design of bolted steel tanks may be found in AWWA Standard D103, *Factory-Coated Bolted Steel Tanks for Water Storage.*

Reservoirs

The term *reservoir* has a wide range of meanings in the water supply industry. A raw water reservoir is typically a pond, lake, or basin that is either naturally occurring or constructed. For the storage of finished water, the term is generally applied to a large storage tank that is larger in diameter than its height and is set on the ground surface or buried.

For equal-sized facilities, the reservoir will have a lower initial cost but must have a pumping station to transfer the water to the distribution system. After the maintenance and operation of the reservoir facilities are taken into account, the long-range costs will be approximately equal.

Reservoirs are usually selected for storage either because a larger quantity of water must be stored than is practical for an elevated tank, or because elevated storage would be visually objectionable to residents. Where a ground-level reservoir would be objectionable, the installation may be made partially or completely underground. Some buried installations have been made into parks with baseball or tennis facilities installed over the reservoirs.

One of the problems with reservoirs is that some water must be removed and replaced daily so that the water does not become stagnant. Most water systems pump out some water every day during peak use periods and refill at night. A reservoir is usually filled through an altitude valve that can be remotely operated and is hydraulically connected to sense the reservoir water level and to automatically close when the reservoir is full. Careful monitoring of water in reservoirs is needed to ensure constant quality.

Early reservoirs were constructed with part excavation and part embankment and paved with riprap, brick, or concrete, and were usually not covered. Among the problems were leakage, freezing, contamination by birds and other animals, algae growth, and potential vandalism. The availability of better construction methods, and state and federal pressure because of the potential health hazards, gradually caused these facilities to be replaced with new reservoirs.

Some ground-level reservoirs are constructed of bolted or welded steel set on a concrete foundation and floor. Although interior and exterior corrosion coatings must be maintained and occasionally replaced, new coating systems currently available last quite a long time.

Concrete reservoirs may be constructed by several methods.

Cast-in-place tanks are constructed about the same as a basement for a building, except that much more reinforcing steel is required to resist the outward pressure of the water. They are generally square or rectangular in shape, and special provisions must be made to prevent leakage at the construction joints. It is difficult to prevent some cracking of cast-in-place concrete tanks, but they can be filled with new types of flexible caulking compound.

Circular prestressed concrete tanks are constructed by beginning with an inner concrete core wall set on a ring foundation. Steel wire is then wrapped around the core under tension, and the wire is protected by a layer of gunite (hydraulically applied concrete), as illustrated in Figure 10-3. Because of the strength of the prestressed concrete, the walls can be made thinner than on a cast-in-place reservoir. The applicable AWWA standards are D110, *Wire- and Strand-Wound, Circular, Prestressed Concrete Water Tanks*, and D115, *Circular Prestressed Concrete Water Tanks With Circumferential Tendons.*

Hydraulically applied concrete-lined reservoirs are constructed much the same as some swimming pools. An earth excavation or embankment is first covered with reinforcing rods and then covered with gunite. A reservoir of this type is typically relatively small and requires special provisions for installing a tight cover.

Hydropneumatic Storage Systems

Hydropneumatic storage tanks are used by very small systems that cannot justify an elevated or ground-level tank. As illustrated in Figure 10-4, a steel pressure tank is kept partially filled with compressed air to provide quantities of water in excess of the pump capacity when required. It also keeps the pump from cycling too often and will provide water for a limited time in the event of pump failure.

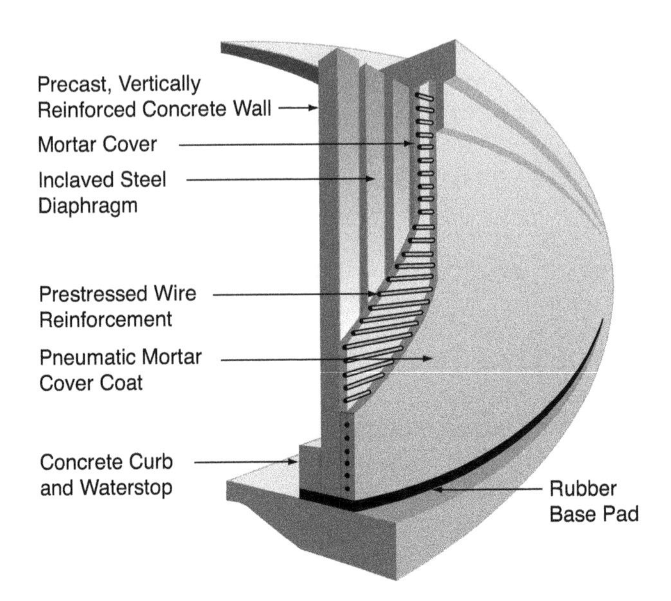

Courtesy of Preload, Inc.

Figure 10-3 Sectional view of a prestressed concrete tank

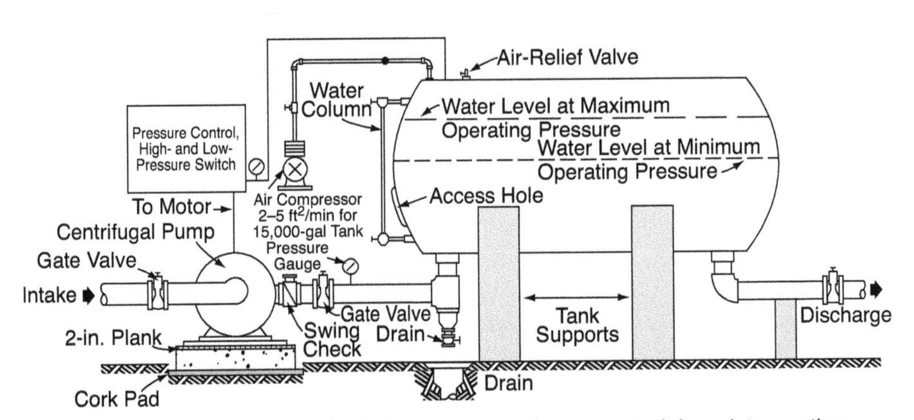

NOTE: Use special rubber hose fitting between pump and pressure tank for quiet operation.

Source: Joseph A. Salvato, Environmental Engineering and Sanitation, *4th ed. Copyright © 1992 John Wiley & Sons, Inc. Reprinted by permission of John Wiley & Sons, Inc.*

Figure 10-4 Hydropneumatic water pressure system

The system provides relatively little fire protection but is a fairly reliable source for domestic water for a few customers. State regulatory agencies usually specify the minimum tank size that must be used, based on the number of persons or houses served.

LOCATION OF DISTRIBUTION STORAGE

The locations of distribution system elevated tanks or ground-level reservoirs are governed by two principal factors: the hydraulics of the system and the availability of appropriate land that is acceptable to the public for construction of the facility.

Hydraulic Considerations

As illustrated in Figure 10-5, there are three primary alternatives in locating storage facilities. It is often most convenient to locate storage at the water source, but as the water system grows, the head loss to the far ends of the system can become excessive. If the only storage for a large system is located at the source, large-diameter transmission mains must be installed to provide adequate flow to all areas. Installing these mains is usually quite expensive if they must be constructed through built-up neighborhoods.

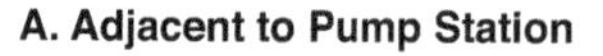

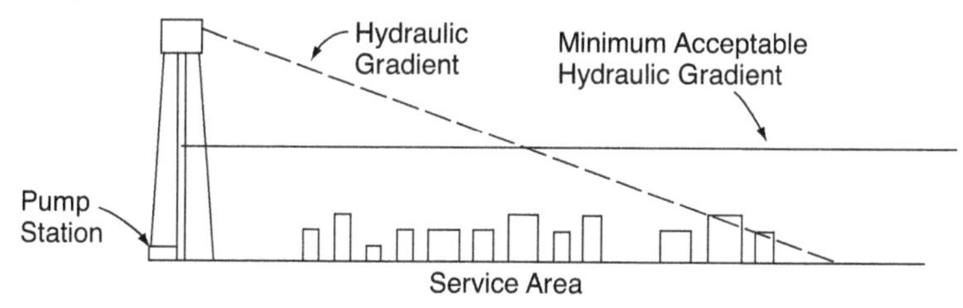

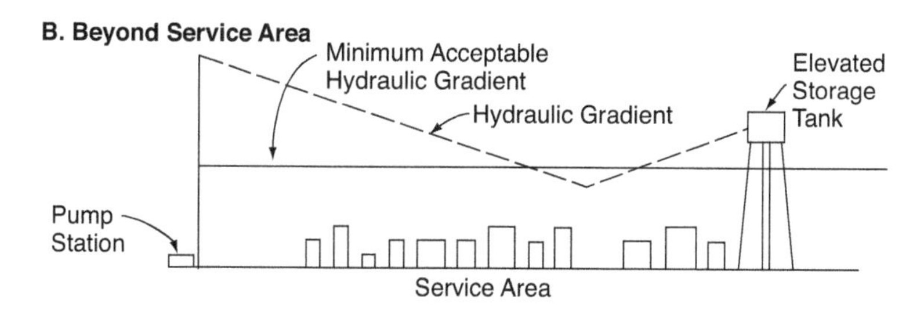

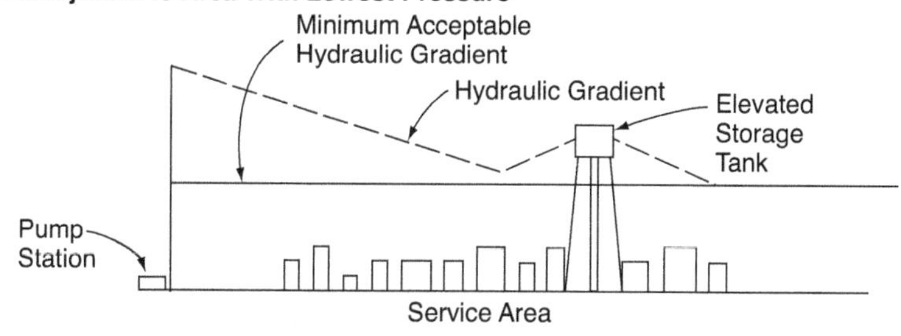

Figure 10-5 Different locations of elevated storage

To avoid the cost of adding transmission mains, storage can be located at the far ends of the distribution system. During high-use periods, water will flow from the tanks to improve the pressure in remote areas. The only problem that must be carefully studied is that there must be adequate main capacity between the source and the remote tanks to refill the tanks at night. If they cannot be completely filled, additional transmission mains may have to be installed just to transport water to the tanks.

Quite often, the best alternative is to install the storage tanks at an intermediate location where the existing distribution system is more likely to support them, but they will still reinforce the far ends of the system.

Installation of one large storage tank will usually require main reinforcement to support it, whereas several smaller tanks at different locations may be able to individually work acceptably on the existing system. The economic factors of the alternatives must be carefully considered during planning.

Aesthetic Considerations

People who live next to an existing elevated water tank rarely notice or are bothered by it. But if a new tank is proposed for an urban area, the residents will usually find it unacceptable. Many new designs and colors are now available that are quite pleasing, but the public reaction is still generally negative. An additional complaint is that the tank might affect television reception in the area.

As a result of these concerns, an ideal piece of land meeting the requirements of a reasonable price, a high elevation, and a good location hydraulically on the distribution system may not be acceptable for the location of an elevated tank. If the tank must be elevated, one of the few alternatives in most communities is to place it in an industrial park, public park, or other area away from homes. In many cases, an acceptable site may not be a very good location for hydraulic flow, elevation, or other reasons, so the installation may cost considerably more than it would at the optimum location.

Another alternative is to install an underground reservoir in the urban area. Many concessions may have to be made with residents, but the installation can usually be designed to gain their acceptance.

ELEVATED TANK EQUIPMENT

The principal accessories to an elevated storage tank are illustrated in Figure 10-6.

Riser Pipe

Usually only one pipe serves as both water inlet and outlet to a tank, and it is called the *riser*. In cold climates, the riser must either be protected from freezing or be large enough in diameter so that it can partially freeze

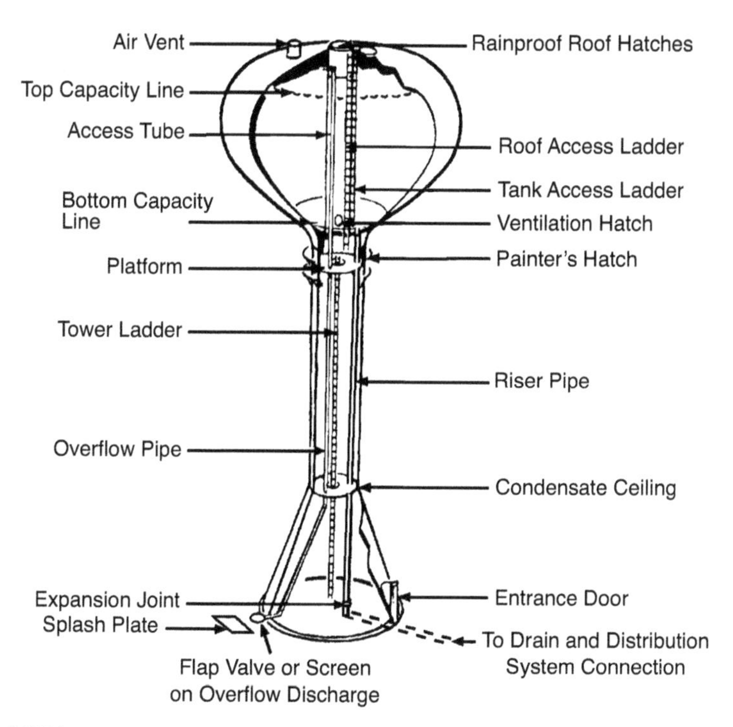

Courtesy of CB&J.

Figure 10-6 Principal accessories for an elevated storage tank

around the outside but still allow flow through the center. The riser of a multicolumn tank is often 6 ft (1.8 m) or larger. An advantage of the single-pedestal tank design is that some heat can be provided inside the column to protect the small-diameter riser from freezing.

Overflow Pipe

An important accessory for all tanks is the overflow pipe, which relieves excess water in the event of excessive system pressure. Proper functioning of the overflow is particularly important in freezing weather. If excess water flows out of the vent at the top of the tank, it will freeze, and the excess weight can be sufficient to cause the tank to collapse. The overflow pipe begins at the maximum tank water level and is brought down to within a foot or so of the ground. The end of the pipe is usually closed by a weighted flap to exclude insects and animals, but it must positively break open if overflow takes place. The overflow discharge should empty onto a splash plate to prevent erosion but should never be directly connected to a sewer or storm drain.

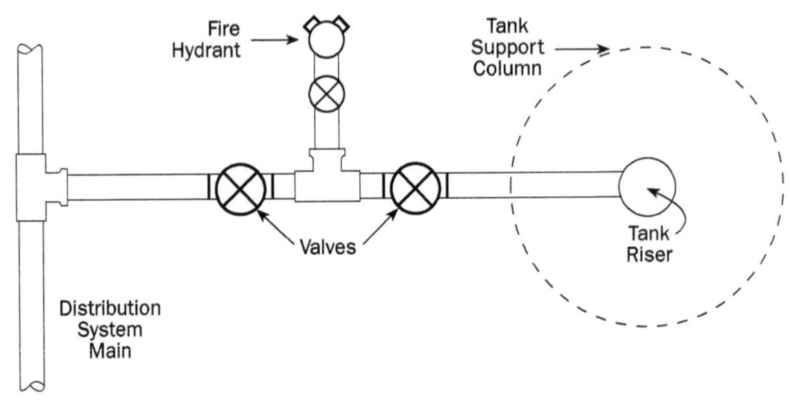

Figure 10-7 Fire hydrant installed for draining an elevated tank

Drain Connection

An elevated tank must be furnished with a drain connection that can be used to empty it for maintenance and repair. Figure 10-7 shows a common method of draining through a hydrant located on the pipe at the base of the tank. Provisions must be made for discharging the flow from the hydrant without causing erosion or creating a cross-connection. If chlorinated water is to be discharged following disinfection of the tank, environmental officials should be consulted about which method of disposal is acceptable.

Air Vent

A tank must be provided with a vent that will allow air to enter and exit as the water level rises and falls. If the vent should become blocked as water is draining from the tank, the vacuum formed could be sufficient to collapse the tank walls.

The vent must be screened to keep out birds, but the mesh cannot be so fine that it might become blocked by freezing moisture. Insects are not usually a problem because of the height of the tank. Most state regulations require a screen with ¼-in. (6-mm) mesh.

Access Hatches

Storage tanks must have hatches installed both for workers to enter and for top and bottom ventilation during maintenance and inspection. Roof vents are designed with a rim under the cover to prevent surface runoff from entering the tank.

A tank having a large-diameter wet riser has a heavy access hatch near ground level that is constructed of steel to withstand the high pressure. This hatch is used for cleaning the riser and to provide ventilation to the tank during repair.

Ladders

A multicolumn tank will usually have three ladders. The tower ladder runs up one leg to the balcony, another ladder goes from the balcony to the roof, and the roof ladder reaches from the balcony ladder to the roof hatch. The tower ladder usually starts at a point about 8 ft (2.5 m) above ground level to deter unauthorized persons from climbing it. A pedestal tank has similar ladders located inside the pedestal.

All ladders must have safety devices in compliance with the federal Occupational Safety and Health Administration requirements. One method is to have a safety cage around the ladder. With another type of ladder safety rail system, a device is fastened to the worker's safety belt and easily rides on the rail while the worker climbs up or down, but if there is a quick downward movement, a locking pawl engages with notches in the rail.

Obstruction Lighting

Depending on the height of a tank and the proximity to an airport, the Federal Aviation Administration (FAA) may require the installation of a red light or strobe light on top of the tank. In locations considered particularly hazardous to aviation, orange and white obstruction painting on the tank may also be required. The FAA should be contacted before construction of any elevated tank to make sure it has no objection and to identify any special requirements.

GROUND-LEVEL RESERVOIR EQUIPMENT

The same general accessories are required for reservoirs as for elevated tanks, with the exception that increased caution must be exercised to prevent vandalism. Vents and hatches in particular are more accessible, so special care must be taken to keep them locked and guarded.

Inlet and Outlet Piping

Reservoirs may in some instances use a single inlet and outlet pipe. Other designs use separate pipelines for the purpose of increasing circulation to maintain uniform water quality and help reduce freezing.

The outlet to a reservoir is usually located a short distance above the floor to prevent any silt that accumulates on the floor from being drawn into the water leaving the tank.

Drains

It is often difficult to install a gravity drain for a reservoir without creating a potential cross-connection, unless the installation happens to be on the top of a hill. The usual practice is to dewater the reservoir for inspection and repair by pumping most of the water to the distribution system. But it is not wise to pump all the way to the outlet opening, because if there is any accumulated sediment on the bottom, it will be picked up and pumped out as cloudy water. The last few feet of water in the tank should be pumped out and wasted using portable pumps. Disposal of reservoir sediment and water may be subject to local regulations that may require dechlorination and/or removal to an approved site.

OPERATION AND MAINTENANCE OF STORAGE FACILITIES

Cold-Weather Operation

Freezing of ground-level reservoirs is typically not a problem because the water picks up some heat from the ground. Some ice may form around the walls of the tank, but it does not harm anything. Some utilities in northern areas drain some of their reservoirs during the winter because they are not needed because of decreased water use. This simplifies winter operation, but it creates a sizable job of having to disinfect the reservoir before returning it to service in the spring.

The problems of ice formation in elevated tanks in northern areas must be clearly understood by system operators. If a tank should be allowed to freeze solid, it not only might damage the structure, but it will be very costly to have it thawed by a professional firm.

It is normal for ice to form around the walls of the tank, and it usually will not damage the structure. What is important is to keep thick ice from forming on the water surface. This is done by keeping the surface of the water moving up and down by continually varying system pressure. Each time the tank level is raised, warmer water is brought in. If system pressure is maintained by automatic controls, they must be set in the winter to purposely vary the pressure over a relatively wide range.

Coatings for Steel Tanks

When steel is exposed to oxygen and moisture, it corrodes; thus, protection is required for both the interior and exterior of a steel tank. The exterior of a tank is exposed to a rather uniform environment, but the interior has more severe conditions, and parts of the interior walls may be severely scraped by ice.

New coatings now available are much more durable than those available just a few years ago, but they will still eventually need to be replaced. If an old tank was originally coated with lead-based paint, it is best removed by sand blasting. This must be done by special means to prevent environmental pollution. State officials should be consulted about the allowable levels of both lead and silica that can be released to the environment.

Care should be taken to ensure that the coatings to be used on the interior of a tank comply with NSF International standards for not imparting any taste or odor or releasing harmful chemicals to the water. When a tank is painted by contract, the work should be inspected to make sure the contractor is fully meeting the specifications. Inspecting the work is both dangerous and beyond the knowledge of most water system operators, so it is suggested that a qualified third party be employed to inspect the work. Competing contractors should not be used for inspection because of possible conflict of interest. Steel tank coating is covered in AWWA Standard D102, *Coating Steel Water-Storage Tanks.*

Cathodic Protection

Even if the interior of a tank is fully coated, there is still some chance for corrosion. If there are any small breaks in the coating (holidays), the corrosion will concentrate in these areas and can do serious damage in a relatively short time. A cathodic protection system can greatly reduce corrosion of a tank's interior.

A cathodic protection system consists of a series of anodes that are suspended in the tank with a direct current impressed on them. Instead of corrosion dissolving iron from the tank surface, the current causes the anodes to corrode (or be sacrificed). In warm climates, the anodes can be suspended from the tank roof and left in place until they require replacement (Figure 10-8). In climates where ice will form in the tank, ice will pull down suspended anodes, so a submerged anode system is used that resists icing damage (Figure 10-9).

Anodes will usually last as long as 10 years, but it is recommended that the system be inspected annually to ensure they are operating properly.

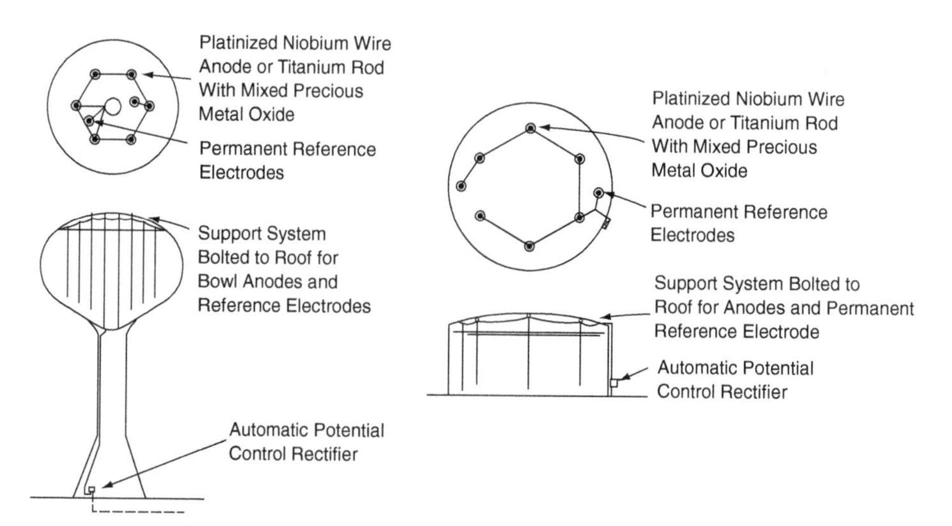

Courtesy of Harco/CPS Waterworks, A Corrpro Company, 1055 West Smith Road, Medina, Ohio 44256.

Figure 10-8 Typical methods of suspending cathodic protection anodes in steel tanks where icing conditions do not exist

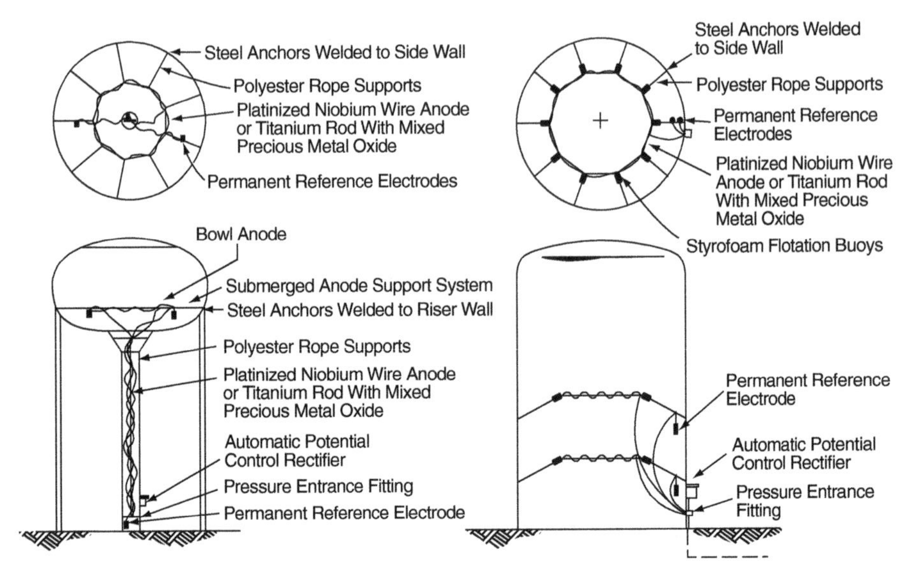

Courtesy of Harco/CPS Waterworks, A Corrpro Company, 1055 West Smith Road, Medina, Ohio 44256.

Figure 10-9 Typical methods of suspending cathodic protection anodes in steel tanks where icing conditions exist

A cathodic protection system can be successfully operated on a tank with no interior coating, but the anodes will disintegrate rather quickly, so it is usually recommended that the system be used in conjunction with a good interior coating. Details of cathodic protection systems are covered in AWWA Standard D104, *Automatically Controlled, Impressed-Current Cathodic Protection for the Interior of Steel Water Tanks.*

Tank Inspection, Cleaning, and Disinfection

Inspection of elevated tanks is both dangerous and specialized, so it is best to employ a qualified firm to inspect the tank and submit a report on the condition with suggestions on any repairs that are necessary. It is generally best to have the inspection done annually or periodically by a firm other than the one that will be doing the repair work.

Annual or periodic inspection and minor repair of ground-level reservoirs is within the capability of most water system repair crews if safety measures are carefully followed. If the reservoir is emptied for interior inspection, special safety measures include provision of adequate ventilation, sufficient lighting with special care to prevent electric shock, proper boots and other gear for workers, and security and fall protection on ladders. In most cases, a confined-space program must be instituted.

Tanks should be inspected for the condition of overflows and vents to make sure they are not blocked and that screens are clean and in place. Leaks in roofs may need to be investigated during wet weather to ensure against intrusion. Concrete tanks should particularly be inspected for cracks. Tanks should be cleaned and inspected at least once every three years, and where there are sediment problems, annual cleaning is recommended. After inspection and cleaning, tanks must be disinfected per AWWA Standard C652, *Disinfection of Water-Storage Facilities.*

Exterior inspection of all tanks and reservoirs should be made frequently (once a month or weekly, if possible). Among the items that should be checked are

- All hatches and locks to make sure they are secure,
- General condition for any signs of vandalism,
- Proper operation of security lights and aircraft warning light, and
- Leaks in the roof where contaminants can enter.

Records of Tank Maintenance

Because most water systems have only a few water tanks or reservoirs, record keeping does not have to be elaborate. Basic information that should be kept in a file includes the tank location and original construction information, copies of all inspection reports, and details of each inspection and repair that is made. A record should also be kept of the names and phone numbers of the tank manufacturer, repair contractors, and equipment suppliers so they are readily available in the event of an emergency.

BIBLIOGRAPHY

ANSI/AWWA C652. *AWWA Standard for Disinfection of Water-Storage Facilities.* Denver, Colo.: American Water Works Association.

ANSI/AWWA D100. *AWWA Standard for Welded Steel Tanks for Water Storage.* Denver, Colo.: American Water Works Association.

ANSI/AWWA D102. *AWWA Standard for Coating Steel Water-Storage Tanks.* Denver, Colo.: American Water Works Association.

ANSI/AWWA D103. *AWWA Standard for Factory-Coated Bolted Steel Tanks for Water Storage.* Denver, Colo.: American Water Works Association.

ANSI/AWWA D104. *AWWA Standard for Automatically Controlled, Impressed-Current Cathodic Protection for the Interior of Steel Water Tanks.* Denver, Colo.: American Water Works Association.

ANSI/AWWA D110. *AWWA Standard for Wire- and Strand-Wound, Circular, Prestressed Concrete Water Tanks.* Denver, Colo.: American Water Works Association.

ANSI/AWWA D115. *AWWA Standard for Circular Prestressed Concrete Water Tanks With Circumferential Tendons.* Denver, Colo.: American Water Works Association.

ANSI/AWWA D130. *AWWA Standard for Flexible-Membrane-Lining and Floating-Cover Materials for Potable Water Storage.* Denver, Colo.: American Water Works Association.

AWWA. 2010. Principles and Practices of Water Supply Operations—*Water Transmission and Distribution*, 4th ed. Denver, Colo.: American Water Works Association.

AWWA. M36—*Water Audits and Loss Control Programs.* Denver, Colo.: American Water Works Association.

AWWA. M42—*Steel Water-Storage Tanks.* Denver, Colo.: American Water Works Association.

Salvato, Joseph A. 1992. *Environmental Engineering and Sanitation*, 4th ed. New York: John Wiley & Sons, Inc.

Chapter 11

Water Services

A water service is the pipeline that carries water from the utility's water mains to the consumer's building or other point of use. Ownership (and therefore maintenance responsibility) of the service line varies for each system.

SERVICE LINE DESIGN

Water services are typically designed in two ways in regard to meter placement: with the meter located either in a pit close to the main or at the building.

Meters Located in Basements

In parts of the country that never experience freezing, meters can be located almost anywhere. Some utilities in the southern United States allow meters to be installed in garages or on the sides of buildings, as long as they are reasonably protected from damage and vandalism.

Where buildings have basements, the meter is usually located in the basement. As illustrated in Figure 11-1, the service pipe is usually placed under the footing so it will be below the frost line. The meter should be located as close as possible to where the pipe enters the building to avoid any temptation to the customer to install an illegal connection ahead of the meter. The curb stop is used for turning off the service for repair or nonpayment of bills. The curb stop with box is normally installed at a point in the parkway or on the property line. It should not be on the customer's property.

In communities where the policy is to install meters in the basements of all buildings having basements, a special problem arises as to where to locate meters in buildings without basements. Some utilities require that meters for these buildings be installed in meter pits or boxes. Some utilities will allow the meter to be installed in a crawl space or utility closet, but this generally creates difficult meter reading and meter replacement problems.

Meters Installed in Meter Boxes

In warmer climates, where the water service does not have to be very deep, it is common to install all meters in meter boxes. The meter box is usually

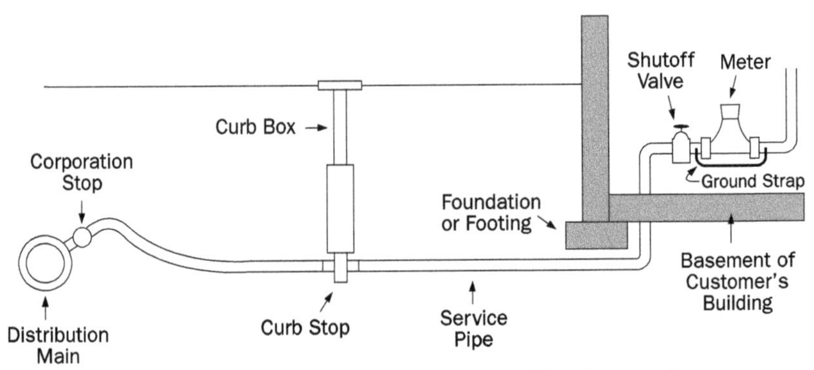

Figure 11-1 Small service connection with the meter located in a basement

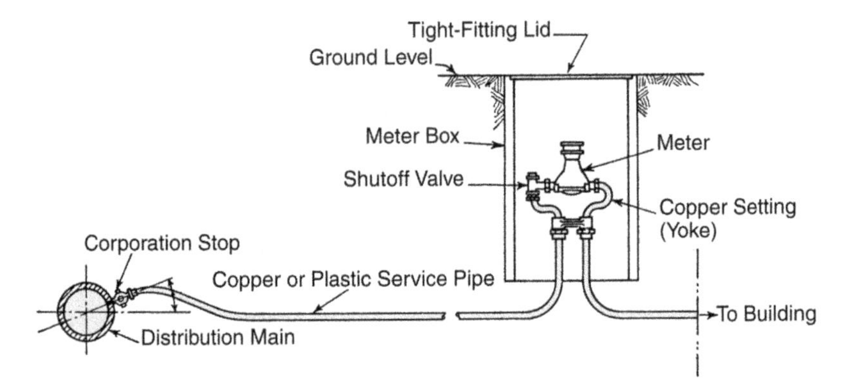

Figure 11-2 Small service connection with shallow meter box

located in the parkway between the curb and sidewalk. The box may be a vitrified or concrete tile, or may be made of cast iron or plastic. Figure 11-2 illustrates a meter box installation.

In northern climates, where the service must be buried quite deep, the meter must be located in a larger enclosure, usually called a meter pit. As long as a pit has a tight-fitting cover, the meter can be raised relatively close to the surface to facilitate reading without danger of freezing because heat from the warmer ground at the bottom circulates within the pit (Figure 11-3).

One of the reasons many water systems originally standardized on installing meters in boxes instead of in buildings was to facilitate meter reading. Manual reading of the meters in pits is much faster than going inside each building for reading. Now, with the various types of electronic meter reading devices that can be installed on meters, this is no longer an issue.

Other advantages of meter boxes are that the meters can be replaced without entering the building and there is less likelihood of customers

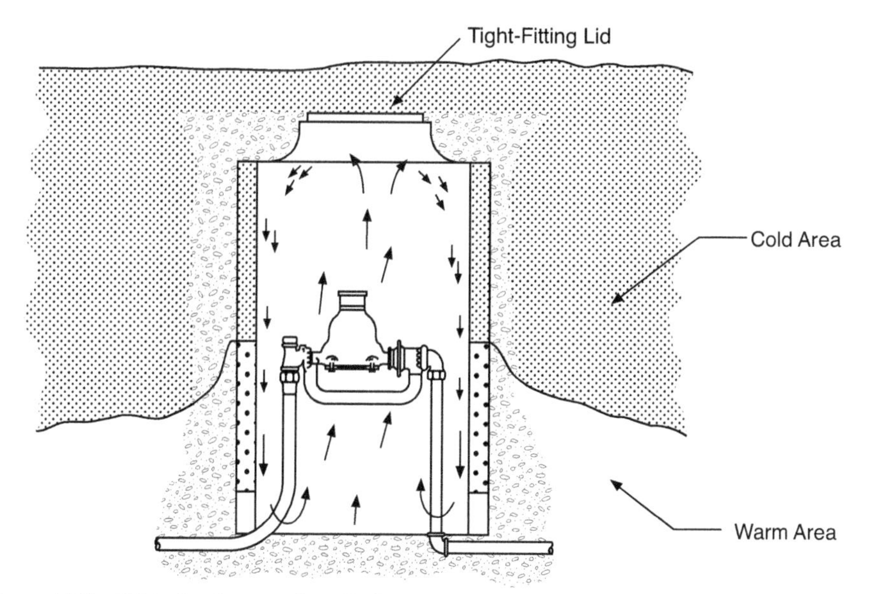

Figure 11-3 Cold-climate meter installation

tampering with them. Two major disadvantages of pits used in colder climates are that they may be difficult to locate and expose when covered with snow, and they are often filled with water that may have to be pumped out before the meter can be read.

Service Line Size

The size of the water service line necessary to properly serve a building generally depends on the following factors: the number of water-using appliances, water pressure, and the distance between the main and the building. Average single-family homes are normally adequately served with a ¾-in. (20-mm) service, and larger homes having several bathrooms with a 1-in. (25-mm) service. If water pressure is low or the building is set far back from the main, the service size should be increased to compensate for friction loss in the piping.

The size of water service for larger buildings depends in part on the use pattern. Hospitals, for instance, use larger quantities of water 24 hours a day. Some industries may have heavy use at some times, and almost none at others. Probably the ultimate in uneven use is in military barracks, where everyone wakes up at the same time in the morning and every shower and toilet is in use simultaneously. Additional information on sizing water services can be obtained from AWWA Manual M22, *Sizing Water Service Lines and Meters*.

SERVICE LINE MATERIALS

The types of material used for small service lines include lead, galvanized iron, copper, and several types of plastic. Services over 2 in. (50 mm) in diameter are usually the same as water main materials.

Lead Pipe

Lead pipe has been used for transporting water since Roman times because it is relatively strong and long lasting. It is also flexible, so it will not be affected by ground movement. The word *plumber* comes from the Latin word for lead, *plumbum*. The connection of lead pipe to fittings was done by partially melting the lead in a process called *wiping*, which required considerable skill by plumbers.

Use of lead for water services gradually decreased as other materials became available and the cost of lead became proportionally higher. Most older water systems still have many lead services in use. Lead services that must be repaired can be connected to newer type materials using a special mechanical fitting, but most utilities have established a policy that a lead service that has broken or is leaking must be completely replaced with new material. Many water utilities have also established a policy that, if a street is to be repaved, all lead services under the street must be replaced as part of the project. One reason for this is that some old lead becomes brittle and may leak if disturbed. Another reason for lead service replacement is new evidence that lead leached from plumbing could present a health risk, particularly to small children.

Galvanized Iron Pipe

Many older water systems allowed the use of galvanized iron pipe as a less expensive substitute for lead as a water service material. Because the pipe is rigid, it was usually connected to the corporation stop with a lead gooseneck a few feet long to provide flexibility at the connection to the main.

The life of galvanized pipe varies widely depending on water quality and soil conditions. Some very old galvanized services continue to provide good service, but if they are disturbed for any reason, they will usually start to leak. Repair of an old galvanized pipe is not usually successful in the long run, so most utilities insist that a leaking galvanized service pipe be completely replaced. There is also concern by many people over the possible adverse health effects of lead leached from the lead goosenecks, so some systems are replacing galvanized services for this reason.

Another problem with galvanized pipe is that direct connection of brass valves and other fittings often creates a galvanic action that hastens corrosion of the pipe at the connection.

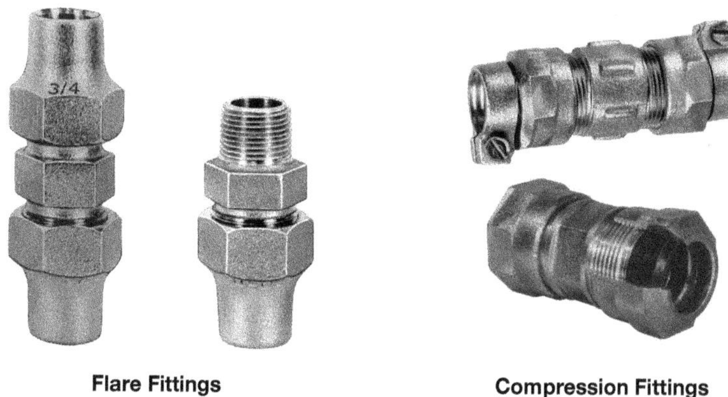

Flare Fittings **Compression Fittings**

Courtesy of The Ford Meter Box Company, Inc.

Figure 11-4 Fittings for water service pipe

Copper Tubing

Beginning shortly before World War II, copper tubing became popular as a replacement for lead and galvanized pipe for water services. The material is flexible, easy to install, corrosion resistant in all but the most corrosive soil, able to withstand high pressure, and not excessively expensive. Other advantages are that there is no serious reaction between the copper and connected brass valves and fittings, and the service lines are easy to locate with an electronic locator.

Very corrosive water may dissolve enough copper from copper tubing to cause stains on plumbing fixtures, but water systems that might have this problem generally must treat the water to reduce corrosivity for other reasons. Although copper tubing used in interior plumbing is usually joined with solder joints, copper that is buried is usually joined by either flare or compression joints (Figure 11-4).

Plastic Tubing

Plastic material used for water services must not soften and swell or become brittle and crack over time. It also must be a material that will not be attacked by underground rodents and will not leach harmful chemicals into the water. Three types of plastic are generally used for water service—polyvinyl chloride (PVC), polyethylene (PE), and polybutylene (PB). PB tubing is not used for new services because of structural problems with older services.

All plastic pipe and tubing to be used for potable water must be tested by an approved laboratory to meet NSF International (formerly National Sanitation Foundation) standards to ensure that chemicals will not leach

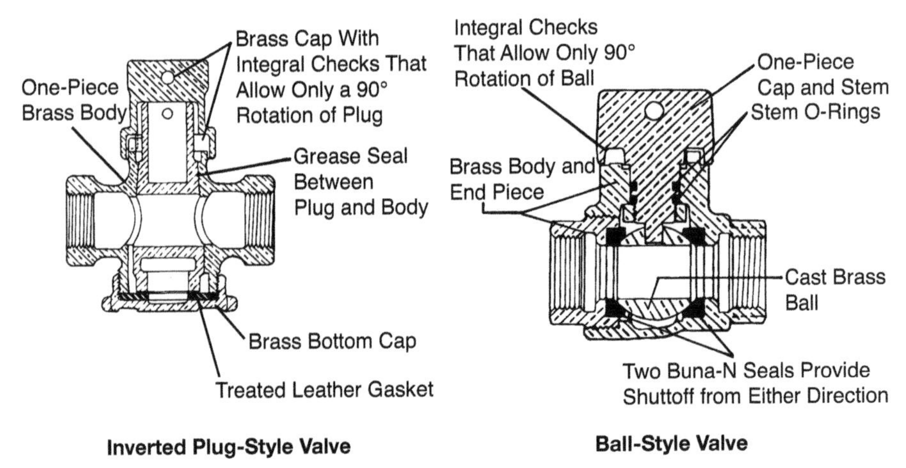

Courtesy of A.Y. McDonald Manufacturing Company, Dubuque Iowa.

Figure 11-5 Principal styles of curb stops

from the plastic and cause tastes or odors, or present a health threat. NSF approval must be printed along the exterior of the tubing.

Plastic tubing has a very smooth interior, and is relatively flexible and very lightweight, which makes it easy to install. Valves and other accessories are connected to plastic tubing using brass fittings similar to those used for copper tubing. Plastic tubing is covered in several AWWA standards in the C900 series.

Plastic tubing must not be used in a location where gasoline, fuel oil, or industrial solvents have contaminated the soil. Petroleum can attack plastic and cause it to eventually fail. Also, low concentrations can pass through the pipe and contaminate the water.

Curb Stops and Boxes

Every water service should have a shutoff valve so that the service line may be easily turned off for repairs or nonpayment of the water bill.

Curb Stops

The valve commonly installed as the main shutoff for a water service is called a *curb valve* or *curb stop*. The type of valve that has been used for this service for many years is the inverted plug style (Figure 11-5). An inverted plug valve that has not been used for many years may be rather stiff but can usually be made to operate. Problems occur in operating a very old valve on a lead or galvanized-iron service because the twisting action on the valve is likely to initiate a leak on the adjoining service pipe.

The newer style curb stop with a ball valve operates very easily, even after many years of nonuse. Curb stops are available with a variety of piping connections. The type with female pipe thread connections requires a

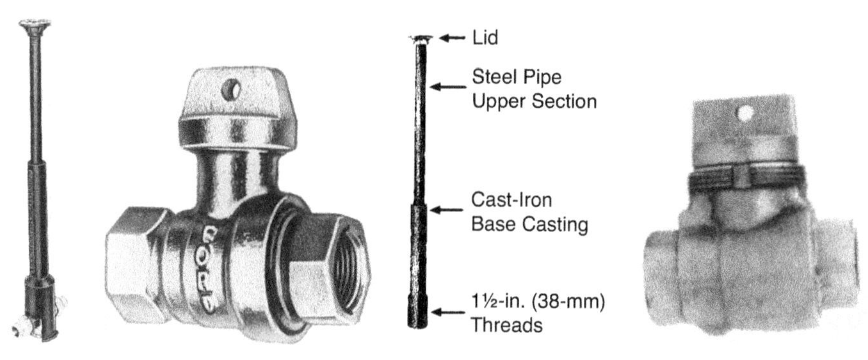

Arch-Pattern Curb Box and Curb Stop

Courtesy of The Ford Meter Box Company, Inc.

Minneapolis-Style Curb Box and Curb Stop

Courtesy of A.Y. McDonald Manufacturing Company, Dubuque Iowa.

Figure 11-6 Styles of curb stops and boxes

connector to adapt to the type of pipe being used. Stops are also available with flare and compression couplings for direct connection to the service line.

If the water meter for a service is located in a box or pit, the curb stop is usually located just ahead of the meter so it will be accessible through the box. A standard curb stop or a special meter valve may be used.

Curb Boxes

If the meter is located in a building, the stop is fitted with a box so that it may be operated with a special valve key. There are two styles of boxes in general use in the United States (Figure 11-6). The arch-style box sits loosely over the top of the stop. The Minneapolis-style box has threads at the bottom that screw onto special threads on the top of a Minneapolis-style curb stop.

Each style has some advantages and disadvantages. If the arch-style box is used in loose soil, some soil may work up into the box from the bottom. It is also possible for the box to shift so that the key will not fit on the valve. On the other hand, if the box is disturbed, the service line will usually be unaffected.

The Minneapolis-style connection eliminates the possibility of filling with soil or misalignment. The one disadvantage is that if there is any serious damage to the box, such as being inadvertently pulled up by construction equipment, the service line will come right along with it.

Various styles of lids are available for curb boxes. Most of them use a pentagon nut that can only be operated by a special wrench to discourage unauthorized persons from removing the lid. It is particularly important to keep children from opening the lid, because they usually drop stones into the box. Water systems generally standardize on a particular style of box and insist that they be used for all new construction and repair.

Valve keys having a slotted end to engage the top of curb stops are available in various lengths to match the standard depth of services. A key is generally easiest to use if it projects to about waist height when it is being used. Most water systems inadvertently end up with some services that are much deeper than normal, usually because fill dirt was added to the properties after the services were installed. It is therefore good practice to have one extra-long key available.

Water service valves and fittings are covered in AWWA Standard C800, *Underground Service Line Valves and Fittings.*

WATER SERVICE TAPS

If the pressure in a water main can be easily turned off, a connection for a water service can be made by drilling a hole in the pipe and either threading the hole to insert a fitting or placing a saddle with fittings over the hole (dry tap). However, it is usually more convenient to make the connection while the main is pressurized (wet tap). Besides being much more convenient, a wet tap is also preferred because there is less chance of contamination entering the main.

Corporation Stops With Different Pipe Connections

Courtesy of The Ford Meter Box Company, Inc.

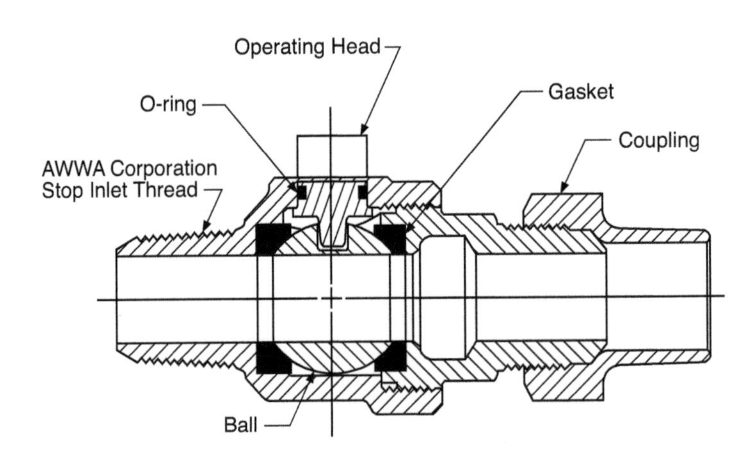

Principal Parts of a Ball-Style Corporation Stop

Courtesy of A.Y. McDonald Manufacturing Company, Dubuque Iowa.

Figure 11-7 Corporation stops

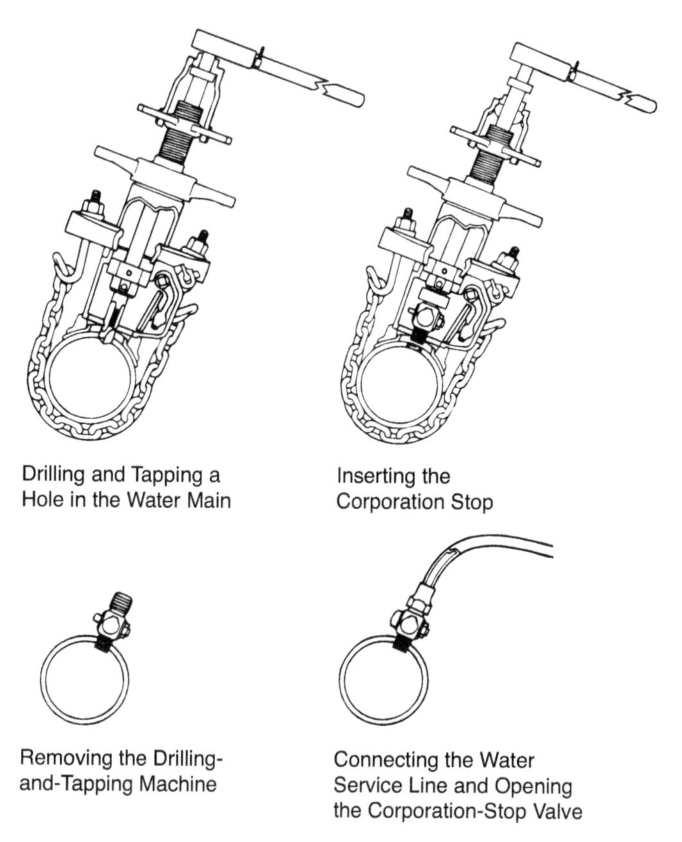

Courtesy of Mueller Company, Decatur, Ill.

Figure 11-8 Steps in making a service tap

Water services have traditionally been connected to the water main through a special brass valve called a *corporation stop* (Figure 11-7). The valve serves two functions. It provides a means of connecting to the main and allows the water to be turned off until the service line is completed and ready for use. It also provides a means of turning off the connection if the service is ever discontinued or is to be completely replaced with new piping.

When small taps are to be made on ductile-iron pipe, they are usually made by direct insertion. A hole is drilled in the pipe, threads are made in the hole, and the corporation stop is screwed into the hole, as illustrated in Figure 11-8. Corporation stops that are to be used for direct insertion have a thread called an *AWWA thread*, which has more of a taper than standard pipe thread so that it will quickly tighten as it is screwed into the pipe. Corporation stops are available with either plug- or ball-type valves, and with various end connections.

Direct insertion may also be used for connecting service to asbestos–cement (A–C) and PVC mains, but the work must be done very carefully.

Courtesy of The Ford Meter Box Company, Inc.

Figure 11-9 Some styles of service clamps

Because of the steep taper in the corporation stop threads, excessive tightening of the valve as it is inserted can either damage the pipe or strip the threads. For this reason, many operators prefer to install all services on A–C and PVC pipe through a saddle.

Pipe manufacturers suggest that all taps larger than 1 in. (25 mm) made on any type of pipe 6 in. (150 mm) or smaller should be made through a saddle. Manufacturer's recommendations should be consulted if a direct tap is to be made in a larger-diameter main.

Various types of service clamps or saddles are available, with each size designed for only a limited range of pipe outside diameter (Figure 11-9). If a saddle tap is to be made on old pipe, the diameter of the pipe should be measured before the clamp is ordered. A saddle tap is made by installing the corporation stop on the clamp and then drilling a hole in the main through the open valve, as illustrated in Figure 11-10. Special service connectors for PVC pipe are also available that incorporate a built-in drill and valve as a complete unit (Figure 11-11).

Tap Location

It is generally recommended that the best location for the tap on a main is at an angle of about 45° down from the top of the main. A tap at the top of the main is more likely to draw trapped air into the service line, and a tap near the bottom of the pipe is likely to draw in sediment. The service pipe should be laid in an S-curve down from the tap so there is plenty of slack to allow for earth settlement and pipe expansion and contraction.

Service Pipe Installation

Water service pipes must be installed below the deepest frost line anticipated for the area. They are particularly prone to freezing at night when water is not moving. The most likely place for a service to freeze is where there is no snow cover, such as under the street, sidewalk, or driveway.

In southern locations, pipe can often be installed deep enough using a trenching machine. In northern areas, a backhoe must usually be used to provide a trench that is deep enough. The pipe can be installed under paved areas by jacking a pilot pipe or by using a portable boring machine.

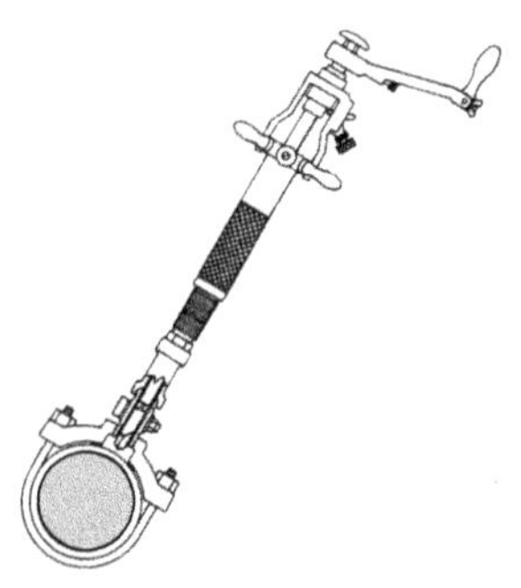

1. With the service clamp attached to the main, the corporation stop is threaded into the clamp. The machine is then mounted on the corporation stop, and the stop is is opened.

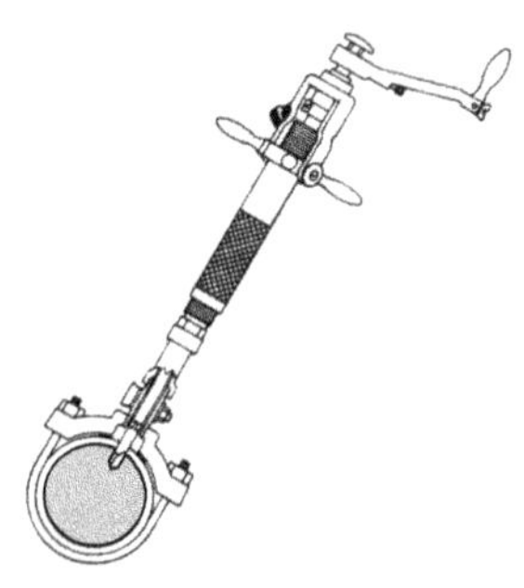

2. The drill penetrates the main without water escaping.

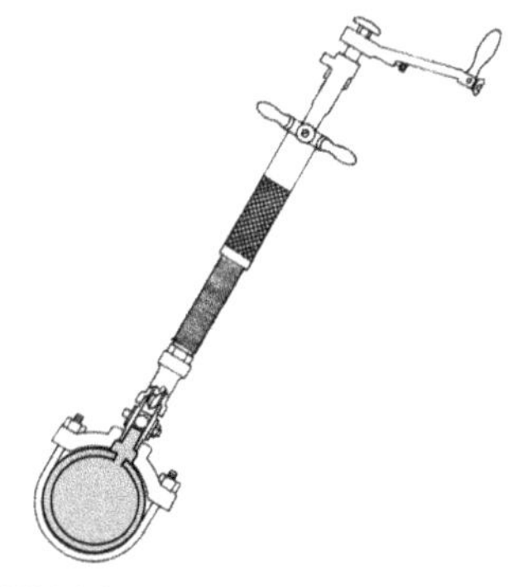

3. The drill bit is retracted, and the corporation stop is closed. The stop now controls the water.

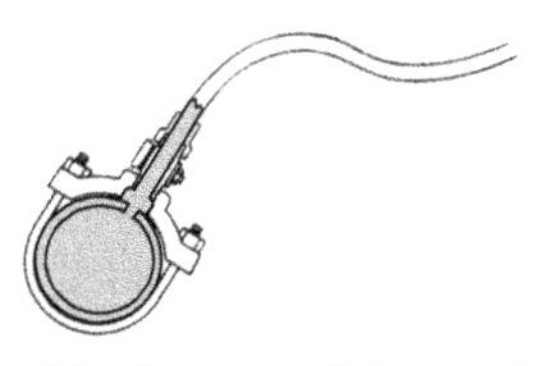

4. The machine is removed, the service line connected, and the corporation stop reopened to activate water service.

Courtesy of Mueller Company, Decatur, Ill.

Figure 11-10 Using a service clamp to install a corporation stop

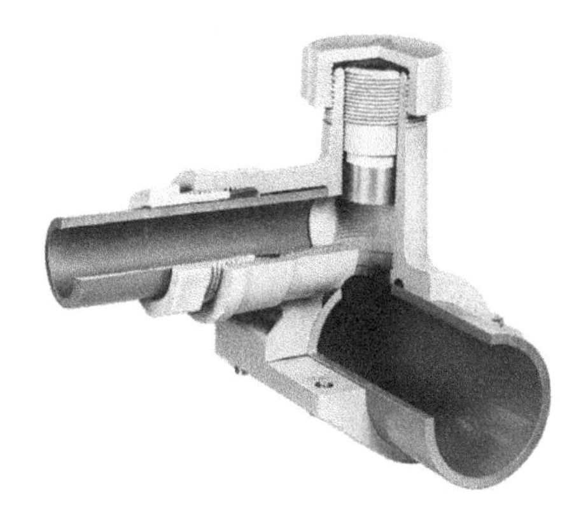

Courtesy of Dresser Piping Specialties.

Figure 11-11 Service connector made for use with PVC pipe

Plastic pipe has a relatively high coefficient of expansion, so the pipe should be installed with ample slack as it is placed in the ditch. If it is installed completely taut, the pipe may eventually pull out of a fitting as it contracts.

WATER SERVICE MAINTENANCE AND REPAIR

Leak and Break Repair

As discussed in chapter 3, water leaks in dense soil will usually either come to the surface or find their way into a sewer. If there is a question of whether or not there is a leak on a water service, the sound can usually be heard by listening on a valve key placed on the curb stop or on the meter. If it appears that the leak may be on the building side or the curb stop, the valve can be closed to determine whether the noise stops and the water stops flowing. In a location with porous soil, small leaks may not come to the surface where they will be noticed, so meter readers should be particularly vigilant in listening for leaks as they make their readings or inspections of the meter installations.

The most common location of leaks on old services is at the connections to the curb stop. This may be due to the twisting of the valve as it is operated or because of pressure on the curb box after heavy equipment has driven over it. An old service may also develop a leak after the adjacent soil has been disturbed, such as in the installation of a new sewer or gas line. Special fittings are available for repairing old lead and galvanized iron pipe, but once the pipe has been disturbed, there is a good chance it will leak again. It is usually best to replace the entire service with new material. If this is not possible, at least the entire leaking section should be replaced.

Thawing Frozen Services

The best way of preventing frozen services is to insist that all service piping be installed to below the recommended maximum frost depth for that part of the country. The most common cause of a shallow service is that the installing contractor did not comply with local requirements. The best policy is to insist on an inspection of the installation by a city or water utility inspector before the trench is backfilled. Some utilities go so far as to make the contractor reexcavate the trench if it is backfilled before inspection, to make sure the proper depth is maintained.

Continuous snow cover typically allows relatively little frost penetration in areas where the snow has not been disturbed. But at the same time, during a very cold winter, there can be several feet of frost in the ground under streets and driveways. The most common services to freeze are those that run under roadways.

A copper, lead, and galvanized-iron service can usually be thawed by running an electrical current through it. A portable source of direct current,

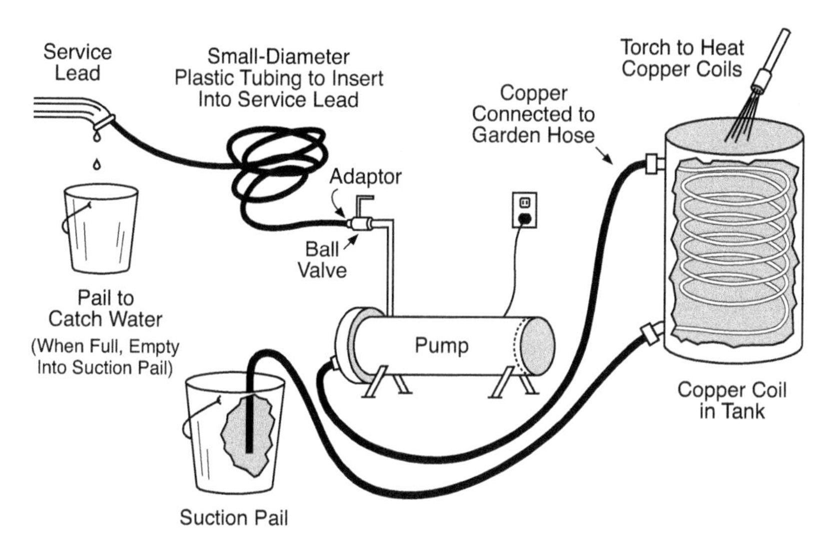

Source: Adapted from drawing by Randal W. Loeslie, manager, G.F.-Traill Water District, Thompson, N.D.

Figure 11-12 System for thawing plastic water line

such as a welding unit, can be connected to the main and the service at the building. This will usually generate enough heat in the line to release the ice blockage. Electrical thawing can be dangerous and should only be performed by someone with experience.

One of the problems that can arise in electrical thawing is a point of poor conductivity in the service line connections. Another potential problem is if the service is in contact with another conductor, such as a gas pipe, which will divert the current and could cause it to enter other adjacent buildings. The current may also damage O-rings or gaskets in the service fittings. Thawing is a service performed by the utility or a contractor on the customer's property, so a waiver should be signed by the property owner, and there must be confirmation by the insurer that the person doing the work has adequate liability insurance in effect to cover any possible consequences of the work.

Hot-water thawing is becoming more common because it is less dangerous and can be used for plastic pipe. As illustrated in Figure 11-12, a plastic tube carrying hot water is fed into the service line and pushed against the ice blockage. The same process can also be used with a steam generator, but it should first be determined if the extreme heat might damage plastic pipe.

If only a meter or small section of service is frozen in a meter pit, it is best thawed using a hair dryer or heat gun. A propane torch should be used with extreme caution for thawing because of the possibility of igniting explosive gases in the pit and the potential of damaging fittings by overheating them. If the line is heated too quickly, there is also the danger of generating steam, which may have no place to escape and could rupture the meter or piping.

After a frozen service pipe has been thawed, the only way to prevent it from freezing again is to open a faucet to allow a small, continuous flow until the ground has thawed. In that this could be several weeks or months, some water utilities will remove the meter or make allowance for the unusual water use when billing the customer.

Service Line Responsibility

Various municipalities and water utilities have different policies regarding responsibility for water services. If a policy does not exist, it is best to formally establish one so there will be no problem of convincing property owners of their responsibility when repairs are necessary. Each policy has advantages and disadvantages. Following are some of the more common policies:

- The entire water service is the responsibility of the property owner, from the main connection to the building. If any part of it leaks or breaks, it is their responsibility to pay for the repair.
- The portion of the service up to the property line is the water utility's responsibility and they will make any necessary repairs. The portion on private property is the responsibility of the property owner.
- The portion up to the curb stop or meter pit is the utility's responsibility, and the remainder is up to the property owner.

Policies also vary on who does the work of installing a new service. A utility with its own tapping machine usually prefers to make the tap on the main to ensure that the work is done correctly, and the property owner is billed for the service. This is often combined with a tapping fee that covers all of the costs of establishing a new service, such as inspection, establishing a new billing account, the initial cost of the meter, and possibly reimbursement for a portion of the cost of the main installation.

BIBLIOGRAPHY

ANSI/AWWA C800. *AWWA Standard for Underground Service Line Valves and Fittings*. Denver, Colo.: American Water Works Association.

ANSI/AWWA C901. *AWWA Standard for Polyethylene (PE) Pressure Pipe and Tubing, ½ In. (13 mm) Through 3 In. (76 mm), for Water Service*. Denver, Colo.: American Water Works Association.

AWWA. 2010. Principles and Practices of Water Supply Operations—*Water Transmission and Distribution*, 4th ed. Denver, Colo.: American Water Works Association.

AWWA. M22—*Sizing Water Service Lines and Meters*. Denver, Colo.: American Water Works Association.

Chapter 12

Water Meters

Water meters can be considered the cash registers of a public water system. It is important that meters be properly chosen, installed, maintained, and read to obtain the necessary revenue for water system operations. Accurate water meters are also an important source of data that are used to determine water use and quantify water losses.

CUSTOMER SERVICE METERS

Almost all water systems in the United States meter water to their customers. A few utilities with relatively abundant water supplies still charge a flat rate to small customers. The advantage of the flat-rate practice is that there is considerable savings in not having to provide, read, and maintain meters for these customers. On the other hand, customers on a flat rate tend to waste water. They commonly water gardens excessively, let water run during the summer to keep it cool, and have no incentive to fix leaks in plumbing fixtures.

Positive-Displacement Meters

The meters used for measuring relatively low flow rates are called positive-displacement meters because they accurately measure all water that passes through them by means of a nutating disk or rotating piston. As illustrated in Figure 12-1, water passing past the disk or piston creates a rotary motion that is transmitted through gears to the register. The register then reads in gallons, cubic feet, or cubic meters.

Positive-displacement meters are usually used for residences and small- to medium-sized commercial establishments. The commonly used nominal sizes are from ⅝ to 2 in. (16 to 50 mm) as listed in Table 12-1.

Although maximum flow rates are shown, positive-displacement meters are not intended to operate at full flow for extended periods of time. Continuous operation of a meter at full flow will quickly destroy it. A meter should ideally be sized so that the maximum flow rate will be one half of its maximum operating capacity. On the other hand, if a meter is greatly oversized,

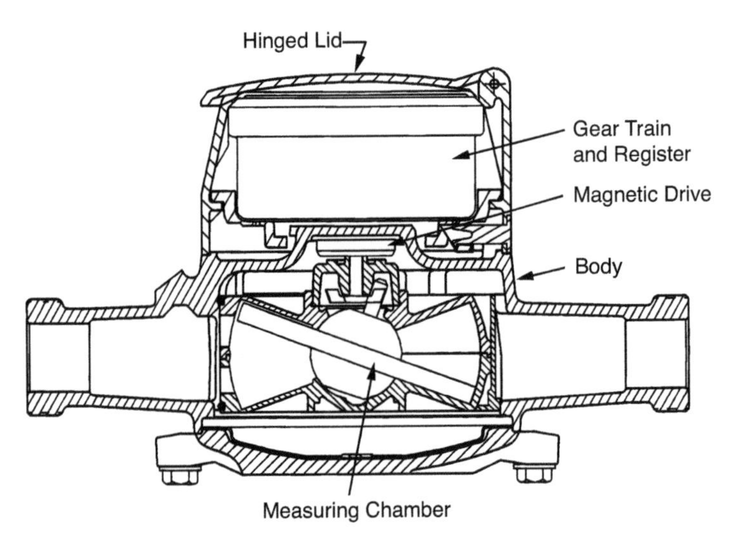

Courtesy of Neptune Technology Group Inc.

Figure 12-1 Nutating-disk meter with a plastic housing

Table 12-1 Characteristics of displacement-type meters

Meter Size		Safe Maximum Operating Capacity		Maximum Pressure Loss at Safe Maximum Operating Capacity		Recommended Maximum Rate for Continuous Operations		Minimum Test Flow		Normal Test Flow Limits	
in.	(mm)	gpm	(m^3/h)	psi	(kPa)	gpm	(m^3/h)	gpm	(m^3/h)	gpm	(m^3/h)
½	(13)	15	(3.4)	15	(103)	7.5	(1.7)	¼	(0.06)	1–15	(0.2–3.4)
½×¾	(13×19)	15	(3.4)	15	(103)	7.5	(1.7)	¼	(0.06)	1–15	(0.2–3.4)
⅝	(16)	20	(4.5)	15	(103)	10	(2.3)	¼	(0.06)	1–20	(0.2–4.5)
⅝×¾	(16×19)	20	(4.5)	15	(103)	10	(2.3)	¼	(0.06)	1–20	(0.2–4.5)
¾	(19)	30	(6.8)	15	(103)	15	(3.4)	½	(0.11)	2–30	(0.5–6.8)
1	(25)	50	(11.0)	15	(103)	25	(5.7)	¾	(0.17)	3–50	(0.7–11.4)
1½	(38)	100	(22.7)	15	(103)	50	(11.3)	1½	(0.34)	5–100	(1.1–22.7)
2	(51)	160	(36.3)	15	(103)	80	(18.2)	2	(0.45)	8–160	(1.8–36.3)

the customer will receive free water at flow rates that are too low to start the meter operating.

The threads on the ends of meters are the same spacing (pitch) as standard pipe thread, but there is no taper; in other words, they are the same diameter for the entire length of the thread. For this reason, they will not screw onto a regular threaded fitting but are intended to be used with a special meter coupling designed to provide a seat against a gasket. The reason for using meter couplings is so that a meter can be simply replaced by

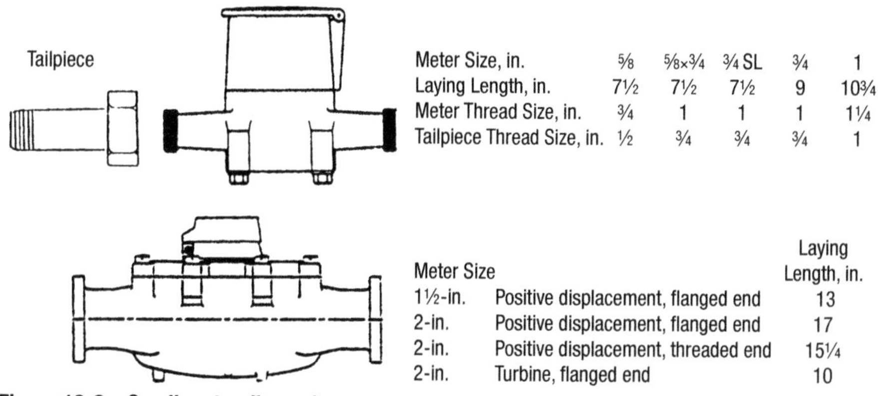

Meter Size, in.	⅝	⅝×¾	¾ SL	¾	1
Laying Length, in.	7½	7½	7½	9	10¾
Meter Thread Size, in.	¾	1	1	1	1¼
Tailpiece Thread Size, in.	½	¾	¾	¾	1

Meter Size		Laying Length, in.
1½-in.	Positive displacement, flanged end	13
2-in.	Positive displacement, flanged end	17
2-in.	Positive displacement, threaded end	15¼
2-in.	Turbine, flanged end	10

Figure 12-2 Small meter dimensions

loosening the nuts and sliding them out of the way. Meters that are 1½ and 2 in. (40 and 50 mm) are most commonly furnished with flanged meter couplings. Meter sizes and nominal dimensions are the same for all meters that conform to AWWA standards (Figure 12-2).

Customers with high water bills often think their meter must be over-registering, but this is almost impossible for a positive-displacement meter. On the contrary, an old, worn meter will usually under-register and may not even start operating at nominal flow rates. Customers who insist on having their meter changed because they think it is running fast usually end up having even larger bills with the new meter. Almost all complaints of higher-than-normal water use can be traced to leaking toilets or other fixtures.

Small Meter Installation

As discussed in chapter 11, meters located in shallow meter boxes are usually installed with a curb stop directly ahead of them. Meter replacement is also greatly facilitated if a valve is installed past the meter. Without the second valve, the installer must enter the building to shut off water at the service entrance before the meter can be removed. Meters in deep pits are usually raised to within about 18 in. (46 cm) of the surface to facilitate reading.

Meters located in basements should ideally be located as close as possible to where the service enters through the floor or wall to discourage customers from installing a connection ahead of the meter. The meter should also be placed where it is relatively accessible for reading and repair. Even if the meter will have a remote reading device installed, it must be accessible for an inspector to periodically make a check reading and determine whether the meter should be repaired or replaced.

Several companies manufacture various styles of meter yokes to facilitate setting meters; examples are shown in Figure 12-3. Some of these devices have valves incorporated with them, and they automatically maintain the correct spacing for easy meter installation and replacement.

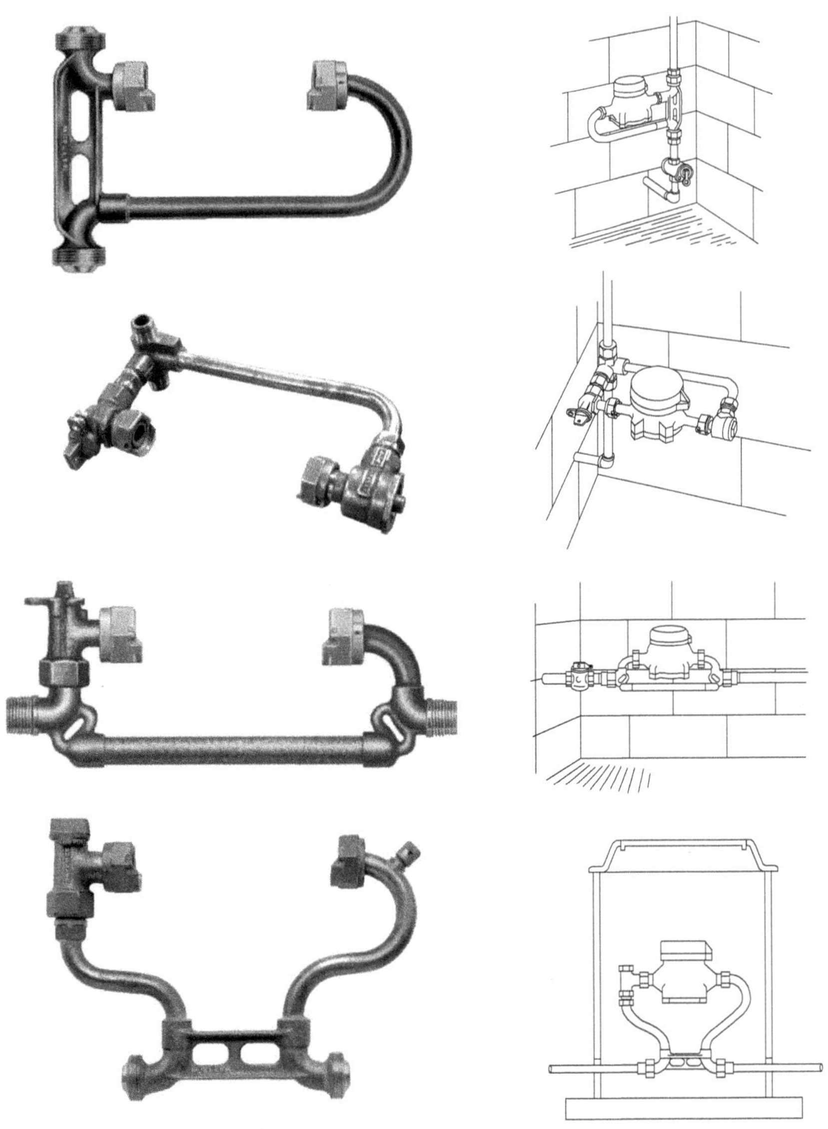

Figure 12-3 Various styles of meter yokes

Although an AWWA policy statement discourages use of a water system as an electrical ground, many homes and buildings have the electric service grounded only to a water pipe. For this reason, many water utilities require an electrical jumper wire from the house piping to the service pipe so any electrical current will bypass the meter. If such jumpers are not installed, it is good practice for personnel replacing meters to carry jumper cables with alligator clips to install temporarily while the meters are changed.

The meter should always be mounted in a vertical position. Installing a meter in any other position will cause premature wear and, in some cases, inaccurate readings.

Water Pilferage

Most water utilities find that there are some customers who, given an opportunity, will attempt to reduce their water bills by altering the meter readings. The common methods are to remove a register for part of the reading period or turn it backward, remove a meter and substitute a pipe nipple, or turn a meter around so it will run backward for part of the billing period.

To prevent this, meter seals should be installed on all meters. Although these will not prevent pilferage, they will show that some unauthorized tampering has taken place, and further action can be taken. Meter registers are generally sealed at the factory with a copper wire and an acrylic seal that cannot be removed without breaking the wire.

A meter coupling has a small hole that can be used for installing a copper wire and seal to prevent unscrewing the coupling nuts without detection. Another device commonly used is a plastic shield that is clipped over the meter nuts and cannot be removed without breaking it.

LARGE WATER METERS

For most large water services, the choice of meters usually lies among types of meters other than the positive-displacement type.

Current Meters

Current meters are commonly called *velocity meters*. The principal types are turbine, multijet, and propeller meters.

A turbine meter has rotors turned by the flow of water. The volume of water passing through the meter is directly proportional to the revolutions of the rotor.

A multijet meter is similar but has a multiblade rotor mounted on a vertical spindle within the measuring chamber. Water enters the chamber through several tangential orifices around the circumference and leaves the chamber through another set of orifices placed at a different level.

A propeller meter has a propeller turned by the flow of water, and the movement is transmitted to the register (Figure 12-4). On larger sizes, the propeller may be small in relation to the diameter of the pipe. Propeller meters are primarily used for main-line measurement where the flow rates do not change abruptly.

Turbine meters will under-register if the blades become clogged or coated with sediment, and multijet meters will under-register if the orifices become partially clogged. Such a meter can also be severely damaged by a

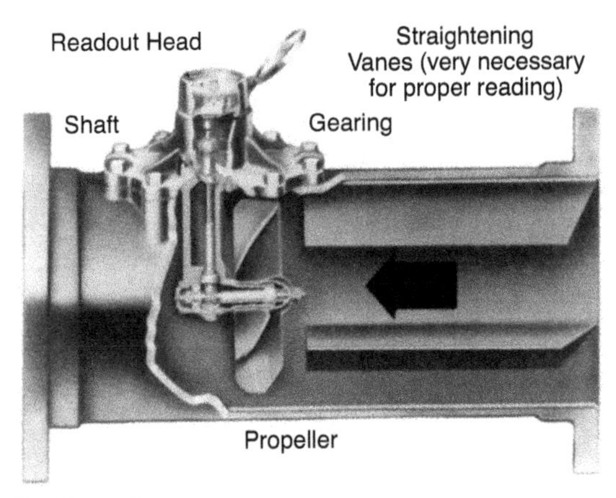

Figure 12-4 Propeller flowmeter

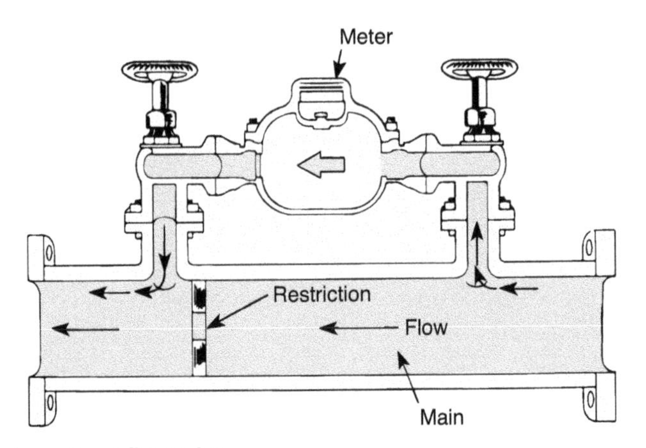

Figure 12-5 Proportional flowmeter

hard object no larger than a pea. If there is a possibility of particles being in the water, a strainer should be installed ahead of the meter.

Proportional Meters

A proportional meter has a restriction in the pipeline that forces a portion of the flow to pass through a small meter, as illustrated in Figure 12-5. The total flow is determined by reading the flow through the small meter and multiplying it by a factor.

Venturi Meters

In the venturi meter, there is a defined throat (constriction) within the meter body. As the flow is constricted by a decrease in the diameter, there is a proportional increase in flow velocity. The difference in pressure before the constriction and at the throat can be measured, as illustrated in Figure 12-6,

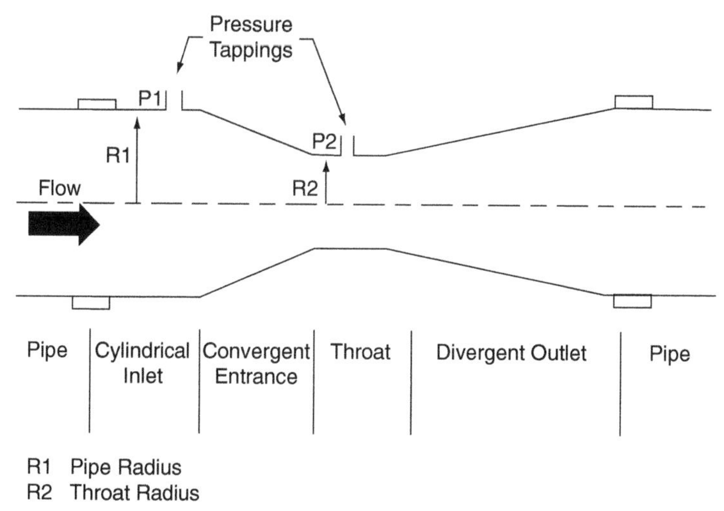

Figure 12-6 Schematic of a venturi meter

and the change in pressure is proportional to the square of the velocity. The quantity of flow can be determined from the flow velocity, using electronic or mechanical instruments.

Venturi meters have the advantages of creating very little friction loss, requiring almost no maintenance, and being particularly useful on large pipelines. Their primary disadvantage is that they are accurate only for a specific flow range.

Orifice Meters

An orifice meter consists of a thin plate with a circular hole that is installed between a set of flanges (Figure 12-7). The flow rate is determined by a comparison between the pressure at taps on each side of the orifice plate in a fashion similar to that used for a venturi meter. An orifice is inexpensive and maintenance-free but creates a relatively high head loss.

Magnetic Meters

A magnetic ("mag") meter measures flow by generating a magnetic field around an insulated section of pipe. Water passing through the field induces a small flow of electrical current that is proportional to velocity, and this is electronically converted to a registration of flow rate. Mag meters are most often used for measuring the flow of dirty or corrosive water that would damage other types of meters.

Ultrasonic Meters

An ultrasonic flowmeter uses transducers to generate and receive sound pulses that are alternately sent in opposite diagonal directions across the

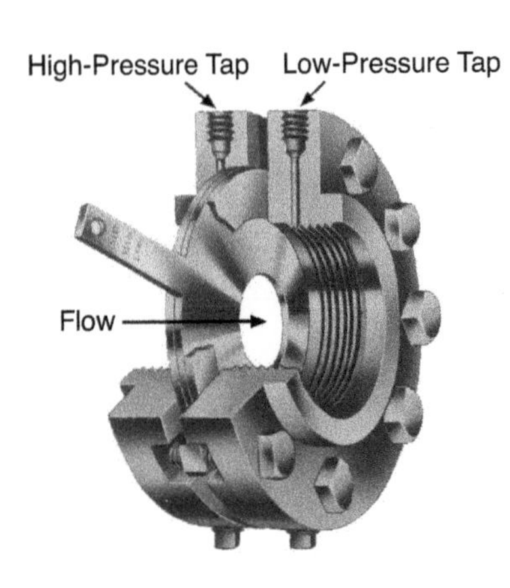

Courtesy of Bristol-Babcock Division, Acco Industries, Inc., Waterbury, Conn.

Figure 12-7 Orifice meter

Courtesy of Great Plains Industries, Inc.

Figure 12-8 Insertion-type meter

flow of water. Because of the Doppler effect, the sound changes with the velocity of water flow, and this can be electronically converted to indicate flow rate.

Insertion Meters

An insertion meter provides a relatively inexpensive way of adding metering to an existing pipeline. The unit consists of a probe having a small rotor at the bottom that is inserted into the pipeline through a mounting on a saddle, as shown in Figure 12-8. Units will mount in pipelines from 1½ in. (40 mm) to 36 in. (0.9 m) or larger and have relatively good accuracy.

The electronic readout can either be mounted directly on the probe or at a remote location, and indicates both flow rate and total flow. Units are also available to provide either a 4–20 mA or pulse signal for use in operating a chart recorder or for pacing chemical feed equipment.

Compound Meters

All meters have specified ranges over which they are acceptably accurate. Positive-displacement meters and propeller meters, for example, will not start to rotate until a certain velocity is reached. Other meters may register at flows below the suggested minimum accuracy point, but the reading may be only a fraction of actual flow.

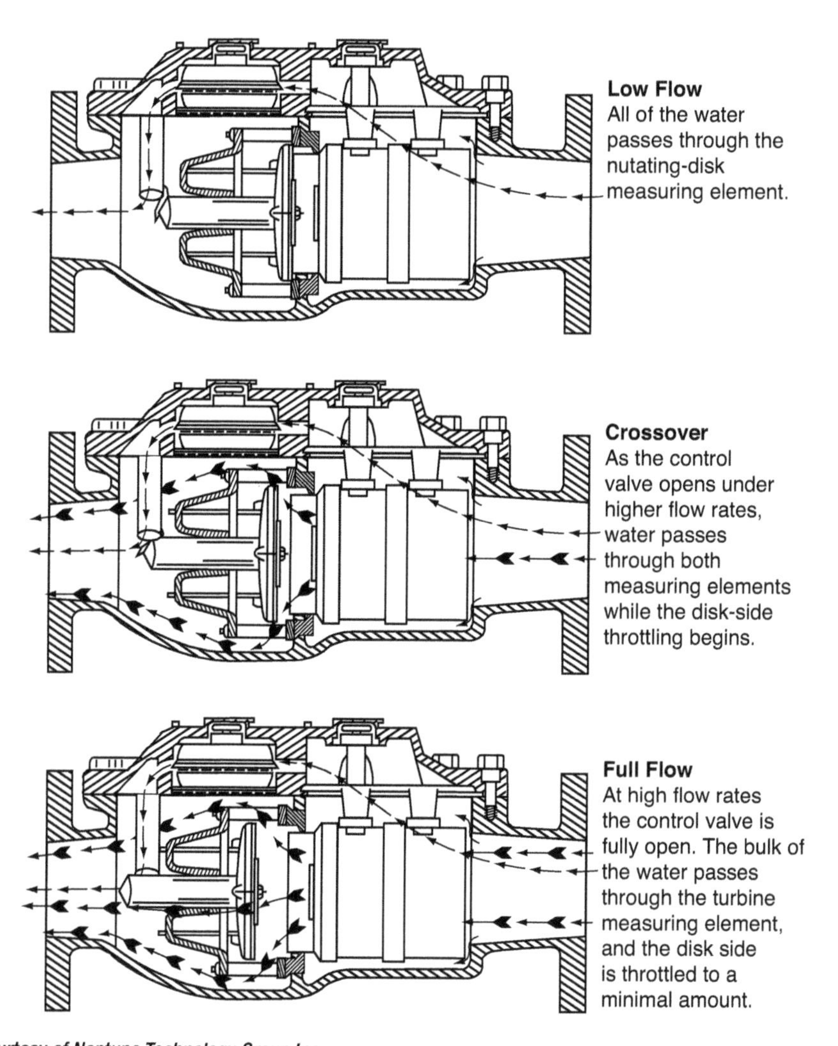

Courtesy of Neptune Technology Group Inc.

Figure 12-9 Compound meter

Occasionally there are situations in which a customer has a wide variation in use and it is desired to meter all of the water accurately. An example would be a factory that uses only small amounts of water during the night but very large quantities during the day shifts. The best method of metering this type of customer is to use a compound meter.

A compound meter consists of both a large turbine meter and a small bypass meter. As illustrated in Figure 12-9, the water flows through the small meter until it reaches a certain velocity, then a valve actuates to divert flow through the turbine. Some compound meters have separate registers

Courtesy of ABB Water Meters, Inc.

Figure 12-10 Compound meter arrangement that uses two standard meters

for each meter and others combine the readings onto one register. Another compound metering arrangement using two standard meters is shown in Figure 12-10.

Meter Accuracy

All meters that operate on the principle of determining flow velocity are designed with the assumption of laminar flow through the meter. Any disturbance, such as pipe bends or valves that might create eddies in the water, either before or after the meter, can cause the meter to be inaccurate. Every meter manufacturer has a recommendation for the minimum straight-pipe distance that should be maintained on either side of a meter. This is usually expressed as pipe diameters because it is dependent on the pipe size. In addition, some flowmeters have vanes ahead of the meter to help ensure straight-line flow as the water enters the meter.

METER READING

Early water meters were furnished with a circular or round *reading register* (Figure 12-11). These were difficult to read, and it was easy for meter readers to make mistakes. It was also difficult for customers to read their own meters. This type of register has gradually been replaced by "straight reading" registers that are like the odometer on a car. Most registers have one or more fixed zeroes, as shown with a black background on the right side in Figure 12-11. Some larger meters also have a note on the register that a multiplier is to be used, such as "10×" or "100×."

Meters used in the United States are generally furnished with registers reading in either gallons or cubic feet, but meters are also available with registers for Imperial gallons or cubic meters.

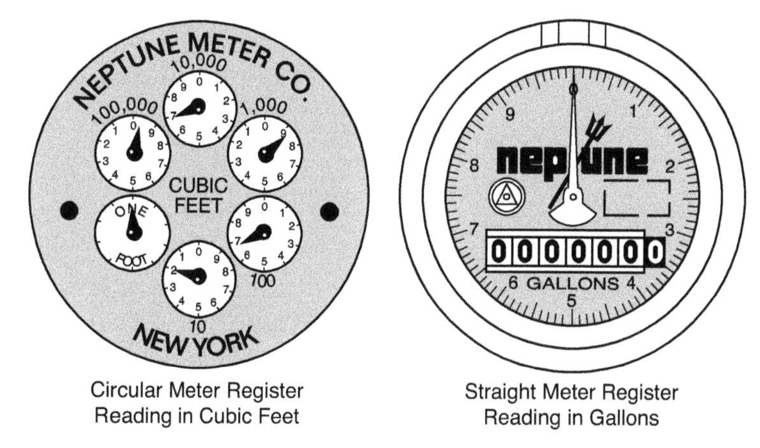

Circular Meter Register Reading in Cubic Feet

Straight Meter Register Reading in Gallons

Courtesy of Neptune Technology Group Inc.

Figure 12-11 Meter registers

Direct Meter Reading

Traditional meter reading involves a meter reader visiting each building and directly reading the meter. One of the few advantages of this system is that the meter reader can make a quick calculation as the meter is read and immediately alert the customer that there may be a leak if use has been greatly in excess of previous billing periods. However, water in the pits or snow cover may make direct reading of meters in pits difficult. If most are located inside buildings, however, there is a long list of problems.

Customers are increasingly reluctant to allow utility workers in their homes because of publicized reports of robbers or attackers gaining access by posing as utility representatives. The best way to combat this is to provide meter readers with identification and distinctive uniforms, and to try to keep the same people on each route so that customers get to know them.

Other reasons customers do not want meter readers entering their homes include meter-reader personnel wearing dirty boots, they are embarrassed about the condition of themselves or their homes at the time, and they do not want to be bothered.

With increasing numbers of homes where everyone is working during weekdays, the only way of obtaining direct readings at these homes is to have a meter reader work on Saturday. There are also areas in some communities where, for a variety of reasons, the meter reader is reluctant to enter to make the reading.

To reduce some of these problems, many water systems are now using a doorknob card that asks the customer to read the meter and drop the postage-free card in the mail (Figure 12-12). Many customers who do not want to be bothered by the meter reader have standing orders to leave the cards. As long as customers promptly make the readings, the system works well.

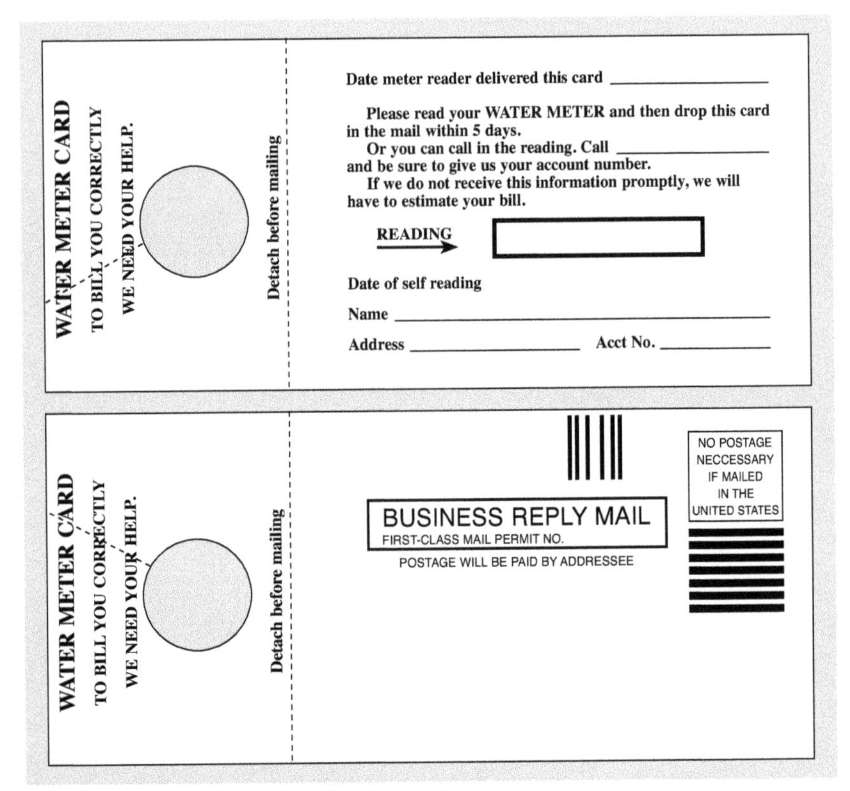
WATER METER CARD
TO BILL YOU CORRECTLY
WE NEED YOUR HELP.
Detach before mailing

Date meter reader delivered this card ______

Please read your WATER METER and then drop this card in the mail within 5 days.
Or you can call in the reading. Call ______ and be sure to give us your account number.
If we do not receive this information promptly, we will have to estimate your bill.

READING →

Date of self reading
Name ______
Address ______ Acct No. ______

WATER METER CARD
TO BILL YOU CORRECTLY
WE NEED YOUR HELP.
Detach before mailing

NO POSTAGE NECCESSARY IF MAILED IN THE UNITED STATES

BUSINESS REPLY MAIL
FIRST-CLASS MAIL PERMIT NO.
POSTAGE WILL BE PAID BY ADDRESSEE

Figure 12-12 Water meter self-reading doorknob card

If one reading is missed, the utility will usually send an estimated bill. If a second self-reading is missed, the meter should be read by the meter reader to verify that water use has been normal. Because of the difficulty and time required to obtain direct meter readings, many water systems read and bill every three months (quarterly).

Remote Reading Devices

Remote reading devices have been developed to eliminate most of the problems associated with making direct readings inside buildings. The basic type uses a meter with a special register that creates a small electrical impulse at each 10 ft^3 (0.3 m^3) or 100 gal (379 L). The register is connected with a pair of wires to the remote register that can be mounted at any convenient location on the outside of the building. The register advances one digit for each pulse it receives (Figure 12-13). The meter reader still carries a meter-reading book and records the numbers by hand.

Another type of remote meter reading uses a similar signal from the meter, but all that is mounted on the outside of the building is a special plug receptacle. The meter reader carries a special electronic unit that is plugged

Courtesy of Badger Meter, Inc., Milwaukee, Wis.

Figure 12-13 Meter with remote-register device

into the receptacle to obtain the reading. The reading may be displayed on a screen and electronically recorded for later direct entry into a computer.

Yet another system is called *electronic meter reading.* It uses an inductive coil on the exterior of the building in place of the receptacle. The meter reader carries an electronic unit (interrogator) with a probe that only needs to be held in the proximity of the coil to obtain a reading. Power from the interrogator is transmitted through the coil to a microprocessor in the meter, and the meter reading is transmitted back to the meter reader. The advantage of this system is that the same unit can be installed on meters in pits with the induction coil mounted under the lid so the meter reader does not have to open the pit to obtain a reading.

Automatic Meter Reading

Rapidly changing advancements in technology are revolutionizing the speed and accuracy of meter reading. These systems, although varied, are included under the general category called *automatic meter reading.* With these systems, each meter has a small transmitter that sends the meter reading electronically on command. The methods used to send the interrogating signal and to receive the reading data include telephone lines, the electric power distribution system, sound transmission through the water mains, cable television systems, and satellite systems. Some of these systems have been coordinated in conjunction with other utilities so that electric and gas meter readings may also be transmitted back to a central location using the same system.

The other alternative is to send the signal by radio. With one system, the meter reader only needs to walk down the street with a handheld computer and enter the codes of adjacent meters to interrogate them. The meter

Courtesy of Neptune Technology Group Inc.

Figure 12-14 Mobile data collector

readings are transmitted back and stored in the computer. An example of a mobile data collector is shown in Figure 12-14.

Another system has the computer mounted in a truck and obtains readings from adjacent buildings as it is driven. If a community is not too large, an antenna can be mounted on a high mast at a central location and obtains all meter readings from that location.

Besides the obvious advantage of saving labor, these systems makes it practical to bill monthly, which has some financial advantages and presents the customers with regular, smaller bills. It also provides a means of easily performing water use studies, such as determining how much water is used by certain customers on high-use days.

The one caution about all remote or automatic reading systems is that the utility must not forget to occasionally inspect the meter. The general condition of the meter should be checked, the seals inspected for any sign of damage, and the register reading checked to be sure it matches the remote reading.

Advanced metering infrastructure (AMI) systems measure, collect, and analyze energy usage, and communicate with metering devices such as electricity meters, gas meters, heat meters, and water meters, either on request or on a schedule. Water utilities that use these systems often refer to them as Smart Grids or using Smart Meters. These systems include hardware, software, communications, customer associated systems, and supplier business systems.

AMI extends current automatic meter reading (AMR) technology by providing two-way meter communications, allowing commands to be sent toward the home for multiple purposes, including "time-of-use" pricing information, demand-response actions, or remote service disconnects. Wireless technologies are critical elements of these systems. Meters that can be effectively used in AMI systems use new solid-state technologies capable of more accurately measuring low flows and providing digital signals.

Smart Meters enable businesses and individual customers to participate in demand control (conservation). Consumers can use information provided by the system to change their normal consumption patterns to take advantage of lower prices. Pricing can be used to control peak consumption. AMI differs from traditional AMR in that it enables two-way communications with the meter.

METER MAINTENANCE

Manufacturers usually test new meters before they are shipped, so many small communities do not retest them before installation. Some state public utility regulatory commissions require that a certain percentage of all new meters be tested. The limits of acceptable accuracy are dictated by AWWA standards.

Customer water meters gradually decrease in efficiency and underregister as a function of both wear from use and water quality. Because of the many variables, it is hard to set specific limits, but AWWA recommends that ⅝-in. (17-mm) meters be tested every 10 years. Larger meters should be tested more often because of the greater loss of revenue if they are not registering correctly. Recommended testing intervals are provided in AWWA meter standards.

When meters are to be periodically tested, there are two procedures in common use. One is to remove the meter and replace it with a space pipe for a few days while the meter is tested and repaired if necessary, and then return to reinstall the same meter. This requires two trips and bothering the customer twice to gain entrance, but it also simplifies meter record keeping because the same meter is used at the address. The other method is to immediately replace the removed meter with a new or reconditioned meter, but this requires that both the meter records and the billing records be changed to show the new meter serial number and initial reading of the new meter.

Testing meters is not very complicated and requires a meter test bench with the components shown in Figure 12-15. Most test benches will test several meters at the same time. Meters are usually tested at high, medium, and low flow rates and checked to be in conformance with AWWA standards. Meters in need of repair must be disassembled and cleaned, and worn parts replaced as necessary. Specific instructions on meter repair should be obtained from the meter manufacturer. General information on meter

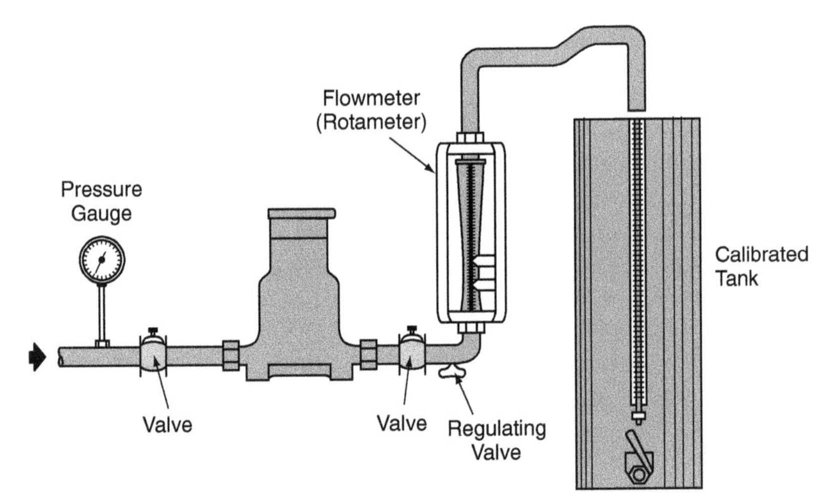

Figure 12-15 Meter-testing equipment

repair is available in AWWA Manual M6, *Water Meters—Selection, Installation, Testing and Maintenance.*

Another alternative to testing meters in-house is to send them to a local firm that specializes in this work. Because of the weight involved, it is not usually practical to ship meters a great distance for repair.

A third alternative that is used by many small water systems is simply to dispose of worn meters and replace them with new ones. This is particularly practical if it is also desired to change the style of meter or to start installing meters with new features, such as remote readout accessories. Some of the replacement cost can sometimes be recovered by selling removed meters to firms that specialize in reconditioning old meters.

It is essential that a water utility maintain good records of each meter installed on the system. Meter record keeping is discussed in chapter 19.

BIBLIOGRAPHY

ANSI/AWWA C700. *AWWA Standard for Cold-Water Meters—Displacement Type, Bronze Main Case.* Denver, Colo.: American Water Works Association.

AWWA. M6—*Water Meters—Selection, Installation, Testing and Maintenance.* Denver, Colo.: American Water Works Association.

AWWA. M22—*Sizing Water Service Lines and Meters.* Denver, Colo.: American Water Works Association.

AWWA. M32—*Computer Modeling of Water Distribution Systems.* Denver, Colo.: American Water Works Association.

AWWA. M33—*Flowmeters in Water Supply.* Denver, Colo.: American Water Works Association.

Berardinelli, B. 2012. Taking Measure of Advanced Water Meter Technologies. *Opflow*, May.

Chapter 13

Cross-Connection Control

The backflow of nonpotable water through a cross-connection can result in contamination of the distribution system and contribute to waterborne disease outbreaks. A carefully designed cross-connection control program is a utility operator's best method of preventing this occurrence. This program and its effectiveness is a primary practice affecting distribution system water quality (chapter 3). System reliability is also affected because suspected contamination from main breaks, for example, can cause interruptions in water service while decontamination procedures are conducted.

CROSS-CONNECTION TERMINOLOGY

Some of the terminology used in describing the problems and conditions relating to cross-connections follows.

- *Backflow* is the flow of any water, foreign liquids, gases, or other substances back into a potable water system. There are two conditions that can cause backflow: backpressure and backsiphonage.
- *Backpressure* is a condition in which the foreign substance is forced into a water system because it is under a higher pressure than the system pressure.
- *Backsiphonage* is the condition under which the water system pressure is less than atmospheric (i.e., it is under vacuum), and the foreign substance is essentially sucked into the potable water system.
- *Cross-connection* is any connection between a potable water system and any source of contamination through which contaminated water could enter the potable water system.

PUBLIC HEALTH SIGNIFICANCE

Water utilities go to great lengths to ensure that the water entering the system is properly treated and the distribution system is operated with care to prevent contamination. In spite of these efforts, many systems have some

Courtesy of USEPA, Region VIII, Water Supply Division.

Figure 13-1 Water tank cross-connection

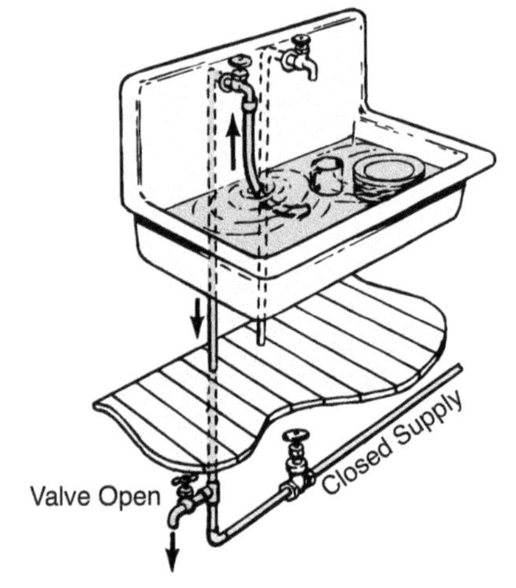

Figure 13-2 Backsiphonage (hose forms cross-connection)

unknown cross-connections that could cause a disease outbreak, poisoning, or degraded water quality if certain conditions should occur.

Diseases and Poisonings Attributable to Cross-Connections

One of the most catastrophic results of a cross-connection occurred in two Chicago, Ill., hotels during the 1933 World's Fair. As a result of backsiphonage during a pressure loss in the system, drinking water was contaminated with amoebic dysentery germs and infected more than 1,400 people. Of this number, at least 98 are known to have died directly from the disease.

Other diseases that have caused the deaths and illnesses of many people as a result of cross-connections are gastroenteritis, hepatitis, and salmonellosis. Many people have also become ill as a result of fertilizer, pesticides, herbicides, and boiler chemicals introduced into a water supply through a cross-connection.

Potential Cross-Connections

Situations in which some of the conditions for a cross-connection exist but require something else to be done to complete the connection are called *potential cross-connections.* In the examples shown in Figures 13-1 and 13-2, the end of the hose must be immersed in liquid for there to be a cross-connection. Although the likelihood is remote that there will be a vacuum on the water system just at the time the hose is submerged in liquid, the tank truck or sink could hold a toxic substance, and consequences of backsiphonage could be very serious.

One common cross-connection that almost everyone has seen is a chemical dispenser connected to a garden hose (Figure 13-3). If a vacuum should

Figure 13-3 Garden hose cross-connection

occur while the unit is in use, the chemical solution would be sucked back into the house plumbing.

Many potential locations for cross-connections exist in factories, restaurants, canneries, mortuaries, and hospitals. Any place where a water fill line is below the rim of a container, a cross-connection can exist. A summary of common cross-connections and potential hazards is presented in Table 13-1.

Figure 13-4 illustrates how heavy use by the fire department in fighting a fire can reduce the pressure on the upper floors of a building, and water could be siphoned from an old-style bathtub with a submerged inlet. Because of this contamination potential, all new bathtubs must have the fill spout located above the tub rim.

BACKFLOW CONTROL DEVICES

When a cross-connection situation is identified, one of two actions must be taken. Either the cross-connection must be removed or some means of defense must be installed to protect the public water supply from possible contamination.

Air Gaps

The least expensive and most positive method of protecting against backflow is to install an air gap. There are no moving parts to maintain or break, and surveillance is necessary only to ensure that it is not altered. The only requirement for an air gap between the supply outlet and the maximum water surface of a nonpotable substance is that it must be at least twice the internal diameter of the supply pipe but no less than 1 in. (25 mm) in any situation.

Typical uses of an air gap are for supplying water to tank trucks, to a nonpotable supply, or a surge tank in a factory as illustrated in Figure 13-5.

Table 13-1 Cross-connections and potential hazards

Connected System	Hazard Level
Sewage pumps	High
Boilers	High
Cooling towers	High
Flush valve toilets	High
Garden hose (sill cocks)	Low to high
Auxiliary water supply	Low to high
Aspirators	High
Dishwashers	Moderate
Car wash	Moderate to high
Photographic developers	Moderate to high
Commercial food processors	Low to moderate
Sinks	High
Chlorinators	High
Solar energy systems	Low to high
Sterilizers	High
Sprinkler systems	High
Water systems	Low to high
Swimming pools	Moderate
Plating vats	High
Laboratory glassware or washing equipment	High
Pump primers	Moderate to high
Baptismal founts	Moderate
Access hole flush	High
Agricultural pesticide mixing tanks	High
Irrigation systems	Low to high
Watering troughs	Moderate
Autopsy tables	High

Reduced-Pressure Zone Backflow Preventers

A device that can be used in every cross-connection situation is the reduced-pressure zone backflow preventer, usually abbreviated RPZ, RPBP, or RPZBP. It consists of two spring-loaded check valves with a pressure-regulated relief valve located between them. As illustrated in Figure 13-6, if there is a potential backsiphonage situation, both check valves will close, and the space between them is opened to atmospheric pressure. If there is backpressure in excess of the water main pressure, both check valves will close, and if there is any leakage in the second valve, it will be allowed to escape through the center relief valve.

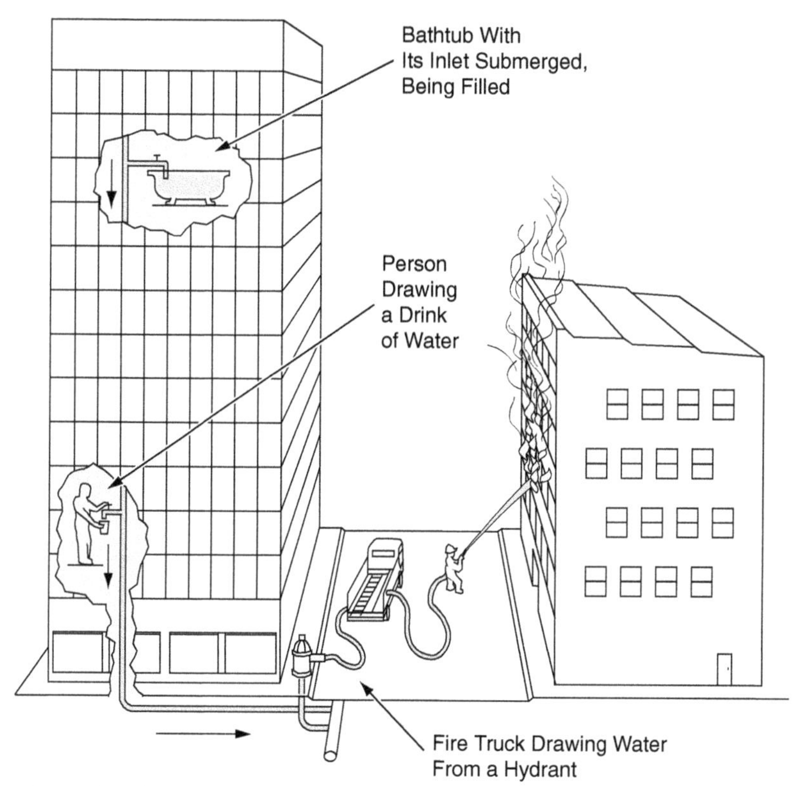

Figure 13-4 Backsiphonage due to pressure loss

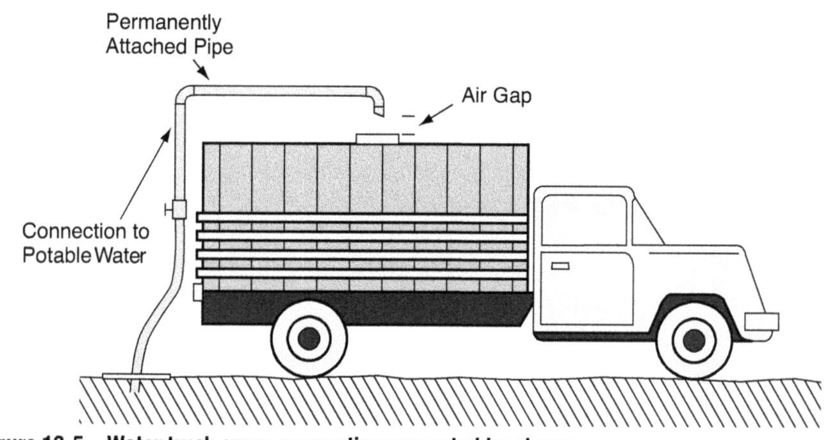

Figure 13-5 Water truck cross-connection prevented by air gap

An RPZ is much safer than one or two check valves because there is always the potential of a check valve leaking. Even though RPZs are designed to be dependable, they are mechanical devices that must be tested and maintained regularly in accordance with the manufacturer's recommendations.

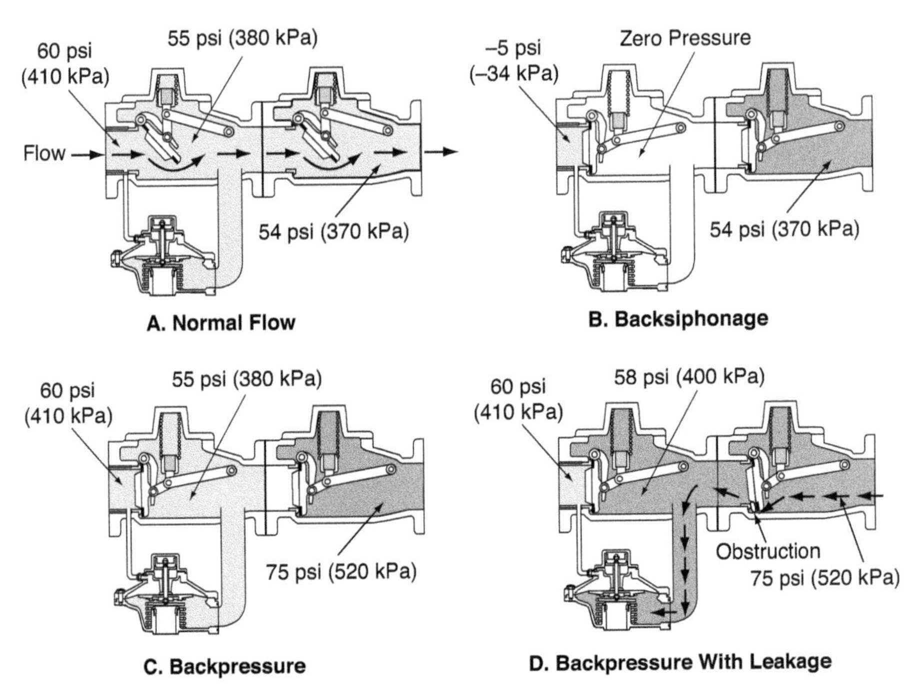

Courtesy of Cla-Val Company Backflow Preventer Division.

Figure 13-6 Valve position and flow direction in an RPZ

They must be installed in locations where the relief port cannot be submerged, and where they are protected from freezing and vandalism.

Double-Check Valves

A double-check-valve backflow preventer is designed similarly to an RPZ except there is no relief valve between the two checks (Figure 13-7). The protection is not as positive as an RPZ because of the possibility of the check valves leaking, so they are not recommended for use in situations where a health hazard may result from valve failure. Local and state officials should be contacted for approval before a double-check valve is installed for cross-connection protection in a potable water line.

Some water utilities install check valves on some or all customer water services. This is particularly desirable for customers with operable private wells because of the potential of well water being forced backward into the utility's system if there is some piping change by the customer. Several manufacturers have developed double-check-valve assemblies for this use, as illustrated in Figure 13-8.

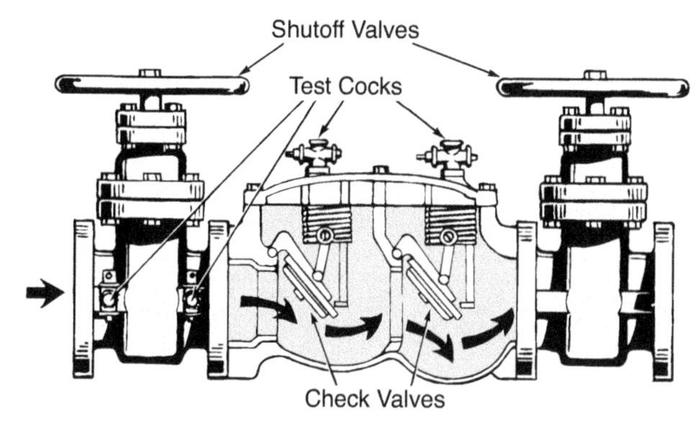

Figure 13-7 Double-check valve assembly

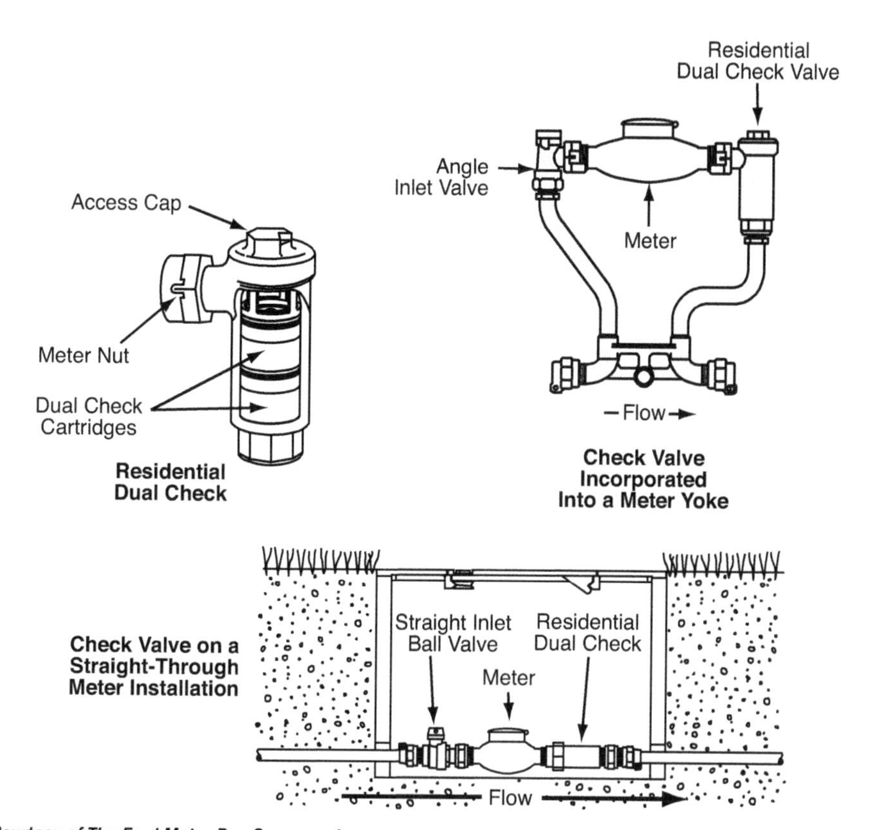

Courtesy of The Ford Meter Box Company, Inc.

Figure 13-8 Examples of residential dual-check valves

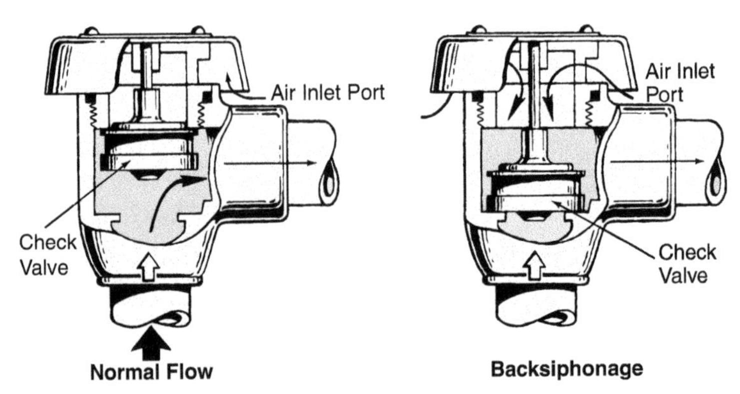

Figure 13-9 Atmospheric vacuum breaker

Vacuum Breakers

There are two general types of vacuum breakers: atmospheric and pressure. Atmospheric vacuum breakers are often called antisiphon valves, backsiphonage preventers, or antisiphon vacuum breakers. They are designed for intermittent use on piping connections where there will be no backpressure. Common uses are on toilet flush valves, on faucets with hose threads such as janitors' sinks, and on lawn sprinkler systems. Atmospheric vacuum breakers must be installed beyond the last valve in the piping system.

As illustrated in Figure 13-9, when the supply pipe is under pressure, the check valve closes against an upper seat to prevent leakage from the valve. When there is no pressure in the supply, the valve drops and allows air to enter the discharge pipe, thus preventing possible backsiphonage. If this device is installed in a situation where it is under continuous pressure, the check valve may stick shut permanently or become unreliable in operation.

The pressure-type vacuum valve is similar but is designed for use under pressure over long periods of time (Figure 13-10). This type of valve, too, should never be used where there is any possibility of backpressure on the discharge pipe. It must, therefore, be installed above the highest fixture on the discharge piping. A typical use is on water lines in industrial plants that should be separated from the potable water supply.

Complete Isolation

The most positive method of preventing connection between piping systems from two different sources is complete separation. When piping systems from two sources are located in the same building, they can be identified by signs and color coding. The need for monitoring continues, however, to ensure that the systems are not inadvertently connected.

Someone who does not realize the potential consequences may install a temporary connection between two systems using a spool piece or a swing connection. Such connections are not recommended for use regardless of

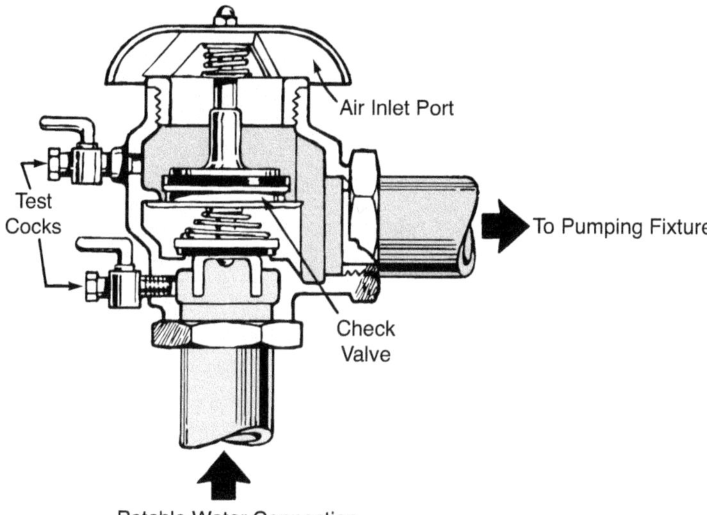

Figure 13-10 Pressure vacuum breaker

the degree of risk involved, and they should be completely removed to eliminate any possibility of a cross-connection.

CROSS-CONNECTION CONTROL PROGRAMS

The Safe Drinking Water Act of 1974 makes each public water supply utility responsible for the quality of water at the consumer's tap. Legal proceedings have also established that the utility is primarily responsible for cross-connection control, and many state regulatory agencies have specifically assigned the responsibility directly to the water supplier.

The problem in administering a cross-connection control program is that the two primary sources of cross-connection problems are usually not directly under the jurisdiction of the water utility. The cross-connections are usually within the customer's premises, which are under the supervision of the local building or health department, or the state health department. Cross-connections between the municipal water system and auxiliary water sources may be under the jurisdiction of a health department or environmental protection agency or its equivalent. When different agencies have an interest, the program must be administered as a cooperative effort.

Most states have established recommended procedures for establishing a cross-connection control program. These recommendations should be reviewed for specific guidance, but the general elements of an effective cross-connection control program are listed here:

- Designation of an organization or agency with overall responsibility and authority for administering the program, with adequate staff
- An adequate plumbing and cross-connection control program

- A program of systematic inspection of new and existing installations
- Follow-up procedures to ensure compliance
- Backflow-prevention device standards, as well as standards for inspection and maintenance of the devices
- Cross-connection control training
- A public awareness and information program

Guidelines for establishing and operating a cross-connection control program are detailed in publications listed in the bibliography.

Device Testing and Repair

To ensure that backflow prevention devices are in working order, they should be tested on a regular schedule. Most state and local jurisdictions require annual testing of approved devices. In addition, most programs require that only certified testers perform these procedures. Any repairs then are only performed by qualified (often certified) personnel.

Regulations usually allow certified testers to be contractors or utility employees. Records are required that document the testing and the results. Repairs or replacement of devices are also documented as required by local regulations. Consult your local agency for the requirements in your location.

BIBLIOGRAPHY

AWWA. M14—*Recommended Practice for Backflow Prevention and Cross-Connection Control.* Denver, Colo.: American Water Works Association.

USEPA. 2003. *Cross-Connection Control Manual.* EPA 816-R-03-002. US Environmental Protection Agency, Office of Water. www.epa.gov/safewater.

USEPA. 2006. *Cross-Connection Control Best Practices Guide.* EPA 816-F-06-035. US Environmental Protection Agency, Office of Water. www.epa.gov/ogwdw/smallsystems/pdfs/guide_smallsystems_crossconnection control.pdf.

Chapter 14

Pumps and Motors

TYPES OF PUMPS

Most pumps used on a public water system are of a type called *velocity pumps*. The pumps move water by a spinning impeller or propeller operating at high velocity. Another type of pump is the positive-displacement pump. Early water systems used piston pumps powered by steam engines to pump water, but these were replaced with centrifugal pumps as higher-speed electric motors became available. About the only positive-displacement pumps used by a modern water utility are some types of portable dewatering pumps (mud pumps) and chemical feed pumps.

Centrifugal Pumps

A centrifugal pump has a rotating impeller within a pump case. Water is drawn in at the center of the impeller and is thrown outward where the high velocity (centrifugal force) is converted to pressure. The most commonly used centrifugal pump is the volute design, in which the impeller discharges into a progressively expanding spiral casing. With this design, the velocity of the liquid is gradually reduced as it flows around the casing, changing velocity into pressure. A diffuser pump operates similarly to the volute type, but instead of having the case designed to convert the energy from velocity to static pressure, the diffuser type has specially designed vanes on the impeller that perform the same task. Volute and diffuser styles of pump casings are illustrated in Figure 14-1.

Smaller centrifugal pumps are usually manufactured with a single suction, as pictured in Figure 14-2. The suction opening is at one end of the pump and the discharge is at a right angle on one side of the casing.

A close-coupled pump has the impeller mounted directly on the motor shaft, whereas a frame-mounted unit has a pump with separate motor bearings and is connected to the motor by a coupling.

Larger pumps are usually manufactured in a style called *double suction* because the inlet water enters on both sides of the impeller (Figure 14-3).

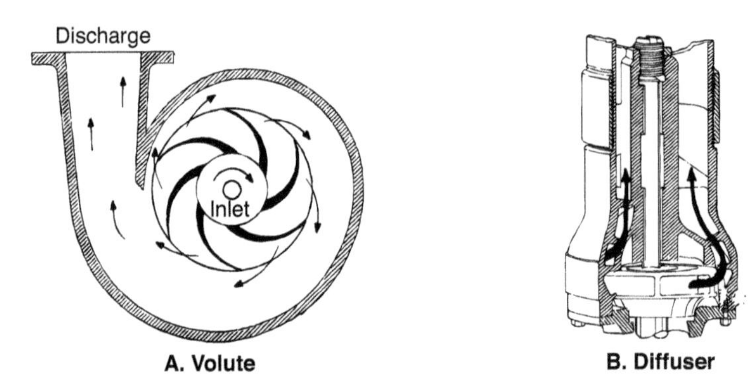

A. Volute

B. Diffuser

Source: Used with permission of John Wiley & Sons, Inc., from Centrifugal Pump Design, *by John Tuzson, © 2000 by John Wiley & Sons, Inc.*

Figure 14-1 Centrifugal pump casings

A. Close-Coupled Pump

B. Frame-Mounted Pump

Figure 14-2 Single-suction pumps

A double-suction pump is also commonly referred to as a *horizontal split-case pump* because the casing is split into two halves along the centerline of the pump shaft, which is normally set in a horizontal position. However, some pumps of this type are designed to be mounted with the shaft vertical.

Centrifugal pumps can be manufactured in a wide range of flow rates and pressures by varying the width, shape, and size of the impeller as well as varying the clearance between the impeller and the casing. Although maximum theoretical suction lift is 34 ft (10 m) at sea level, the maximum practical lift is generally between 15 and 25 ft (4.6 and 7.6 m). Pumps can develop heads up to 250 ft (76 m) for a single stage, and higher heads can be achieved by using multiple stages as illustrated in Figure 14-4. Pump efficiencies can be as high as 75 to 85 percent.

The advantages of centrifugal pumps include simple construction, moderate initial cost, small space requirements, low maintenance, and ability to operate against a closed discharge head for short periods of time. One disadvantage is that maximum efficiency is achieved only over a rather limited pressure range. If a pump is operating under conditions somewhat different than the design conditions, efficiency may be considerably decreased.

Axial-Flow Pumps

Axial flow pumps are often called *propeller pumps.* As illustrated in Figure 14–5, they have neither volute nor diffuser vanes and achieve flow by the lifting action of a propeller-shaped impeller. These pumps handle very high volumes of water but can achieve only limited head. They must be installed so that the impeller is submerged at all times.

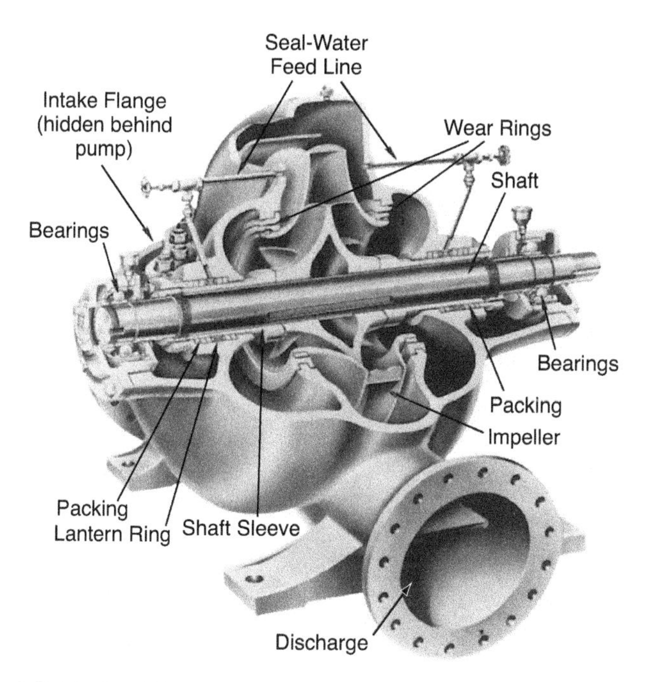

Courtesy of Ingersoll-Dresser Pump Company.

Figure 14-3 Double-suction pump

Courtesy of Aurora Pump.

Figure 14-4 Section of a two-stage pump

Mixed-Flow Pumps

The mixed-flow pump illustrated in Figure 14-6 is a compromise between centrifugal and axial-flow pump designs. The impeller is shaped so that centrifugal force will impart some radial component to the flow. This type of pump is particularly useful for moving large quantities of water containing some solids, such as in raw water intakes.

Vertical Turbine Pumps

Vertical turbine pumps are really a type of centrifugal pump. They are usually furnished in multiple stages to obtain very high pressure at efficiencies as high as 90 to 95 percent. However, to achieve these efficiencies, the impellers are very close fitting, so sand, silt, or other grit in the water will quickly wear them and decrease the efficiency. Turbine pumps have a higher initial cost and are considerably more expensive to maintain than regular centrifugal pumps of equal capacity.

A deep-well pump has a turbine pump suspended below water level in a well, and the pump is driven through a long shaft from the motor located at the ground surface (Figure 14-7). Deep-well turbines have been installed with lifts of more than 2,000 ft (610 m).

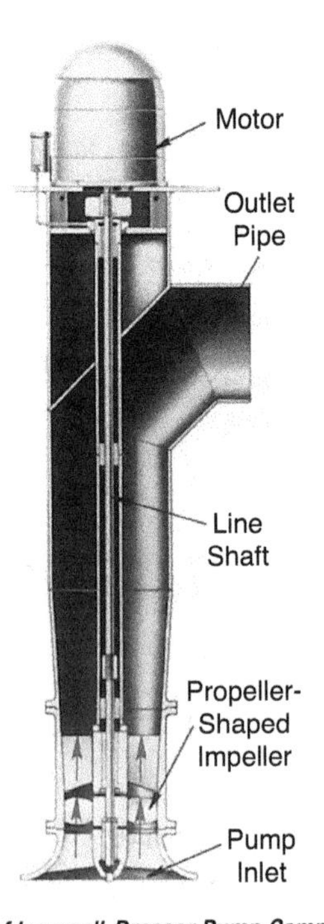

Courtesy of Ingersoll-Dresser Pump Company.

Figure 14-5 Axial-flow pump

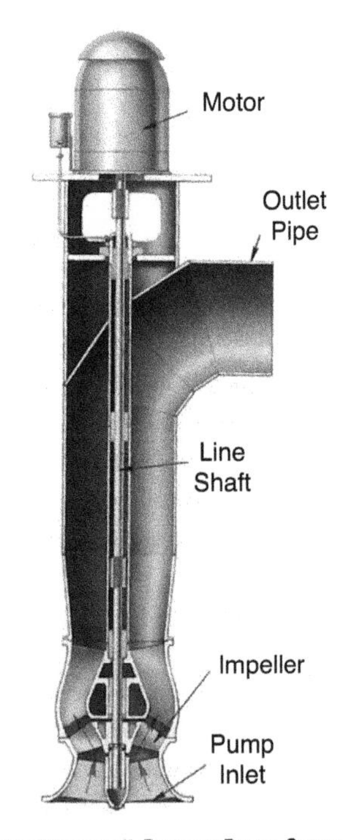

Courtesy of Ingersoll-Dresser Pump Company.

Figure 14-6 Mixed-flow pump

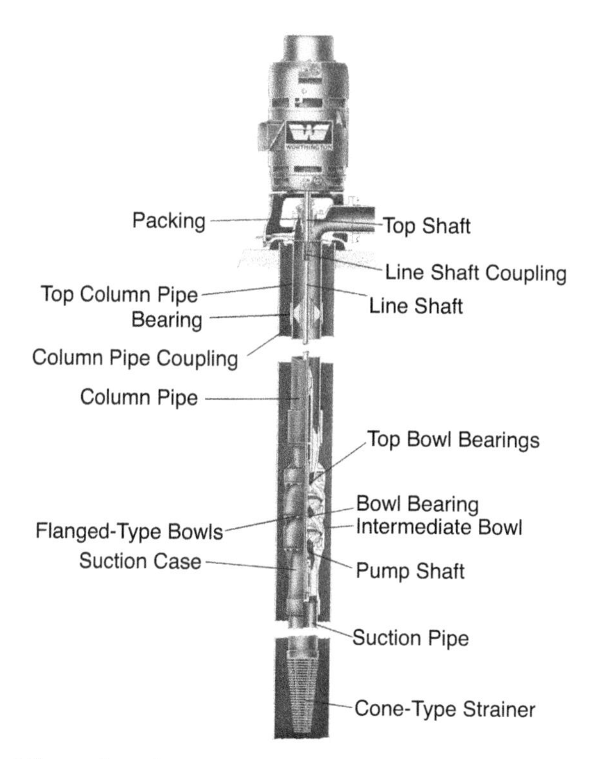

Courtesy of Ingersoll-Dresser Pump Company.

Figure 14-7 Deep-well pump

A submersible pump combines a turbine pump with a submersible motor that is close-connected in a single unit (Figure 14-8). Submersible pumps are generally replacing deep-well pumps for deeper installations because they eliminate the maintenance and problems associated with the long drive shafts of the deep-well pumps.

Turbine pumps are also used as in-line booster pumps. An example of this use is in a distribution system where pressure must be raised for a service area that is at a higher elevation than the rest of the system. The pump is essentially a turbine pump mounted in a container (Figure 14-9), thus it is commonly called a *can pump.*

Jet Pumps

As shown in Figure 14-10, a jet pump uses a centrifugal pump at the ground surface to generate high-velocity water that is directed down the well to an ejector. The partial vacuum created by the ejector then raises additional water to the surface. The discharge of the pump is split, with part of the water going to the distribution system and part being returned to the ejector.

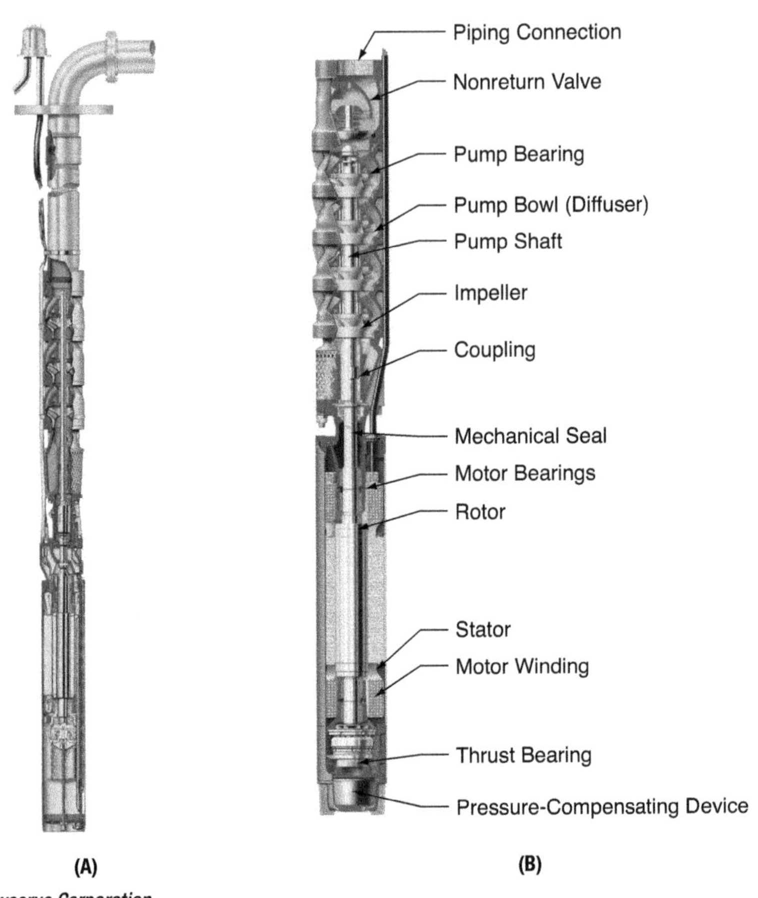

Courtesy of Flowserve Corporation.

Figure 14-8 (A) Vertical turbine pump driven by a submersible motor, and (B) cross-sectional view of a submersible pump

Jet pumps are widely used for private wells because of their low initial cost and low maintenance. The pumps have relatively low efficiency, however, and are seldom used for public water systems.

PUMP OPERATION AND MAINTENANCE

Reading Pump Curves

A pump curve is a graph showing the four characteristics of a particular pump. The four characteristics of capacity, head, required power, and efficiency are interrelated.

The graph furnished by the manufacturer for each type and style of pump generally has the following three curves:

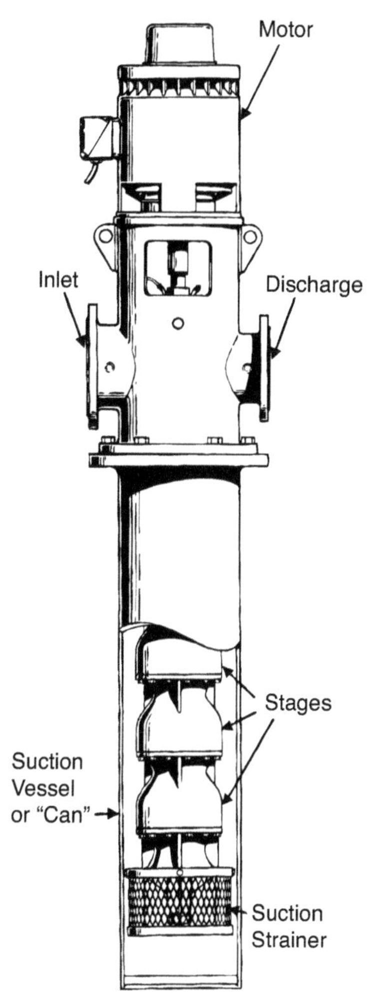

Figure 14-9 Turbine booster pump

1. The H-Q curve is the relationship between the head (H), usually expressed in feet of water, and the capacity (Q, for quantity) in gallons per minute (gpm). The highest possible head that the pump can attain is when it is not pumping at all, and head drops at an ever-increasing rate as the quantity increases.
2. The P-Q curve shows the relationship between power required (P) and capacity (Q). Power is in brake horsepower, so motor efficiency must be known to determine the exact motor horsepower required.
3. The E-Q curve provides the relationship between pump efficiency (E) in percent and capacity (Q). In sizing a pump, a model should be selected that provides the desired flow rate at or near the peak pump efficiency. The more efficient a pump is, the less costly it will be to operate.

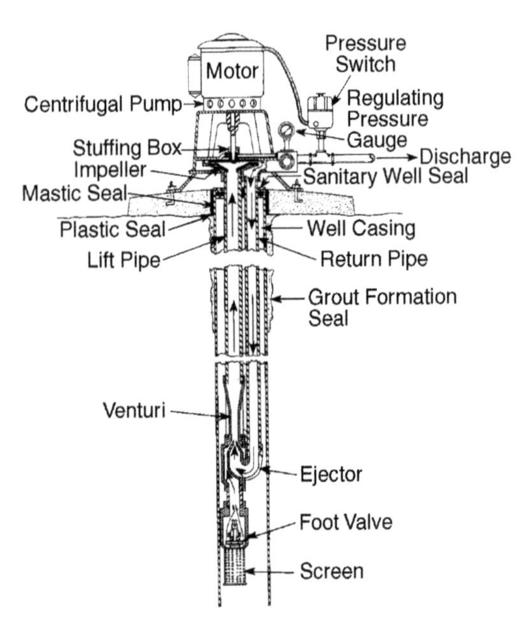

Figure 14-10 Jet pump

Use of a typical set of curves is illustrated in Figure 14-11. In this case, suppose one wanted to determine the pump head, power, and efficiency when the pump is operating at 1,600 gpm. A line is first drawn upward from 1,600 gpm, and the three graphs are marked at the intersections. A line drawn to the left of the intersection with the E-Q curve indicates that the pump efficiency is about 83 percent. A line drawn to the left of the H-Q curve shows that the pump will be operating at a head of about 122 ft under these conditions. And a line to the right of the P-Q curve shows the brake horsepower required is about 60.

Another factor that must be figured into pump efficiency is the efficiency of the electric motor. Motors are available with various degrees of efficiency, with the more efficient ones being more expensive. Consultants often specify the overall efficiency of the pump and motor, usually called the *wire-to-water efficiency*. In this way, a manufacturer whose pump does not operate at maximum efficiency at the specified head and capacity may still meet the overall specified efficiency by supplying it with a very high-efficiency motor.

Pump Starting and Stopping

If the suction side of a pump is supplied with water under some pressure (flooded suction), the pump will always be primed and ready for immediate use. The only thing that is necessary is to provide a means of releasing any air that may accumulate at the top of the pump casing.

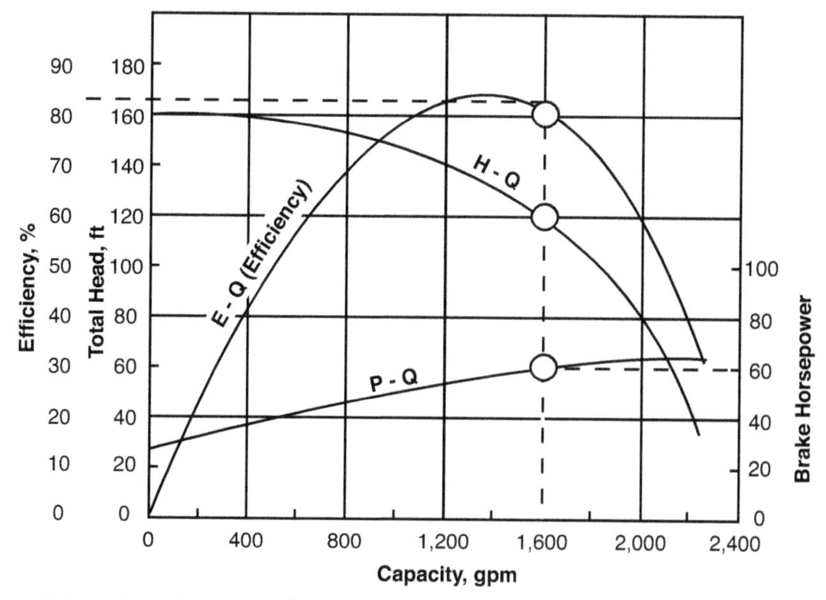

Figure 14-11 Example pump performance curve

Priming Systems

Most pump installations have the pump located above the water source, so some provision must be made for priming the pump before use. One method of maintaining prime is to install a foot valve at the bottom of the suction pipe. This will usually hold enough water in the pump after each use to allow it to be started again. There must also be an emergency method of filling the pump with water in the event the foot valve does not hold or the pump must be drained for repairs. So a priming pump must be available to create a vacuum on the highest point on the pump casing to draw water up into the pump (Figure 14-12).

The method of maintaining prime used by most pumping stations today is to have a central priming system connected to all pumps.

As illustrated in Figure 14-13, each pump is fitted with a priming valve. This valve has a float that raises and closes a small valve to the priming vacuum line as long as there is water in the pump. If there should be air in the pump, the float drops and opens the valve, which will apply a vacuum to pull water into the pump.

Start and Stop Operations

A small pump is usually provided with only a check valve on the discharge, which will automatically open when the pump is running and close when the pump is stopped. If there is some water hammer created by the stopping and starting, it can sometimes be relieved by installation of a pressure relief valve or a surge chamber filled with air.

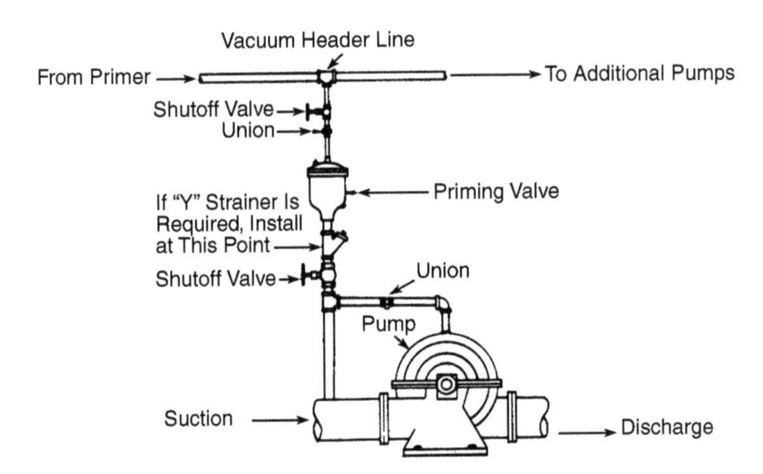

NOTE: Use pipe sizes recommended by the priming valve manufacturer.

Source: Reprinted with permission of APCO/Valve & Primer Corp., 1420 S. Wright Blvd., Schaumberg, IL 60193, from APCO Valve Index *by Ralph DiLorenzo, Executive VP, © 1993.*

Figure 14-12 Vacuum priming system

For a larger pump, the surge is so great that other means must be used to slowly bring the pump on line and slowly remove the flow as it is stopped. The general procedure requires special controls and a power-activated discharge valve that operate as follows:

1. When a pump is to be activated, power is first applied to the motor.
2. A pressure sensor on the pump discharge then determines that the pump is producing pressure before it will allow the discharge valve to open.
3. If pressure is detected, a contact closes that allows the power-operated discharge valve to open slowly. Hydraulically operated valves have a needle valve for adjusting operating speed, and should be set to give the motor plenty of time to come up to speed and to apply the additional supply of water slowly to the system.
4. On shutdown of the pump, the signal first goes to the discharge valve, which closes slowly while the motor continues to run.
5. As the valve reaches the end of travel and completely closes, a limit switch is activated that finally shuts off the motor.

Special provisions must also be made for closing the discharge valve in the event of power failure. If the motor suddenly stops with the valve open, the rush of water from the system will quickly begin to spin the motor backward with possibly disastrous results. Controls must therefore be provided to shut pump discharge valves quickly in the event of power failure.

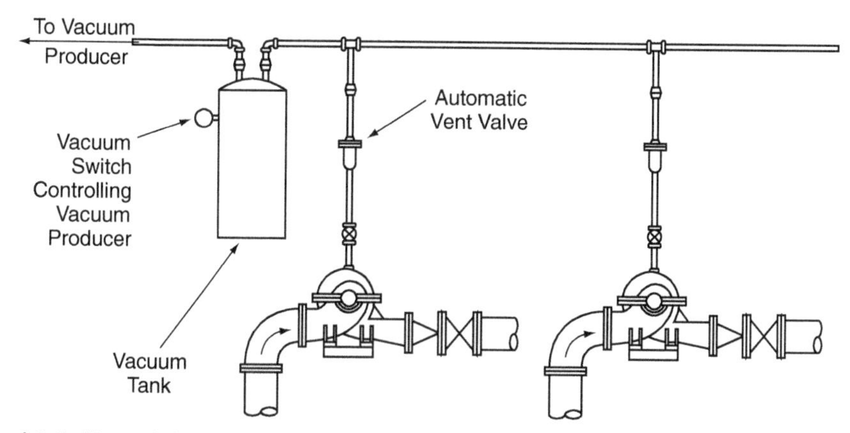

Reprinted with permission of the McGraw-Hill Companies from Pump Handbook *by Krassik et al., © 2001. Published by McGraw-Hill.*

Figure 14-13 Vacuum-controlled central automatic priming

Flow Control

Most pumps used in water system operations are constant speed, and varying demands are met by turning pumps on and off in different combinations. The discharge valve of a centrifugal pump can be partially closed to throttle flow without harm to the pump, but it should only be done in emergency situations. Throttling flow with a gate valve will eventually damage the valve because the gates are hung loose when the valve is in a partially open position. A butterfly valve can be used for throttling flow for short periods of time, but if used long-term, the gate will begin to flutter and eventually will be damaged.

If continual varying of pump discharge is required, several types of variable-speed drives are available, including stepped-speed and variable-speed motors, as well as hydraulic and mechanical variable-speed drives used with constant-speed motors. Pumps operated at variable speed may not have very good overall efficiency because all centrifugal pumps have a relatively narrow range at which they operate at maximum efficiency.

Monitoring Pump Operation

The following actions should be performed regularly to monitor pump operation and to determine whether a pump requires maintenance:

- Suction and discharge head should be monitored by reading gauges mounted on the suction and discharge sides of the pump. The initial readings for a new pump should be recorded so that comparisons can be made over the years.
- The bearing and motor temperatures should be regularly monitored on both pumps and motors. Many experienced operators make a habit of putting their hands on each surface as they make an inspection of

operating pumps. A more accurate way is to use an infrared heat gun to record the exact temperatures periodically to determine whether there are changes. If there is a sudden increase in heat on any surface, the unit should be shut down until the cause is determined.

- As with temperature, experienced operators learn the normal sound and vibration levels of a pump. A sudden increase in vibration can be caused by misalignment of the motor coupling, imbalance of the pump impeller, or other problems that could become serious if the pump is allowed to continue operation. Automatic vibration sensors are also available that can be mounted on the pump unit and will alarm or shut down the motor if higher than normal vibration occurs.
- The pump output can be easily monitored if the discharge of individual pumps is metered. The pumpage rate at specific pressure conditions should be recorded for new pumps and then periodically compared. If the output is dropping, it is generally a sign of pump wear or damage that should be checked.
- Packing leakage should be observed. If the pump has mechanical seals, there should be no leakage, and if there is, it is a sign that repairs should be made relatively soon. If the pump has packing glands, water should be slowly dripping from them. If they are not dripping, the packing gland has been pulled too tight or there are other problems that should be investigated.
- The amperage drawn by the motor should be initially recorded and then periodically checked. If the amperage increases over a period of time with the pump running under the same operating conditions, the cause should be investigated.

Pump Maintenance

Every water system should have a regular program of major pump inspection and maintenance. It is better to take a pump out of service in winter and take all the time that is necessary to perform a thorough maintenance job than to have it break down in summer and do a rush repair.

Besides a thorough cleaning of the pump and motor, other things that should be checked are the condition of the impeller, bearings, seals, and alignment of the coupling. The pump manufacturer should be consulted for a complete list of maintenance items and methods of repair.

Bearing Maintenance

Pump bearings may be either oil lubricated or grease lubricated. The bearing housing of oil-lubricated bearings should be kept full with oil of the grade recommended by the equipment manufacturer. The oil should be changed in new pumps after the first month of operation. After the initial change, most manufacturers recommend that the oil be changed every 6 to 12 months, depending on the operating frequency and environmental conditions.

For grease-lubricated bearings, bearing temperature on new pumps should be closely monitored for the first month. Bearings that are not operating properly can be touched with the bare hand. After the first month of service, the bearings should be regreased at the frequency recommended by the manufacturer.

It is important that bearings not be provided with too much grease. If the grease is packed too tightly, the bearing will run hot. Bearings should be regreased according to the following procedure:

1. Open the grease plug at the bottom of the bearing housing.
2. Fill the bearing with new grease until new grease flows from the drain plug.
3. Run the pump with the drain plug open until the grease is warm and no longer flows from the drain.
4. Replace the drain plug.

MOTORS

Electric motors are used to power 95 percent of the pumps used in water supply operations. Internal combustion engines are primarily used for standby service, although some utilities operate engines during peak demand periods to reduce electrical costs.

The alternating current (AC) electricity furnished by electric utilities for water utility use is generally in the form of three-phase current. This is then usually reduced to single-phase current within the plant for operating lights and small equipment. Principal motor components are labeled in Figure 14-14.

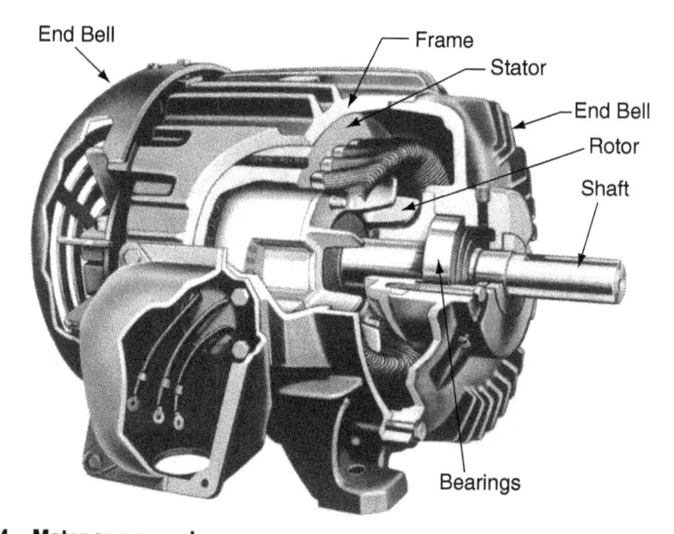

Figure 14-14 Motor components

Single-Phase Motors

Single-phase motors are typically used only in fractional-horsepower sizes, but if there is a special requirement, they can be furnished with ratings up to 10 hp (7.5 kW) at 120 or 240 V. A single-phase motor has no power to bring it up to speed (starting torque), so it must be started by some outside device.

A starting winding is usually built into the motor to provide initial high torque. Then as the motor comes up to speed, a centrifugal switch changes connections to the running winding. Single-phase motors are of the following three basic types:

1. *Split-phase motors* use a rotor with no windings. They have a comparatively low starting torque, so they require a comparatively low starting current.
2. *Repulsion-induction motors* are more complex and expensive than split-phase motors and also require a higher starting current.
3. *Capacitor-start motors* have high starting torque and high starting current. They are used in applications where the load can be brought up to speed very quickly and infrequent starting is required.

Three-Phase Motors

Motors used in water treatment or distribution systems that are more powerful than ½ hp (4 kW) are generally three phase, and may be operated at 230, 460, 2,300, or 4,000 V. The three main classes used are squirrel-cage induction, synchronous, and wound-rotor induction.

- A *squirrel-cage induction motor* is the simplest of all AC motors. The rotor windings consist of a series of bars placed in slots in the rotor and connected together at each end (which has the appearance of a squirrel cage). The stator windings located in the frame are connected to the power supply, and the current flowing through them induces a rotating magnetic field. Simple starting controls are usually adequate for most of the normal- and high-starting-torque applications of these motors.
- *A synchronous motor* has power applied to the windings in such a way that a revolving magnetic field is established. The rotor is constructed to have the same number of poles as the stator and they are supplied with direct current so the rotor's magnetic field is constant. A slip-ring assembly (commutator) and graphite brushes are used to connect power to the rotor. Synchronous motors are used where the motor speed must be held constant and, because the motor has a power factor of 1.0, in areas where the power company has a penalty for low power factor conditions.
- A *wound-rotor induction motor* has a stator similar to a squirrel-cage motor, except that the resistance of the rotor circuit can be controlled

while the motor is running, which varies the motor's speed and torque output. The starting current required for a wound-rotor motor is seldom greater than the full-load operating current. In contrast, squirrel-cage and synchronous motors generally have starting current requirements between 5 and 10 times their full-load current.

Motor Temperature

Motors convert electrical energy into mechanical energy and heat. About 5 percent of the energy is lost in heat and it must be removed quickly to prevent the motor temperature from rising too high. Motors are designed for an external (ambient) temperature of 104°F (40°C), so ventilation air should never have a higher temperature. The useful life of a motor is considerably shortened by being run at high temperatures. Care must be taken not to obstruct air flow around motors.

Mechanical Protection

The design of the motor housing must be considered in relation to where the motor will be located. A motor powering a pump inside a building can generally be of the simplest design because it will be in a clean environment. A motor to be installed outside must have protection from rain, dust, and wind-driven particles.

Motor housing designs commonly available include open, drip-proof, splash-proof, guarded, totally enclosed, totally enclosed with fan cooling, explosion-proof, and dust-proof.

MOTOR CONTROL EQUIPMENT

Motor Starters

Small motors are usually started by directly connecting line voltage to the motor. Motors larger than fractional horsepower are typically started and stopped using a motor starter. As illustrated in Figure 14-15, a typical motor starter includes a main disconnect switch, fuses or a circuit breaker, motor protection by temperature monitors on each of the phases, and provisions for remote operation.

Reduced-Voltage Controllers

When the starting current of a motor is so high that it may damage the electrical system or deprive other operating motors of sufficient current, a reduced-voltage controller is used. The controller supplies reduced voltage to start the motor, then applies full voltage when it is about up to speed.

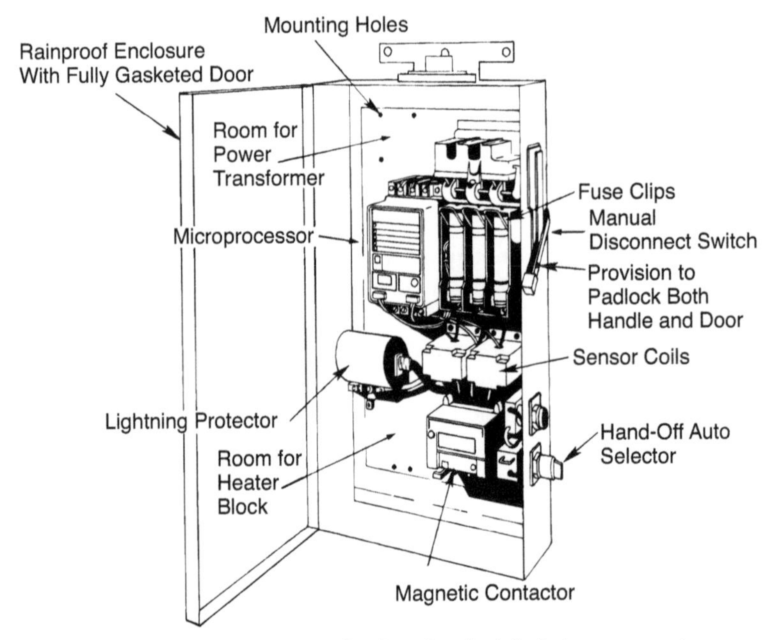

NOTE: Lightning protector is more effective when installed above power, in the area where transformer can be placed.

Figure 14-15 Combination motor starter

Motor Control Systems

Remote and automatic controls eliminate the need for an operator to be near a pump to operate it. Manual remote controls for pumps are now usually located in a central control room.

Figure 14-16 illustrates how controls can start and stop pumps on several levels to automatically meet customer demands. The system is also furnished with both high- and low-level alarms that will alert the operator of trouble if any of the preset limits are exceeded.

INTERNAL-COMBUSTION ENGINES

Internal-combustion (IC) engines are used by water utilities to power pumps during emergencies and for portable applications. Also, emergency electric generators are often powered by gas (natural gas, propane, or methane), gasoline, or diesel engines. Remote locations may use IC engines to provide power for the entire utility operation. Some utilities use IC engines as part of an energy (cost) efficiency strategy to avoid peak-demand electrical charges.

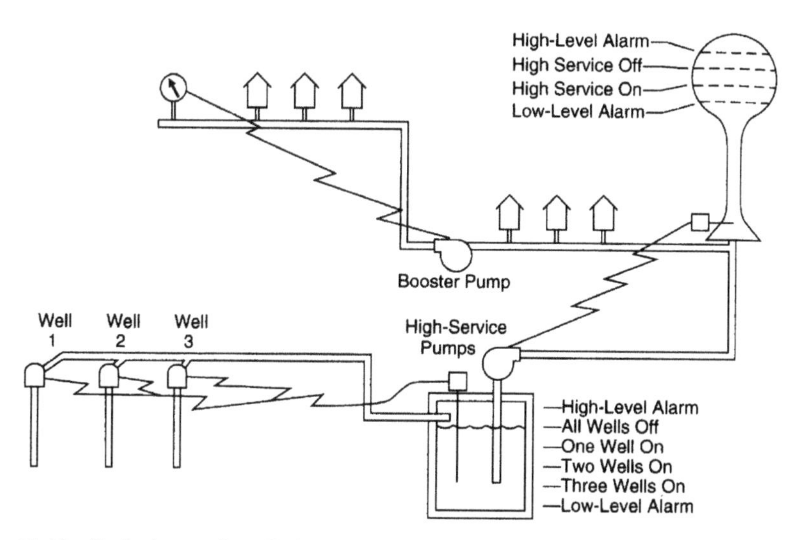

Figure 14-16 Typical pump installation with a number of controls

Gasoline Engines

Most gasoline engines are used for standby or emergency purposes. The cost of the engines is low, but fuel and maintenance can be relatively high. These engines are available in sizes ranging from 1 hp (0.7 kW) to hundreds of horsepower.

Diesel Engines

Diesel engines are reliable and often used as emergency generator drives that supply power to pump stations and water treatment plants. Initial cost can be higher than similar size gasoline engines, but operating costs are usually lower.

Gas Engines

Gas (natural gas, propane, or methane) engines can be converted from gasoline engines or original equipment designed for these fuel sources. Methane fuel is usually only available from wastewater treatment facilities when it can be produced. Gas engines are selected where an inexpensive source of fuel is available or where there may be concerns about air emissions.

Steam Engines

Although historically significant, these are not often used in water utilities today. Where they are in use, the steam turbine engine is most often employed to power very large pumps.

Operation and Maintenance

Start-up and operating procedures and maintenance practices for engines are supplied by the equipment supplier. These procedures should be posted near the engine location. The following general procedures are recommended for most engines.

- *Service prior to operation.* Inspect the engine and components before starting. Check fluid levels, belts, and hoses. Make sure that cooling water is flowing and the clutch is disengaged.
- *Initial operation service.* Soon after starting the engine, check the idle speed, oil pressure, water temperature, and all other operating indicators. After the recommended warmup period, check all operating indicators again.
- *Service during operation.* Check all gauges and other operating indicators periodically during operation. Fuel levels should be closely monitored for diesel engines.
- *Service after operation.* Any issues observed during operation should be corrected. Lubricant should be examined.
- *Routine preventive maintenance.* Engines that are used to provide emergency service, including power backup, should be exercised under load according to a strict schedule. Many systems operate emergency power engines at least 15 minutes a week. Routine inspection and preventive maintenance should be performed according to manufacturer's recommendations. All maintenance procedures should be recorded for future reference. Diesel fuel filtering that may be required for injection-type systems must be performed by qualified personnel who have been specially trained.

PUMP, MOTOR, AND ENGINE RECORDS

Detailed records should be maintained on all equipment for use in scheduling maintenance, evaluating operation, and performing repairs. Most water systems maintain a quick-reference data card or notebook sheet for each piece of equipment, or a computer file with similar information. Typical information that should be recorded for each piece of equipment includes

- Make, model, capacity, type, serial number, and warranty information;
- Date and location of installation and name of installer;
- Part numbers of special components likely to require replacement;
- Results of initial tests and of all subsequent tests of the equipment;
- Manufacturer's suggested inspection and maintenance schedules; and
- Names, addresses, and phone numbers for the manufacturer and local representative.

A file folder should also be maintained for each piece of equipment, containing the original manufacturer's literature, operating and repair manuals, and copies of correspondence, purchase orders, and other pertinent information.

BIBLIOGRAPHY

ANSI/AWWA E102. *AWWA Standard for Submersible Vertical Turbine Pumps.* Denver, Colo.: American Water Works Association.

ANSI/AWWA E103. *AWWA Standard for Horizontal and Vertical Line-Shaft Pumps.* Denver, Colo.: American Water Works Association.

Beverly, R. 2009. *Pump Selection and Troubleshooting Field Guide.* Denver, Colo.: American Water Works Association.

DiLorenzo, Ralph. 1993. *APCO Valve Index.* Schaumberg, Ill.: APCO/Valve & Primer Corp.

Jones, G.M., R.L. Sanks, G. Tchobanoglous, and B.E. Bosserman II. 2008. *Pumping Station Design.* Boston: Butterworth-Heinemann.

Karassik, I.J. 2008. *Pump Handbook,* 4th ed. New York: McGraw-Hill.

Rishel, J.B. 2002. *Water Pumps and Pumping Systems.* New York: McGraw-Hill.

Chapter 15

Water Wells

Groundwater sources are usually relatively simple to develop and often require little or no treatment. About 95 percent of the rural population in the United States is served by wells.

AQUIFERS AND CONFINING BEDS

Groundwater is the result of the infiltration and percolation of water down to the water table through the soil and cracks in consolidated rocks. Referring to Figure 15-1, the water table that is immediately below the surface is not confined and often rises and falls due to variations in rainfall. When a well is drilled into the water table, water rises in the well just to the level of the top of the water table.

A confined aquifer is a permeable layer of rock that is confined by upper and lower layers that are relatively impermeable. As illustrated in Figure 15-1, if the recharge area for the confined aquifer is elevated, the water in the aquifer will be under pressure. If there is sufficient pressure for the water to flow from a well that taps the aquifer, it is called a *flowing artesian well.* More often, there is pressure in the aquifer so that water will rise in the well but not sufficiently to flow out of the well. In this case, it is called a *nonflowing artesian well.* The height to which water will rise in wells located in an artesian aquifer is called the *piezometric surface.*

On occasion, there are several aquifers available, each with different water quality. In this case, the well is drilled and cased through the aquifers of poor quality so as to tap only the best aquifer.

The mineral content of water from shallow wells will vary with the mineral content of the soil in the area. Water from deep wells almost always contains concentrations of dissolved minerals and may have considerable hardness, but it is usually clear and colorless.

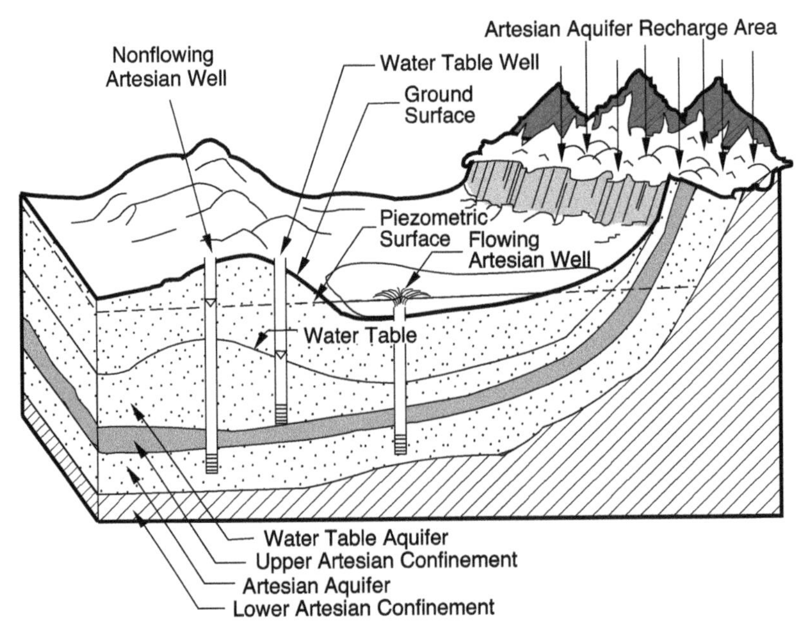

Figure 15-1 Cross-section of an artesian aquifer

GROUNDWATER SOURCES

Groundwater used for supplying public water systems is obtained principally from springs and infiltration galleries or from drilled wells.

Springs and Infiltration Galleries

It is rare for springs to have adequate capacity to serve a public water system, but if a spring is to be used, the collection point must be enclosed in a concrete box constructed around the spring to prevent insects, animals, and surface drainage from entering the water. It is usually difficult to tell where spring water originates. It may come from a distant point and travel through the ground for many years, or it could come from a nearby septic tank. A thorough analysis of the water should be conducted before a spring is used for a public water supply source.

Infiltration galleries are often constructed by burying a perforated pipe under or along the edge of a lake or stream and conducting the water to a central collection box. The intent is to obtain surface water that is partially filtered, thus eliminating the problems of a surface water intake. If the proper soil conditions are present, infiltration galleries work quite well. But remember that the water collected by the perforated pipe may actually be surface water from the adjacent river banks that is migrating toward the stream. If this is the case, care must be taken that there are no sources of surface or buried contamination in the surrounding area.

Under provisions of the federal Surface Water Treatment Rule (SWTR), state regulatory agencies must require that treatment of water from springs and infiltration galleries be the same as for a surface water source.

Drilled Wells

Drilled wells are the most common type of well installed for public water supply use. They may be up to 4 ft (1.5 m) in diameter and anywhere from relatively shallow to several thousand feet (meters) deep. Common drilling methods include the cable tool, rotary hydraulic, rotary air, and down-the-hole hammer methods. All methods use some means of breaking through the soil or rock and bringing the removed material to the surface.

If the water-bearing stratum is sand or gravel, the usual practice is to drill the well and install the casing to the bottom of the sand layer. The well screen is then lowered into the casing and the casing is withdrawn to expose the screen, as illustrated in Figure 15-2. The seal between the top of the screen and the bottom of the casing is called a packer.

When the source of water is from fractures or openings in consolidated rock, the casing is usually seated firmly on top of the rock layer, and the hole is then left open through the rock, as illustrated in Figure 15-3.

If the water-bearing stratum is composed of very fine sand, it is usually difficult to hold back the sand with a well screen while still admitting an acceptable flow of water. In this situation, a gravel-wall (or gravel-packed) well may be installed. One common installation method is to first install a large-diameter construction casing into the sand layer. The working casing and screen is then dropped into place, and selected coarse gravel is placed around it. As illustrated in Figure 15-4, the construction casing is then withdrawn, and the gravel holds back the fine sand.

WELL TERMS

Some important terms for the hydraulic characteristics of a well are illustrated in Figure 15-5 and/or defined as follows:

- *Static water level* is the natural water surface with the well not in operation. It is measured from the ground surface to the water surface. In some areas of the country, the water surface may be just below the ground surface; in other locations, it may be near the level of a deep aquifer.
- *Pumping water level* is the level of the water in the well after the pump has been operating for a period of time. The intake of the well pump must be located below this level, otherwise it will begin to draw in air.
- *Drawdown* is the drop in water level from the static level to the pumping level. A well located in very porous gravel may have almost no drawdown because water flows to the well screen as fast as it is pumped. A well located in an aquifer with low porosity will have a very large drawdown.

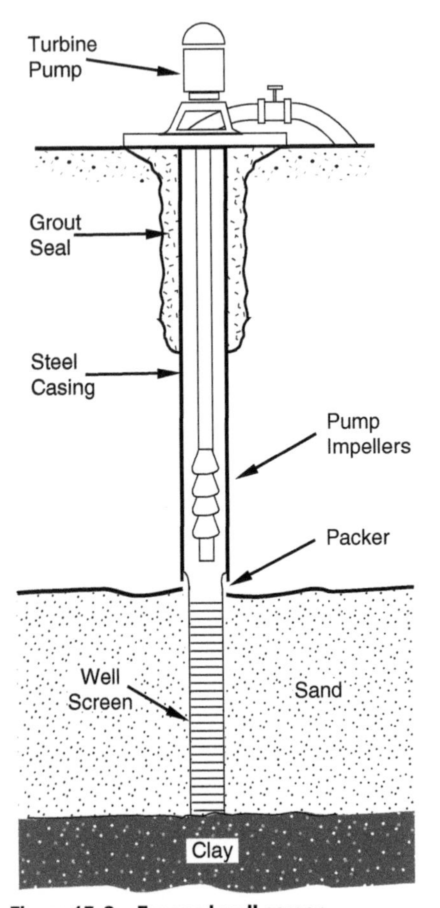

Figure 15-2 Exposed well screen

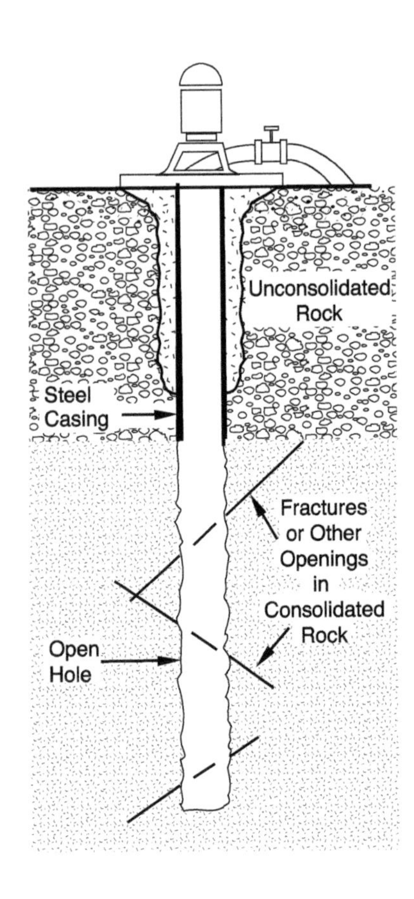

Figure 15-3 Casing seated at top of rock layer with an open hole underneath

- *Cone of depression* is the depressed water surface surrounding the well when the well pump is in operation. The water surface takes the shape of an inverted cone.
- *Zone of influence* is the distance that the cone of depression affects the normal (static) water level. It may be from a few feet (or meters) to hundreds of feet (or meters), depending on the porosity of the aquifer and other factors. In general, it is not good practice to install wells so that their zones of influence overlap. As illustrated in Figure 15-6, the cone of depression of closely situated wells will significantly lower the required pumping level of two adjacent wells.
- *Residual drawdown* is a lowered water level, below the original static level, that remains after pumping has been stopped for a period of time.
- *Well yield* is the rate of water withdrawal that a well can supply over a long period of time. In other words, it is the recharge rate that the aquifer can continuously sustain to the well. The yield of small wells

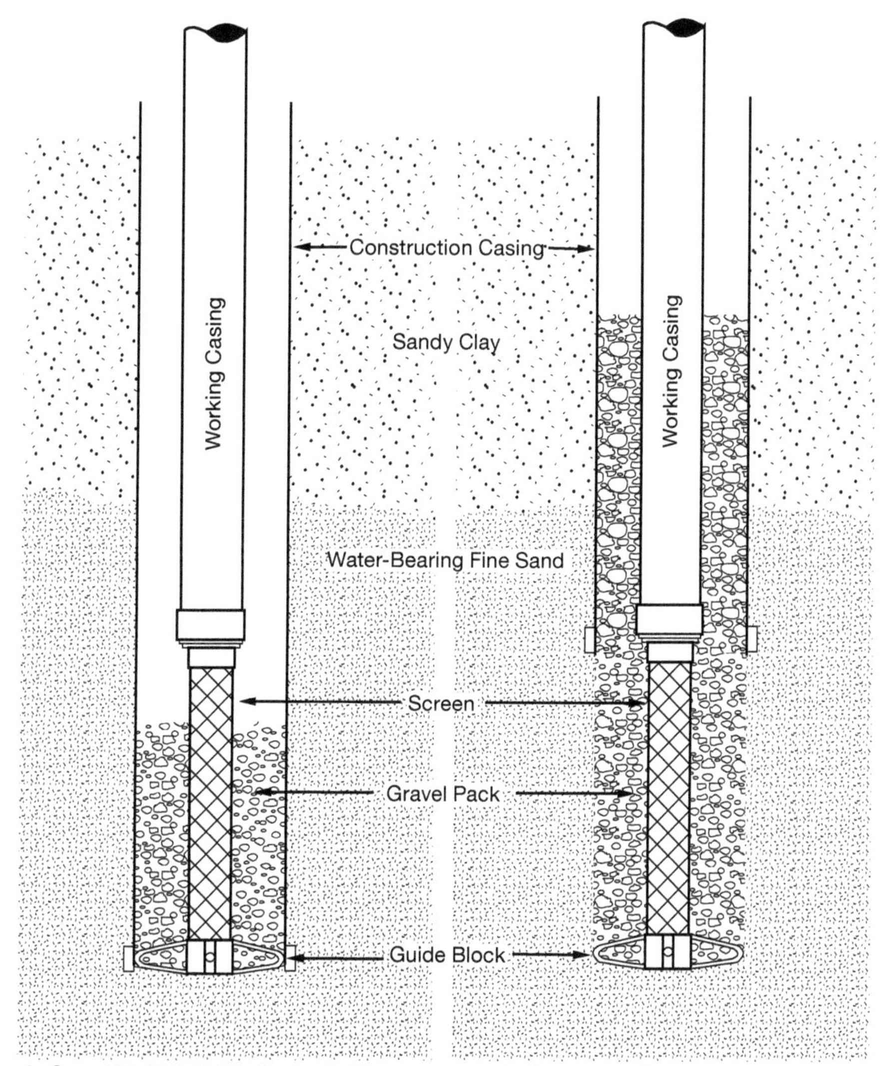

Figure 15-4 Gravel-wall well construction

is usually expressed in gallons or liters per minute. For large wells, it may be expressed in cubic feet or cubic meters per second.

- *Recovery time* is the time it takes after pumping has stopped for the water level to return to the static water level.
- *Specific capacity* is the well yield per unit of drawdown, or

$$\text{specific capacity} = \frac{\text{well yield}}{\text{drawdown}}$$

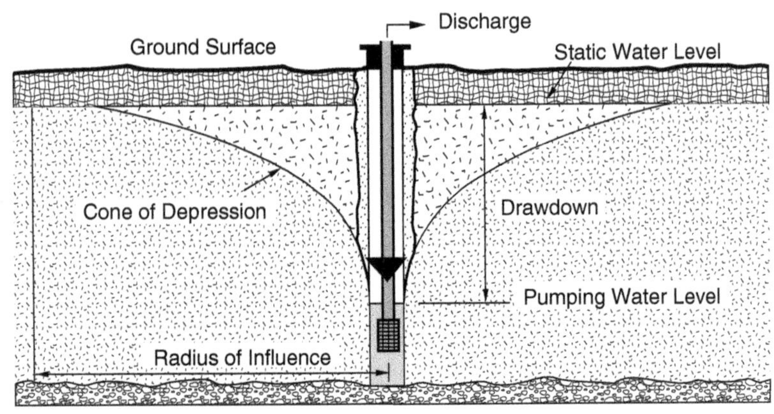

Figure 15-5 Hydraulic characteristics of a well

Example 1: A well has been tested by pumping at various rates, and it has been found that the maximum amount it can produce while operating for an extended period of time is 500 gpm. When operating at that rate, the drawdown is 10 ft. The specific capacity is therefore

$$\frac{500 \text{ gpm}}{10 \text{ t}} = 50 \text{ per ft drawdown}$$

WELL PROTECTION

Until about 1985, all groundwater was considered pristine because the water was thought to be cleansed as it passed through the soil. Then, as new chemical analysis techniques were developed and more wells were tested, it was found that many aquifers were affected by both natural and human chemical contamination that could be a danger to health.

Many shallow wells are at risk of contamination from surface water. Under provisions of the federal SWTR, the state primacy agency must make a ruling on each well, and those that are considered vulnerable are labeled "groundwater under the direct influence of surface water." These wells must be treated as surface water sources. In general, wells less than 50 ft (15 m) deep are likely to be considered vulnerable.

Contamination Sources

Under federal and state regulations, all wells serving public water systems must be tested for a variety of contaminants that might pose a threat to public health. In some cases, contamination originates from sources that were not anticipated or known when the well was constructed. Potential sources of groundwater contamination include landfills, liquid-waste storage ponds, septic systems, agricultural activities, and illegal dumping activities.

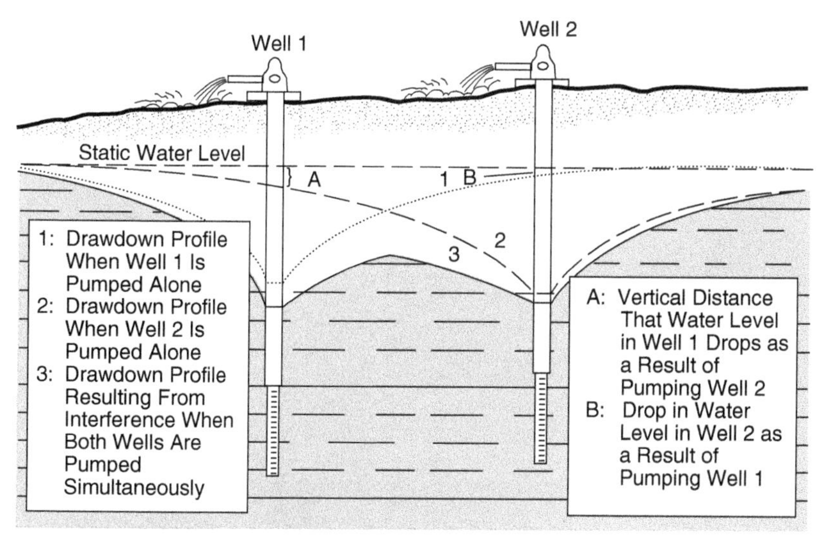

Figure 15-6 Overlapping cones of depression of two wells

Problems caused by groundwater contamination include disagreeable tastes and odors, the presence of disease-causing organisms, and chemical contamination above recommended health limits for drinking water. Unfortunately, there is no set limit on how far a well must be from a contamination source to be safe. It depends on groundwater flow, soil conditions, and many other factors. Some organic chemicals in aquifers have been tracked to sources more than 0.5 mi (0.8 km) away. As illustrated in Figure 15-7, when siting a new well, it is necessary to consider the possibility that the groundwater flow may be reversed when the well is put into operation.

Well Grouting

An important installation procedure in well construction is that the space between the walls of the drilled hole and the well casing must be cemented or grouted, as illustrated in Figure 15-8. The primary purpose is to prevent surface water pollution from traveling down along the outside of the casing and causing contamination of the water being drawn into the well. Secondary purposes of grouting are to protect the casing from corrosion, to restrain unstable soil and rock formations, and to seal off any aquifers with poor water quality. Also illustrated in Figure 15-8 is a well slab that is poured around the top of the well to pitch rain and surface water away from the well.

Sanitary Seal

The top of the well casing must always be sealed to prevent the entrance of water, insects, and animals. The seal has openings for the discharge pipe,

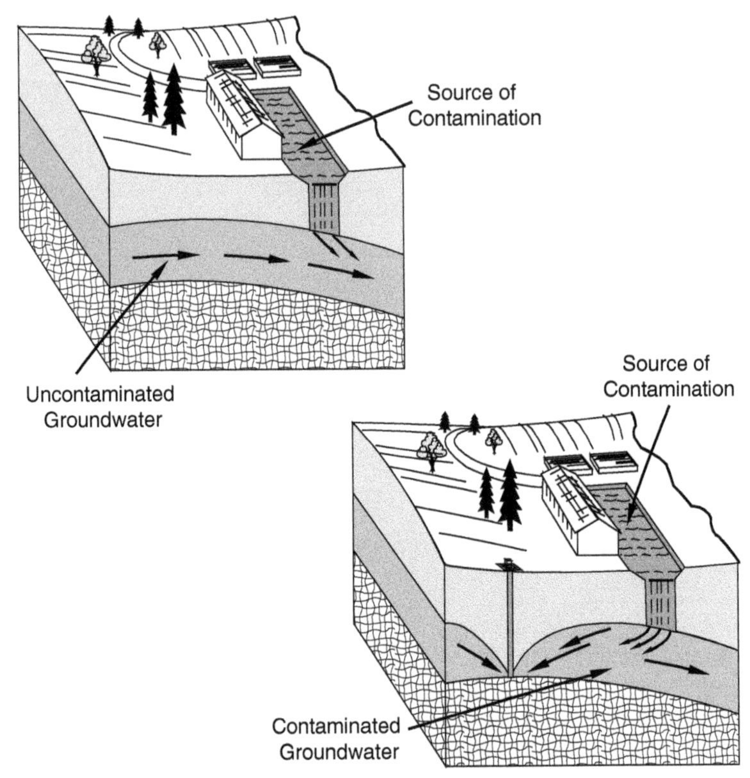

Figure 15-7 Reversal of flow in an aquifer due to well drawdown

pump controls, and a screened air vent that allows air to enter and leave the casing as the water level rises and falls. Figure 15-9 illustrates a sanitary well seal for a small well having a submersible pump.

Measuring the Water Level in a Well

Three common methods are used for measuring the depth of the water in a well: chalked steel tape, electronic, and air-pressure tube. The methods are illustrated in Figure 15-10.

Chalked Steel Tape Method

With the chalked steel tape method, a weight is attached to the end of a steel tape and chalk is applied for a distance of several feet (meters) at the end of the tape. The tape is lowered into the well until the end is sure to be below the water level, and the tape reading is recorded at ground level. When the tape is withdrawn, the wetted portion shows up because of the chalk, and the wetted length can be deducted from the reading at ground level to arrive at the water depth.

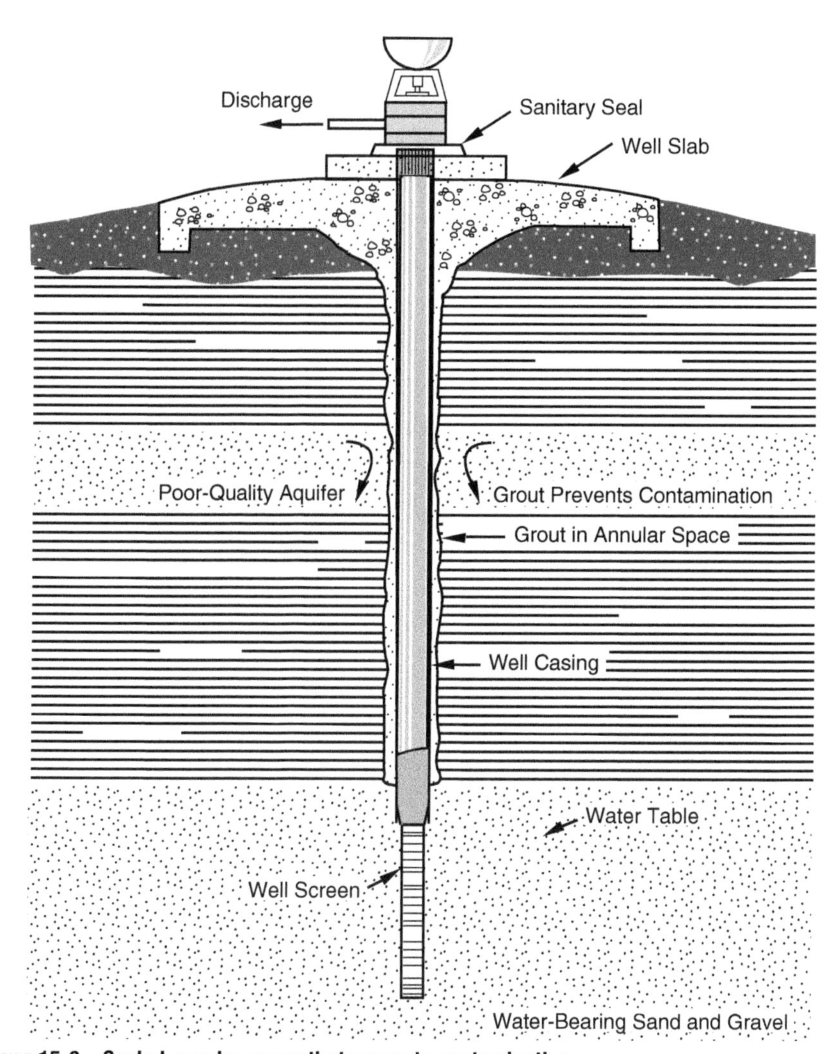

Figure 15-8 Sealed annular space that prevents contamination

Electronic Method

Electronic depth-measuring devices are the quickest and easiest to use. A probe on the end of a wire or tape is lowered into the well, and a light or sound signal is activated when the probe touches the water. The depth is then read from markings on the wire or tape.

Air-Pressure Tube Method

The method of measuring well water level used by most water systems before the electronic units became available included a small-diameter air tube permanently installed in the well. The end of the tube was suspended so that it would always be below the lowest drawdown of the water; then

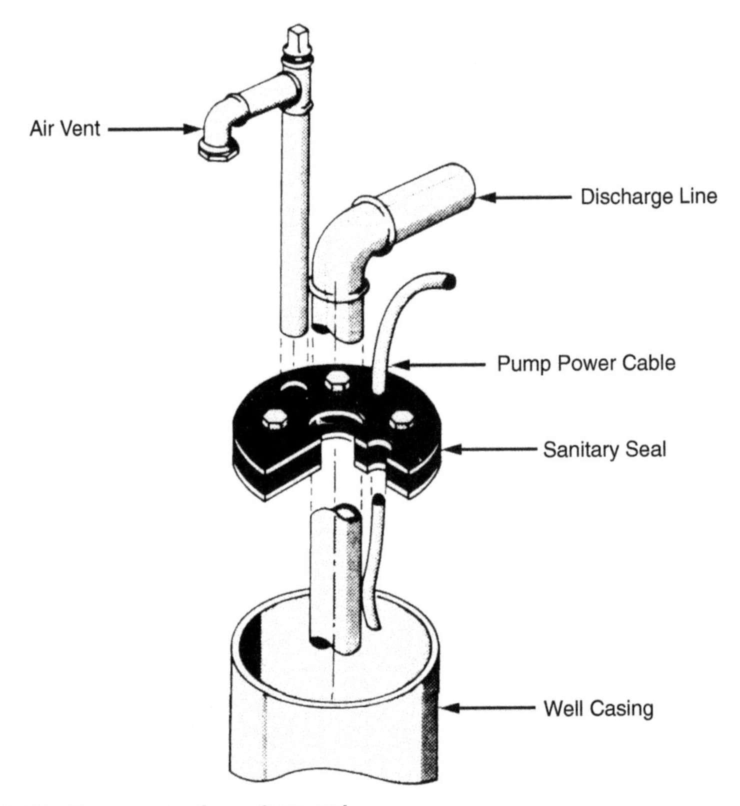

Figure 15-9 Components of a sanitary seal

the exact distance from the end of the tube to the surface was measured before installation.

The top of the tube was fitted with a pressure gauge and a source of air, such as a tire pump. The water depth was determined by forcing air down the tube until air bubbles formed at the end of the tube, and at that point, the gauge pressure was read. For each pound of pressure shown on the gauge, 2.31 ft (0.70 m) of water has been forced from the end of the tube, so the length of tube below the water surface could be computed. This amount could then be subtracted from the length of the tube to arrive at the water depth. This is illustrated in the following example.

Example 2: The air tube in a well is 90 ft long. When the tube is pumped with air, the gauge reading stabilizes at a reading of 20 psi. What is the water level in the well?

20 psi × 2.31 ft/psi = 46.2 ft (the water level is 46 ft above the end of the tube)

90 ft (the length of the tube) – 46.2 ft = 43.8 ft (the distance from surface to water level)

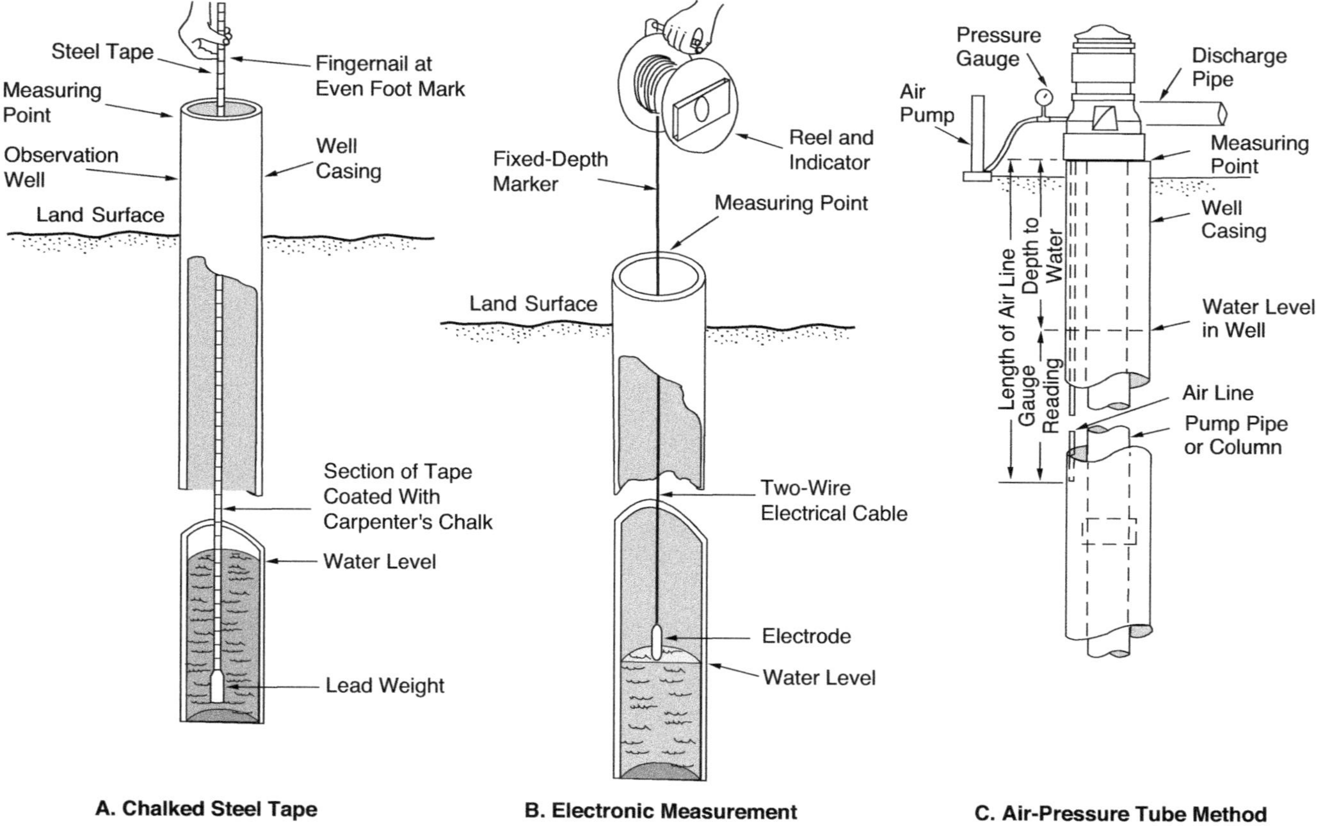

Figure 15-10 Methods of measuring water level depth in wells

WELL RECORDS

It is important that the water system operator maintain good records of the performance of each well. When properly analyzed, the records can indicate problems such as blocking of the well screen, decreasing capacity of the well pump, and drop in water level in the aquifer. Many well problems occur slowly, so by anticipating the problem, arrangements can be made to take corrective action before complete failure occurs.

It is suggested that the following data be recorded at regular intervals:

- Static water level
- Drawdown
- Well yield
- Specific capacity
- If the well is relatively shallow, the correlation between water level and rainfall
- Pumping water level
- Well production
- Recovery time
- Water quality analysis reports

Analysis of these data over a period of time can provide information about the aquifer, pump, and well. For example, if the static level drops but the drawdown remains the same, the operator knows that the water table is falling. This could mean that the aquifer is being pumped in excess of the recharge, so the well could eventually become useless if use of the aquifer is not reduced.

If the static level is unchanged but the drawdown is decreasing, it usually indicates that the pump is losing efficiency. If the static level is unchanged but the drawdown is increasing, it may indicate that the screen is clogged and water is not flowing freely into the well.

BIBLIOGRAPHY

ANSI/AWWA A100. *AWWA Standard for Water Wells.* Denver, Colo.: American Water Works Association.

AWWA. 2010. Principles and Practices of Water Supply Operations—*Water Sources*, 4th ed. Denver, Colo.: American Water Works Association.

AWWA. M21—*Groundwater.* Denver, Colo.: American Water Works Association.

Bloetscher, F., A. Muniz, and J. Largey. 2007. *Siting, Drilling and Construction of Water Supply Wells.* Denver, Colo.: American Water Works Association.

Houben, G., and C. Treskatis. 2007. *Water Well Rehabilitation and Reconstruction.* New York: McGraw-Hill.

Sterrett, R.J., ed. 2007. *Groundwater and Wells.* New Brighton, Minn.: Johnson Screens.

Chapter 16

Distribution System Safety

PERSONAL SAFETY CONSIDERATIONS

Everyone is responsible for maintaining safe working conditions, including water utility upper management, supervisors, and operations personnel. Management is responsible both for maintaining safe working conditions and for supporting a policy that encourages safe performance of work duties. Supervisors at all levels are responsible for the direct control of work conditions. This means that each crew chief has a responsibility to see that all work is done in compliance with safety practices and regulations. Supervisors have a duty to correct safety violations.

The employees have a special position with regard to safety. Safety practices and safe equipment must be used or the safety program will not be successful. The employees must help guard against unsafe acts and conditions. Supervisors are required to discipline employees who do not comply with safety policies.

Regulatory Requirements

The obvious reason for exercising safe practices is to eliminate suffering, injury, and possible death of individuals. Beyond that is the cost to the utility from lost time, medical costs, and possibly even legal judgments. Other considerations are damage to equipment and property with their resulting repair costs.

Another reason for safe practices is the federal Occupational Safety and Health Act. Under this act, the federal government has established minimum health and safety standards that are applicable to every industry. The act mandates that every employer must furnish employees with a workplace that is free from recognized hazards that are likely to cause death or serious physical harm. The act provides for six months in prison and/or a $10,000 penalty on conviction for violations.

Causes of Accidents

An accident can almost always be traced to either an unsafe act or to an unsafe condition. All supervisors and workers should learn to recognize unsafe conditions and unsafe acts.

Personal Protection Equipment

Most water utilities now provide basic personal protection equipment for workers. The equipment should be issued to all workers who will need it, thus they will have no excuse for not using it. Each person must be responsible for maintaining his or her equipment in good condition and having it available when it is needed.

A hardhat should be worn whenever a worker is in a trench or has someone working above him or her, or when he or she is near electrical equipment. Metal hardhats should never be used near an electrical hazard.

Gloves are necessary for protection from rough, sharp, hot materials and cold weather. Special long gloves are available to provide wrist and forearm protection. Workers should also wear gloves when handling oils, solvents, and other chemicals. Gloves should not, however, be worn around revolving machinery because of the danger of a glove being caught and pulled into the machine.

Respiratory equipment must also be available for use in some situations. For example, masks should always be worn when working with asbestos–cement pipe because of the danger of inhaling asbestos fibers. A self-contained breathing apparatus should be available for emergency use wherever chlorine gas is being used.

Other personal safety equipment that should be used when a related danger exists includes safety goggles, face shields, steel-toed shoes, aprons, and ear protection.

EQUIPMENT SAFETY

Material Handling

One of the most common and most debilitating types of injury related to material handling is back injury. Lifting heavy objects can be done safely and easily if common sense and a few basic guidelines are followed. These recommended procedures should be followed when lifting heavy objects:

1. Bend at the knees to grasp the weight, keeping the back straight.
2. Get a firm hold on the object.
3. Maintain good footing with feet about shoulder-width apart.
4. Keeping the back as straight and upright as possible, lift slowly by straightening the legs.
5. If the object is too heavy to lift alone, do not hesitate to ask for assistance.

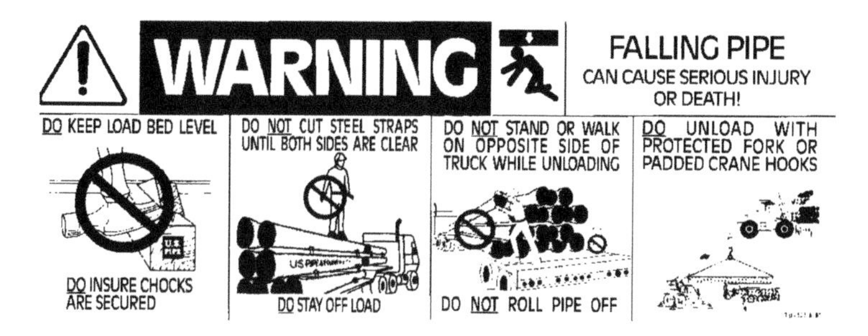

Courtesy of US Pipe and Foundry Company.

Figure 16-1 Warning of the dangers of injury during pipe unloading

Other general safety guidelines for workers who are handling material or doing manual labor are listed here.

- Do not lift or shove sharp, heavy, or bulky objects without the help of other workers or the use of tools.
- To change direction when carrying a heavy load, turn the whole body, including the feet, rather than just twisting the back.
- Never try to lift a load that is too heavy or too large to lift comfortably. Use a mechanical device to assist in lifting heavy objects.
- Even though pipes and fittings look tough, they should be handled carefully. They should be lifted or lowered from the truck to the ground—not dropped. In addition to potentially damaging the equipment, dropping it can be dangerous in several ways.
- When pipe is being unloaded from a truck, there is considerable danger of it rolling. All workers must be warned to stay clear of the load at all times. Figure 16-1 shows a sign warning of the dangers involved in unloading pipe.
- If several people are moving or placing a pipe, they must all work together. Only one person should give directions and signals.
- When a crane is handling pipe, only one person should direct the machine operator. No one should ever stand or walk under the suspended pipe or crane boom.
- If pipe is to be lowered by skids, it is important to make sure the skids are strong enough to hold the weight and are firmly secured. Snubbing ropes should be both strong enough to support the weight and large enough for workers to maintain a good grip.
- Individuals working with ropes should wear gloves to prevent rope burns.
- For lifting and moving large valves and fire hydrants, slings placed around the body should be carefully secured so as not to slip. Special lifting clamps for valves and hydrants should be used if possible.
- Equipment for transporting objects by hand, such as wheelbarrows and hand trucks, should be properly maintained and not overloaded.

- Workers using jackhammers must wear goggles, ear protection, and protective foot gear.
- Horseplay has been the cause of many serious accidents and should be absolutely prohibited on the job.

Trench Safety

Trenches can be made safe if the excavation is properly made and appropriate equipment is used. Proper trench shoring cannot be reduced to a standard formula. Each job is an individual problem and must be considered in relation to local conditions.

If an excavation is 5 ft (1.5 m) or more deep, cave-in protection is required under any soil conditions. Where soil is unstable, protection may be necessary in shallower trenches.

Confined-Space Safety

Dangers of confined spaces include injury, acute illness, disability, and death. The National Institute for Occupational Safety and Health estimates that, until recent years, an average of at least 174 confined-space deaths were occurring each year. Many of the incidents are extremely tragic because they involve multiple deaths. A common scenario is that a worker enters a confined space without proper safety preparations and equipment and is overcome, then co-workers attempt rescue and are also overcome. A great number of those who have died were water and wastewater system workers. Among the dangers that may cause injury or death in a confined space are oxygen-deficient or oxygen-enriched atmosphere, toxic gases, flammable gases, temperature extremes, flooding potential, slick or wet surfaces, falling objects, and electrical hazards.

The Occupational Safety and Health Administration (OSHA) established standards for confined-space entry (29 CFR Part 1910.146) in 1993. The number of yearly deaths is now decreasing but is far from acceptable. A *confined space* is defined by OSHA as any space that

- Has limited or restricted means of entry or exit,
- Is large enough for an employee to enter and perform work, and
- Is not designated for continuous work occupancy.

Examples of confined spaces in the water and wastewater industries include access holes for valves, meters, and air vents; sewer access holes; tanks; wet wells; digesters; and reservoirs.

Confined spaces, as further defined by OSHA, fall into two categories: permit-required and nonpermit-required. Permit-required (permit space) means a confined space that has one or more of the following characteristics:

- Contains, or has a potential to contain, a hazardous atmosphere
- Contains a material that has the potential for engulfing an entrant

- Has an internal configuration such that an entrant could be trapped or asphyxiated by inwardly converging walls or by a floor that slopes downward and tapers to a smaller cross-section
- Contains any other recognized serious safety or health hazard

Nonpermit confined space means a space that does not have any of the above hazards.

OSHA requires a written program for any permit-required, confined-space entry. This includes identifying locations and making preparations prior to entry. Before an employee enters a permit space, the internal atmosphere must be tested with a calibrated, direct-reading instrument for oxygen content, flammable gases and vapors, and potentially toxic air contaminants. In addition, the permit space must be periodically tested during work inside the space to ensure that acceptable conditions exist.

The employer is required to supply the following equipment and to be sure that employees use them:

- Testing and monitoring equipment
- Ventilation equipment needed to obtain acceptable entry conditions
- Communications equipment
- Personal protective equipment
- Lighting equipment needed to enable employees to see well enough to work safely
- Barriers and shields as required
- Ladders needed for safe entry and exit
- Rescue and emergency equipment (Figure 16-2)

The OSHA standards include many other requirements that must be observed. These may include a requirement for a trained attendant outside the confined space that can assist with a rescue or operate emergency equipment. Additional information should be obtained from state and federal offices and is summarized in publications listed at the end of this chapter.

Hand Tool Safety

Some of the basic rules for safe use of hand tools are as follows:

- Always use an appropriate tool for the job. A very large percentage of on-the-job injuries are caused by the use of an improper tool. A screwdriver is not the same as a crowbar, a wrench should not be used as a hammer, and so forth.
- Check the condition of tools frequently. Repair or replace them if they are damaged or defective.
- Avoid using tools on machinery that is moving. It is best to shut off the machine and lock it out before making adjustments.
- Check clearance at the workplace to make sure there is sufficient space to recover a tool if it should slip.

Courtesy of DBI/SALA.

Figure 16-2 Confined space rescue and retrieval system

- Maintain good support underfoot to reduce the hazard of slipping, stumbling, or falling when working.
- Wearing rings while doing mechanical work should be discouraged.
- Carry sharp or pointed tools in covers and pointed away from the body. Do not carry sharp-edged tools in trouser pockets.
- Wear eye protection when using impact tools.
- After using tools, wipe them clean and put them away in a safe place. Keep the workplace orderly.
- Do not lay tools on tops of stepladders or other elevated places from which they could fall on someone below.
- Learn and use the correct way of using hand tools.
- Try not to hurry unduly under emergency conditions. When hurrying, one tends to forget good safety practices and often takes dangerous shortcuts.

Portable Power Tools

An electric power tool should be grounded unless it has an all-plastic case. A ground-fault interrupter circuit should always be used whenever a power tool is used outdoors or in a damp situation. Power cords should be inspected frequently and replaced if any breaks are noted. Power tools should not be lifted by the cords. A worker must make sure to have a firm footing before starting to use a power tool to avoid being thrown off balance when it starts.

Air tools can be dangerous if the hoses and connections are not properly maintained. Workers should not point air tools at anyone and should not use compressed air to clean off their bodies or clothing.

Traffic Control

Barricades, traffic cones, warning signs, and flashing lights should be used to warn the public of construction work that is taking place. These devices should be placed far enough ahead of the work site so that the public has ample opportunity to stop or avoid the obstructions. If necessary, one or more flaggers should be used to slow and direct traffic. Everyone involved in work near roadways should wear bright, reflective vests.

Approved traffic safety control devices and procedures for various classes of roadways are detailed in *Manual on Uniform Traffic Control Devices for Streets and Highways*, prepared by the US Department of Transportation, Federal Highway Administration. Most states have also prepared simplified booklets describing work area protection.

Water utility operators should be aware that they could be liable for damages if an accident occurs as a result of work on a street or highway that was not guarded in conformance with state-directed procedures. Figure 16-3 illustrates the guarding procedures required for construction work that will obstruct one lane of a low-traffic-volume, two-lane roadway.

Chemical Safety

Three chemicals are commonly used to disinfect mains and other facilities. *Liquid chlorine* is used with a gas-flow chlorinator and ejector to disinfect water mains. This type of disinfection should only be used by experienced personnel with specific training in chlorine safety and emergency response for chlorine release. *Sodium hypochlorite*, a liquid, may be used to disinfect mains and other facilities. *Calcium hypochlorite* is generally found in tablet or powder form, and is the most common means of disinfecting mains. It is a strong oxidizer, and while it is not flammable by itself, its heat of reaction with other materials may cause fires. Hypochlorites are corrosive, and personal protective equipment (such as eye goggles and gloves, at a minimum) should be used for worker safety.

Vehicle Safety

Records indicate that most accidents in the water works industry involve vehicles. Workers should be particularly made aware of the potential dangers to themselves, fellow workers, and the public while they are operating large trucks and heavy construction equipment. Many utilities require all workers to periodically attend a safe-driving school.

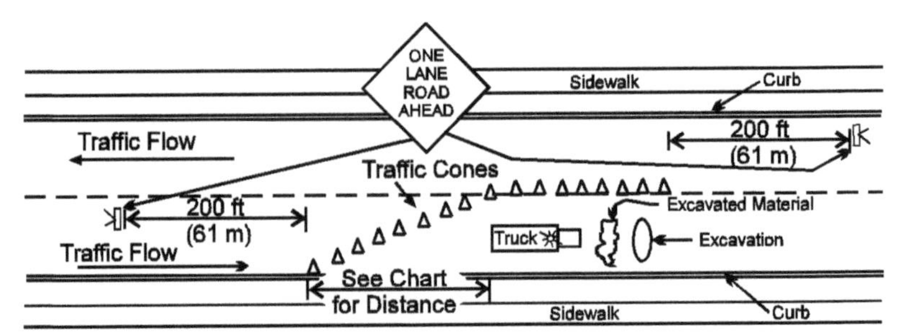

(If traffic is heavy or construction work causes interference in the open lane, one or more flaggers should be used.)

Speed Limit *mph (km/h)*	10 ft (3 m) Taper Length, *ft*	*(m)*	11 ft (3.5 m) Taper Length, *ft*	*(m)*	12 ft (3.7 m) Taper Length, *ft*	*(m)*	Number of Cones Required
20 (32)	70	(21)	75	(23)	80	(24)	5
25 (40)	105	(32)	115	(35)	125	(38)	6
30 (48)	150	(46)	165	(50)	180	(55)	7
35 (56)	205	(62)	225	(69)	245	(75)	8
40 (64)	270	(82)	295	(90)	320	(98)	9
45 (72)	450	(137)	495	(151)	540	(165)	13
50 (81)	500	(152)	550	(168)	600	(183)	13
55 (89)	550	(168)	605	(184)	660	(201)	13

Figure 16-3 Recommended barricade placement for working in a roadway

BIBLIOGRAPHY

AWWA. 2013. *Let's Talk Safety.* Denver, Colo.: American Water Works Association.

AWWA. M3—*Safety Practices for Water Utilities.* Denver, Colo.: American Water Works Association.

OSHA. *Code of Federal Regulations.* Occupational Safety and Health Standards, 29 CFR Part 1910.146. 2011. Permit-Required Confined Spaces. Washington, D.C.: Occupational Safety and Health Administration.

US Department of Labor, Occupational Safety and Health Administration. *Confined Space Regulations.* Washington, D.C.: US Government Printing Office. http://www.osha.gov/Publications/osha3138.pdf.

US Department of Transportation, Federal Highway Administration. 2009. *Manual on Uniform Traffic Control Devices for Streets and Highways.* Washington, D.C.: US Government Printing Office.

Chapter 17

Security, Emergency Preparedness, and Response

The goal of distribution system operations is to provide an uninterrupted supply (system reliability from chapter 3) of potable water to all customers. In spite of this goal, natural or human-made events may jeopardize the integrity of the water supply system. A comprehensive emergency plan is an essential tool that is used before, during, and after a disaster to meet the objectives of a utility's critical mission. An emergency plan

- Assesses the vulnerability of the water system to potential threats,
- Identifies mitigation measures to reduce or eliminate vulnerabilities and protect system assets,
- Provides complete incident response and crisis communications procedures, and
- Includes a description of a post-event recovery process.

A good emergency plan is coordinated with other agencies. Ensuring a continuous water supply may be part of these agencies' own emergency plans. The plan must be updated and tested on a periodic basis. This is critical to ensure that all elements of the plan are still valid.

Although distribution systems personnel should have significant input to the emergency plan, operators are primarily responsible for implementing the plan in an emergency. Make sure you understand the plan and what you need to do. Here are the major elements of the plan and what is included in each.

VULNERABILITY ASSESSMENT

All major system components are identified in the vulnerability assessment. It also includes the power, transportation, and communications requirements needed to reliably operate the system. Critical security-related physical components (alarms, lighting, access control) are listed for each major item.

Threats (Figures 17-1 through 17-3) and potential effects on the distribution system are evaluated. The types of threats that may disrupt water service are ranked by probability in this list:

1. Accident
2. System malfunction
3. Vandalism
4. Human error
5. System aging
6. Natural disaster

Courtesy of Ghasson Al-Chaar.

Figure 17-1 Downed power lines at Homestead Air Force Base, Fla., after Hurricane Andrew

Courtesy of D.B. Ballantyne.

Figure 17-2 Flocculator/clarifier center mechanism damaged from sloshing water in Loma Prieta earthquake (Rinconada Water Treatment Plant, San Jose, Calif.)

7. Extortion
8. Sabotage
9. Waste leakage/seepage
10. Terrorism
11. Civil unrest

This section of the emergency plan will have the performance goals for the system established by utility leadership. These are the level-of-service minimums that should be maintained even in an emergency. This part of the plan lists the critical system components, which receive priority during an emergency.

SUPPLY CONTINGENCIES

In the event of an emergency that interrupts water service, contingencies should be included in the emergency plan to restore critical water supply. These contingencies may include

- Use of standby power or auxiliary power feed,
- Use of temporary portable pumps,
- Connecting to another water system (system interconnection),
- Water tanks or tankers,
- Bottled water, and/or
- Emergency water treatment facilities.

Emergency water supply contingencies should be tested and approved by state regulatory agencies as required by regulations.

Courtesy of D.K. Sander.

Figure 17-3 Road washed out during flooding in Washington State

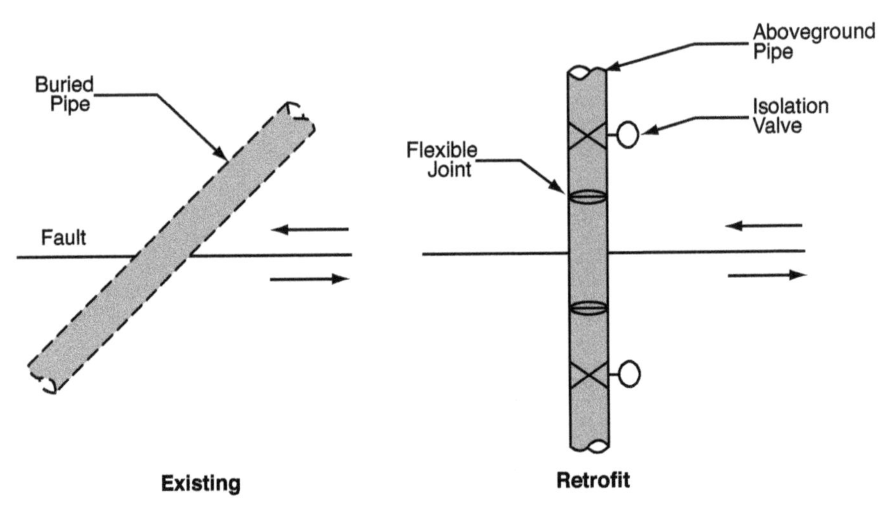

Courtesy of American Society of Civil Engineers, Technical Council on Lifeline Earthquake Engineering.

Figure 17-4 Mitigation options for pipelines at fault crossing

MITIGATION MEASURES

Once the critical assets and the potential threats to those assets are identified, measures must be taken to protect them. These measures include both physical facility improvements (Figure 17-4) and changes to policies and procedures. Access control is one of the most important measures to protect facilities. Included in this measure is the control of keys and locks. Detection devices such as motion sensors and video surveillance are useful in controlling access to facilities. Delay systems like fences, gates, and locks can provide time for a trained response. The emergency plan includes a listing of who to contact if a chemical or biological threat is encountered.

RESPONSE, RECOVERY, AND COMMUNICATIONS

This section of the plan is your guide of what to do if there is an emergency situation in spite of all your efforts to prevent one. It will tell you to analyze the type and severity of the emergency. You will need to take immediate action (Figure 17-5) to save lives and prevent additional injuries or damage to facilities. The plan will help you prioritize any repairs to minimize effects on your critical system components. Following the initial emergency response, you will then take steps to return service to normal levels.

Communications are a primary tool needed to deal with an emergency. You must have a reliable communications system within the utility to be able deal with the incident effectively. Customers will need information on the status of the water supply. In some cases, the news media may need to be informed so that they can help with this task. Distribution system operators

Figure 17-5 A 400,000-gal (1.5 million-L) steel tank was uplifted and underwent "elephant's-foot" buckling on impact during the Landers, Calif., earthquake

are not usually the point of contact with the news media, but employees in the field need to be well informed so they can respond properly to inquiries from customers and others whom they may encounter. A consistent, accurate message needs to be delivered by the utility to all who may be affected.

After the emergency has ended, the plan needs to be re-evaluated. Distribution system operators should provide feedback so the plan can be strengthened.

WATER UTILITY SECURITY INITIATIVES

The US Department of Homeland Security (DHS) has taken steps in partnership with all public and private stakeholders to ensure the protection and resilience of water services. These Water Sector partners collaborate to be better prepared to prevent, detect, respond to, and recover from terrorist attacks and other intentional acts, and natural disasters, otherwise known as the "all-hazards" approach.

Homeland Security Presidential Directive 7 (HSPD-7) identifies 18 critical infrastructure and key resources sectors and assigns an agency lead for each. The US Environmental Protection Agency (USEPA) is the designated agency for the Water Sector.

All-Hazards Approach

Water utilities have taken steps to protect critical water supply facilities from theft and sabotage, and have planned for response and recovery from events that interrupt water service. An all-hazards approach to security and emergency preparedness mirrors the time-tested multibarrier approach used in water treatment.

The Water Sector Coordinating Council consists of representatives from a cross section of water agencies and professionals. The council interacts with federal agencies regarding homeland security issues. In this role, the council has developed several critical resources for water utilities:

- *Water Sector-Specific Plan: An Annex to the National Infrastructure Protection Plan* (USEPA 2010)
- *All-Hazard Consequence Management Planning for the Water Sector* (DHS CIPAC 2009)
- *Roadmap to Secure Control Systems in the Water Sector* (AWWA and DHS 2008)
- *Recommendation and Proposed Strategic Plan: Water Sector Decontamination Priorities* (DHS CIPAC 2008)

In addition, standards have been developed as part of the ANSI/AWWA process to assist water systems in their efforts to improve facility security and emergency preparedness:

- G430, *Security Practices for Operations and Management*
- G440, *Emergency Preparedness Practices*
- J100, *Risk and Resilience Management of Water and Wastewater Systems*

Distribution system operators should be aware of the requirements for the storage and transportation of hazardous substances. Specifically, chlorine gas and large quantities of sodium hydroxide (caustic soda) have extensive requirements for emergency response and security. Many systems have elected to abandon using chlorine gas because of these requirements and concerns. Onsite generation of sodium hypochlorite has gained popularity as a way to avoid these hazards.

Distribution System Security Measures

Protecting a public utility's assets to ensure an uninterrupted, safe water supply is a fundamental concern for all levels of management. Operating staff should participate in the development of a master plan that will establish a level of protection for their areas of responsibility. A utility can implement a fully developed master plan to provide for overall security of the utility. These plans are site-specific but contain common components. By breaking down these components into areas of responsibility, a utility can ensure an integrated approach to plant security development. The major features of a security master plan are described in the following paragraphs.

Physical Requirements

The physical requirements of the plan focus on actions that terrorists, saboteurs, or criminals might take to damage or contaminate the water supply or sabotage system infrastructure and the electric transmission system.

Preventing acts of sabotage at water treatment plants, electric substations, maintenance facilities, other utility facilities, and in the distribution system requires physical barriers to prevent intruders from entering a site and reaching vital areas or equipment.

Integrated Approach

An integrated systems approach that combines people, equipment, and procedures is used to protect against possible threats and intrusions. The security plan has multiple layers of protection that provide several ways to monitor sites and provide deterrence measures for a multitude of possible threats. It also monitors employee activities and documents the presence of contractors, visitors, and other invitees to the facilities through the use of swipe cards and proximity card readers.

Detection Systems

Electronic detection systems, which are more reliable and predicable than systems that depend on people, provide alarms to alert security and plant personnel that the plant or facility is being approached or an intrusion attempt is underway.

Delaying Tactics

Barriers, such as chain-link fences around perimeters, gates, and proximity card readers, are arranged in layers to create physical and psychological deterrents and to delay intrusion long enough for security to detect the intrusion and respond. Barriers channel entry and exit through specified control points, facilitating efficient identification and control of authorized people and equipment.

Assessment Systems

Closed-circuit television cameras monitor facility entrances and special secured areas, and allow security to assess an alarm received while limiting exposure until assistance arrives.

Communications and Response Systems

Security assessment and monitoring devices are linked to communications systems to transmit information to a local monitoring site, a central monitoring site, and the security force. As the security officer or facility personnel respond, they remain in communication with a central monitoring site and other site personnel at all times by several means of alternate communication, such as 800-Mhz radios, cell phones, walkie-talkies, pagers, and other communication devices available at the site.

Training and Qualifications

Security and facility personnel effectiveness must be maintained through training. Training exercises against mock adversaries should be conducted periodically to add realism to the training. In addition, training on system maintenance, and reporting and policy enforcement, must be conducted annually for effective system management.

Access Control

Employees are issued identification badges that must be carried at all times. Access control systems at the entrance gate and within buildings will confirm a person's identity and authorization before entrance is granted. Visitors are also issued badges upon entry to the facility.

Reliability of Personnel

Security awareness will become part of each employee's initiation to the work site. Thorough background investigations should be conducted in accordance with department policies as the primary method for learning about any past actions that may indicate problems with individual employees.

Testing, Maintenance, and Auditing

Periodic testing of all security equipment is required to make sure it is operating correctly. If problems are found, corrective measures are required to ensure that the security function provided by the equipment is immediately restored. Physical security programs should be audited periodically to be certain requirements are being met.

Contingency Planning

Plant operators and facility managers must plan and provide for all reasonable contingencies, such as strikes, natural disasters, and onsite emergencies that require quick support and coordination from outside agencies and security personnel.

BIBLIOGRAPHY

ANSI/AWWA G430. *AWWA Standard for Security Practices for Operation and Management*. Denver, Colo.: American Water Works Association.

ANSI/AWWA G440. *AWWA Standard for Emergency Preparedness Practices*. Denver, Colo.: American Water Works Association.

ANSI/AWWA J100. *AWWA Standard for Risk Analysis and Management of Water and Wastewater Systems*. Denver, Colo.: American Water Works Association.

APHA, AWWA, and WEF (American Public Health Association, American Water Works Association, and Water Environment Federation). 2012. *Standard Methods for the Examination of Water and Wastewater*, 22nd ed. Edited by E.W. Rice, R.B. Baird, A.D. Eaton, and L.S. Clesceri. Washington, D.C.: APHA.

AWWA. M3—*Safety Practices for Water Utilities*. Denver, Colo.: American Water Works Association.

AWWA. M19—*Emergency Planning for Water Utilities*. Denver, Colo.: American Water Works Association.

AWWA. Utility Security. Website resources. Denver, Colo.: American Water Works Association. http://www.awwa.org/legislation-regulation/issues/utility-security.aspx.

AWWA. Water/Wastewater Agency Response Network (WARN). Denver, Colo.: American Water Works Association. http://www.awwa.org/resources-tools/water-knowledge/emergency-preparedness/water-wastewater-agency-response-network.aspx.

AWWA and DHS. 2008. *Roadmap to Secure Control Systems in the Water Sector.* Water Sector Coordinating Council Cyber Security Working Group. Denver, Colo.: American Water Works Association and Department of Homeland Security.

DHS CIPAC. October 2008. *Recommendation and Proposed Strategic Plan: Water Sector Decontamination Priorities.* Final Report. Arlington, Va.: Department of Homeland Security, Critical Infrastructure Partnership Advisory Council Working Group. http://www.awwa.org/Portals/0/files/legreg/Security/Decontaminate.pdf.

DHS CIPAC. November 2009. *All-Hazard Consequence Management Planning for the Water Sector.* Arlington, Va.: Department of Homeland Security, Critical Infrastructure Partnership Advisory Council Working Group.

Homeland Security Presidential Directive 7: Critical Infrastructure Identification, Prioritization, and Protection. December 17, 2003.

OSHA. *Code of Federal Regulations.* OSHA Regulations. Occupational Safety and Health Standards, Part 1910. Washington, D.C.: Occupational Safety and Health Administration.

OSHA. *Code of Federal Regulations.* OSHA Regulations. Safety and Health Regulations for Construction, Part 1926. Washington, D.C.: Occupational Safety and Health Administration.

USEPA. 2010. *Water Sector-Specific Plan: An Annex to the National Infrastructure Protection Plan.* EPA 817-R-10-001. Washington, D.C.: US Environmental Protection Agency.

Chapter 18

Instrumentation and Control

Instruments allow a distribution system operator to monitor and control flow rates, pressures, and water levels. The main categories of instrumentation and control are

- *Primary instruments*, which measure flow, pressure, level, and temperature;
- *Secondary instruments*, which respond to and display the information from the primary instruments; and
- *Control systems*, which either manually or automatically operate equipment such as pumps and valves.

MONITORING SENSORS

The sensors that measure process variables are called *primary instrumentation* because they are basically necessary to obtain the information required to operate the monitoring system.

Flow Sensors

The various types of flow sensors (meters) used in a water distribution system are discussed in chapter 12. The types of meters primarily used for measuring flow in mains are differential pressure meters (such as venturi meters) and velocity meters (such as propeller meters). For monitoring flow at remote locations and for control purposes, meters must be provided with either a pulse or an electrical output that is proportional to the flow rate.

Pressure Sensors

Pressure sensors are commonly used to measure suction and discharge pressure on pumps and at points on the distribution system. The three common types of direct-reading pressure gauges that have been used for many years are illustrated in Figure 18-1. The bellows sensor uses a flexible copper can that expands and contracts with varying pressure. The helical sensor has a spiral-wound tubular element that coils and uncoils with changes

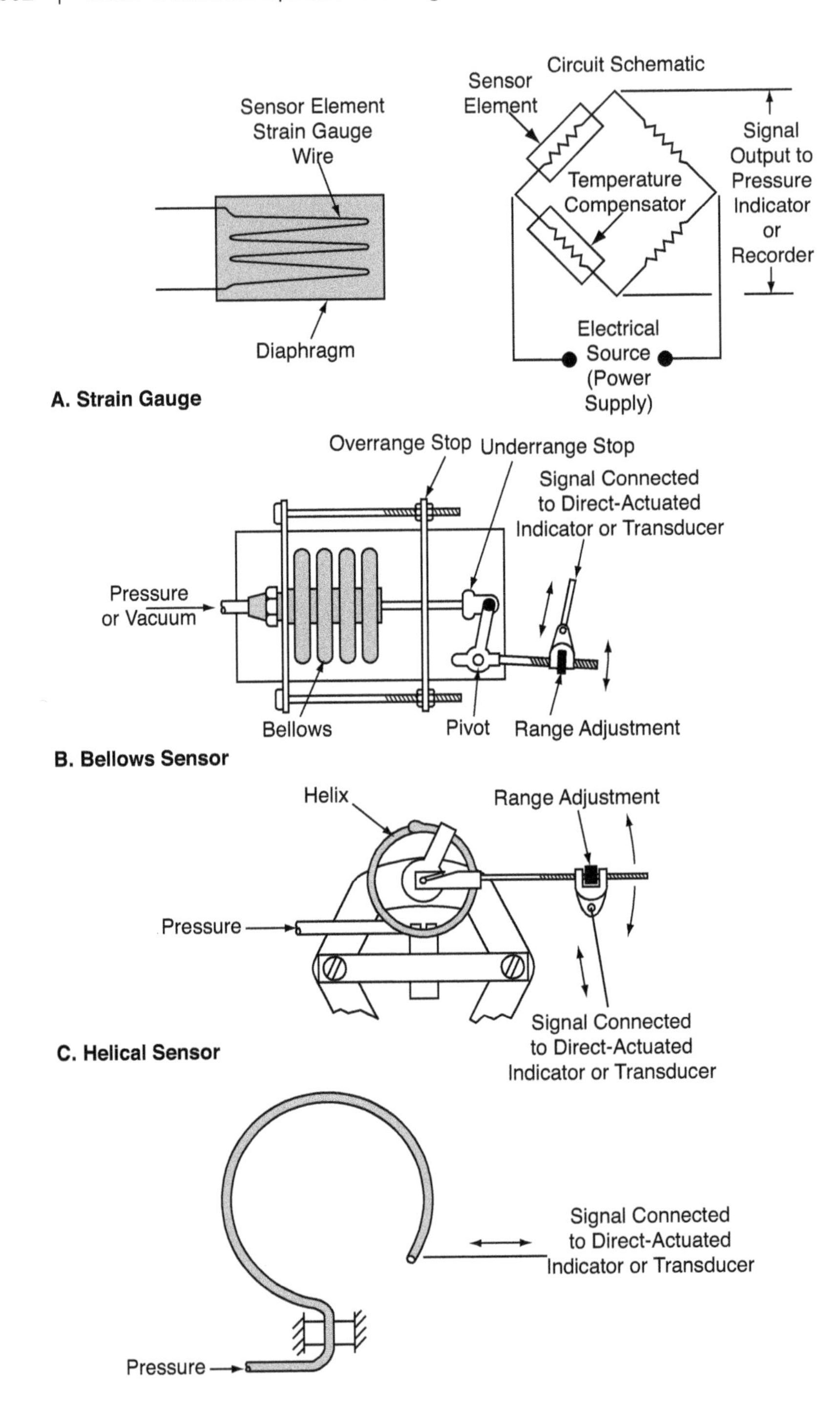

Figure 18-1 Types of pressure sensors

in pressure. The Bourdon tube uses a semicircular tube with a C shape that opens under increasing pressure.

Most pressure sensing for operating electronic recorders and controls today is done with pressure transducers that produce an electrical current in proportion to pressure. Transducers are very accurate, require essentially no maintenance, and can be adjusted to register pressure over either a narrow or wide range.

Level Sensors

Several methods of measuring the depth of water in a tank or the elevation of a water surface are illustrated in Figure 18-2. One of the oldest types uses a float mechanism attached by a wire to a pulley on a shaft, which is rotated as the water level rises and falls.

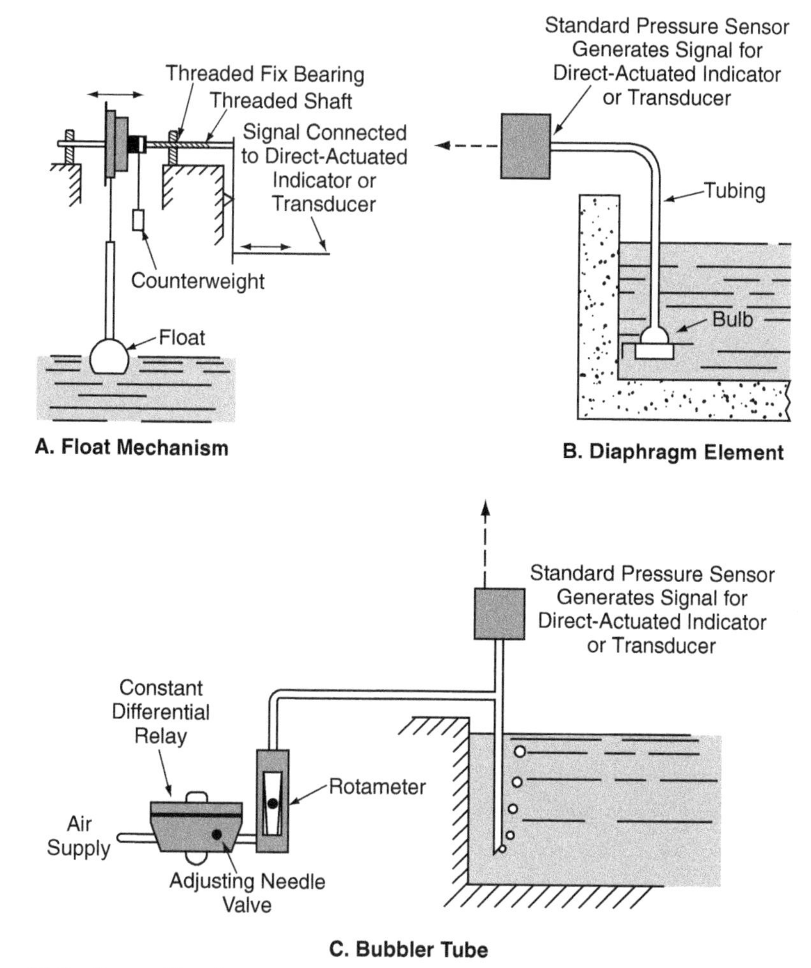

Figure 18-2 Types of level sensors

The diaphragm element type of sensor operates on the principle that the confined air in the tube compresses in relation to the head of water above the diaphragm. Changes in pressure are detected by a transducer and electronically translated into water depth.

Bubbler tubes were once widely used for determining water depth, particularly in dirty water. A constant flow of air must be maintained in the tube, which is suspended in the water. Bubbler tubes work on the principle that the pressure required to discharge air from the tube is proportional to the head of water above the bottom of the tube. Although the bubbler assembly is essentially maintenance free, maintaining a continuous air supply presents some maintenance problems.

Ultrasonic units bounce radio signals off the water surface and translate the return times into distances. Knowing the elevation of the water surface, one can then convert the figure into water depth.

Transducers are now widely used for measuring water depth. The pressure of the water over the unit can be translated directly into feet of head or pressure. In a similar fashion, a transducer can be installed at the base of a water tower to indicate the elevation of water in the tower.

Electronic probes may also be used for sensing water depth. An insulated metallic probe suspended in the water has an electronic circuit that detects a change in capacitance between the probe and water and electronically converts the information into depth. If a probe is installed in a nonmetallic tank, a second probe is required.

A variable-resistance level sensor consists of a wound resistor inside a semiflexible envelope. As the liquid level rises, the flexible outer portion of the sensor presses against the resistor and gradually shorts it out. The resistance is then converted to a liquid-level output signal.

Temperature Sensors

Two types of temperature sensors that are commonly connected to instrumentation are thermocouples and thermistors. Thermocouples use two wires of different material, as illustrated in Figure 18-3. The wires are joined at two points. One is called the *sensing point* and the other is the *reference*

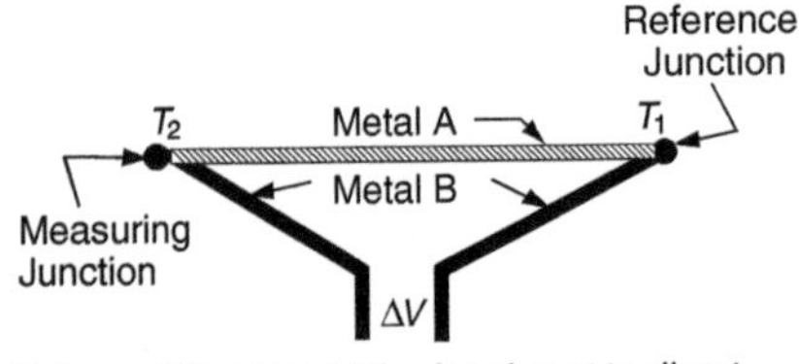

Voltage difference ΔV is signal sent to direct-acting indicator (voltmeter) or transducer.

A. Thermocouple

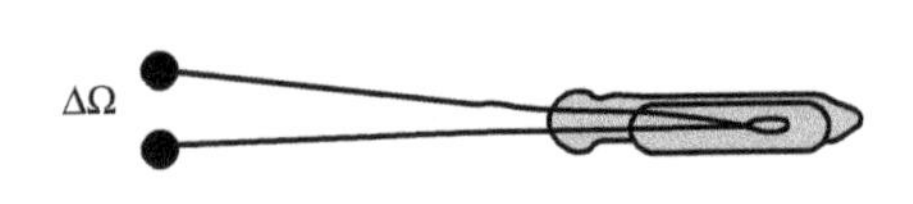

Change in electrical resistance $\Delta\Omega$ is the signal sent to direct-acting indicator (ohmmeter) or transducer.

B. Thermistor

Figure 18-3 Types of temperature sensors

Courtesy of Fluke Corporation.

Figure 18-4 Digital multimeter

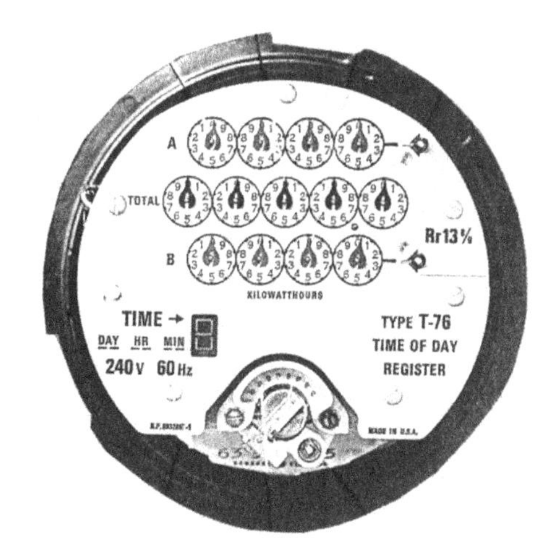

Figure 18-5 Totalizing wattmeter

junction. Temperature changes between the two points cause a voltage to be generated, which can be read directly or amplified.

Thermistors are also called *resistance temperature devices* (RTDs). They use a semiconductive material, such as cobalt oxide, that is compressed into a desired shape from the powder form, and then is heat treated to form crystals to which wires are attached. Temperature changes are reflected by a corresponding change in resistance as measured through the wires.

Electrical Sensors

It is frequently necessary to monitor the status of operating electrical equipment. The variables that may be measured are voltage (in volts), current (in amperes), resistance (in ohms), and power (in watts). The measurement of volts, amperes, and ohms can all be made with the same instrument, the D'Arsonval meter. Electric current passing through the meter's coil creates a magnetic field, which reacts with the field of the permanent magnet and causes the indicator needle to move. The parameter measured by the meter depends on how it is connected in the circuit. The D'Arsonval meter is now rapidly being replaced by digital equipment that indicates the electrical value directly (Figure 18-4).

To measure power, an instrument must combine the measurements of volts and amperes (watts = volts × amps, assuming the power factor is 1). The electrical energy used is measured and totalized by a wattmeter that reads in kilowatt-hours (kW·h). One thousand watts (1 kW) drawn by a circuit for 1 hour results in an energy consumption of 1 kW·h. Meters that register kilowatt-hour usage are essentially totalizing wattmeters (Figure 18-5).

A rotating disk can be seen through the glass front on older kilowatt-hour meters, and the rotations can be timed with a stopwatch to determine power consumption. Newer meters have other types of dial and digital readouts. A local power utility representative can be consulted on how to read power consumption on a particular meter.

Equipment Status Monitors

Some of the more common equipment status monitors are vibration sensors, position and speed sensors, and torque sensors. Vibration sensors are one of the most commonly used types in the water industry and are often mounted on pumps and motors that are located in unsupervised locations. The sensors are usually wired into the power circuit to disconnect power to the motor and activate an alarm if the vibration in the equipment exceeds a specified value.

Process Analyzers

Because of increasingly stringent water quality requirements, many water systems are installing online monitoring equipment to continuously analyze water quality. Typical equipment of this type includes turbidity, pH, and chlorine residual monitors; particle counters; and streaming current meters. These monitors provide continuous analyses of water quality parameters and will activate alarms if values go above or below preset limits. They may also feed into a computer control system that will automatically make corrective adjustments of the operating system.

Turbidity Monitors

Online turbidity monitors are critical in water treatment plant operation and they are required by water quality regulations. However, they are also useful for distribution system monitoring. By carefully placing these analyzers, they can be effectively used to detect changes in water quality.

pH Monitors

These analyzers provide another sensitive measure of changes within the distribution system. The pH of water entering the distribution network can change unacceptably because of water reactions of pipe walls and storage tank coatings. Also, if pH modification is being used to provide corrosion protection, this can be monitored to ensure effectiveness.

Disinfectant Residual Monitors

Chlorine and chloramine residuals (and chlorine dioxide) can be reliably measured by online analyzers within the distribution system. Strategically placing these monitors near storage tanks and in areas of concern can provide early warning of disinfectant decay. This information can be used by operators to adjust water flow and respond to avoid the possibility of problems developing.

Other Analyzers Used in Distribution Systems

Several other types of analyzers are commonly used to monitor water quality in distribution systems. These are often employed along with those previously mentioned to provide a "suite" of analyses where related results can provide the following important information:

- Conductivity
- Ultraviolet absorbance
- Temperature
- Dissolved organic carbon
- Online gas chromatography or mass spectrometry

SECONDARY INSTRUMENTATION

Secondary instruments display information provided by sensors. The display may be mounted adjacent to the sensor, in a nearby control room, or in a distant control center.

Signal Transmission

Many older plants have instrumentation that transmits signals and control equipment using pneumatic (compressed) air run through small-diameter tubing. The standard operating range of this type of system is 3 to 15 psig (20 to 100 kPa gauge).

Because of improvements in electronic circuitry, most new equipment operates with electrical transmission using either current or voltage. The most common is 4–20 mA direct current (DC). It is also possible to convert a pneumatic signal to an electrical signal (P/I converted), or an electrical to a pneumatic signal (I/P converted).

Receivers and Indicators

Receivers and indicators convert signals from sensors for use by the water system operator or to be fed into the control system. Alternative methods include the following:

- Direct-reading value of the parameter (e.g., gpm, volts, or pressure)
- Recording of the information (as on a strip chart)
- Total accumulated value since the unit was last reset (e.g., 5,000 gal)
- Some combination of the above methods

Two types of instrument display are analog and digital. An example of an analog display is a dial indicator. The values range smoothly from the minimum to the maximum, and it is easy to see the relative position of the reading to the entire range. An analog display makes it easy to estimate readings that fall between the primary divisions on the dial.

A digital display shows decimal numbers. The numbers may be a mechanical readout or an electronic display like a digital watch. Digital indicators are generally more accurate than analog displays because they are not subject to the errors associated with mechanical systems. They are also easier to read correctly. A disadvantage is that there is no way of estimating the exact value when it is between the divisions provided on the display.

Telemetry

When the distance between a sensor and the indicator is relatively short, the information can be transmitted between them by using variations in current or voltage. But if there is an appreciable distance between them, telemetry must be used because the signal must be a type that will not vary in spite of variations in the wiring or radio signal.

Early telemetry equipment used audio tones or electrical pulses. Most new systems are digital and use a binary code. The sensor signal feeds into a transmitter that generates a series of on–off pulses that represent the exact numerical value of the measured parameter. For example, off–on–off–on represents the number 5. The receiver then translates the code to number or letter readings.

The transmitting device in a digital system is called a *remote terminal unit* (RTU) and the receiver is called the *control terminal unit* (CTU).

Multiplexing

Several methods are available for sending signals from more than one sensor over the same transmission line.

Tone-frequency multiplexing sends several signals over one wire or radio signal by having tone-frequency generators in the transmitter and sending each parameter at a different frequency. Filters in the receiver then sort out the signals and send them to the proper indicator. As many as 21 frequencies can be sent over a single voice-grade telephone line.

Scanning equipment transmits the value of each of several parameters one at a time in a set sequence. The receiver decodes the signal and displays each one in turn. Scanning can be used with all types of signals and all types of transmission. Scanning and tone-frequency multiplexing can be combined to allow even more signals over a single line. For example, a four-signal scanner combined with a 21-channel, tone-frequency multiplexer would yield 84 signal channels.

Polling is another method of sending several different signals over a single line. In this system, each instrument has a unique address (identifying number). A system controller, usually located at the central control center, sends out a message requesting a specific piece of equipment to transmit its data.

The controller can poll the instruments as often as necessary, which may be every few minutes to every hour or so. In more sophisticated units, the controller regularly polls each piece of equipment to determine whether there is any new information. If the status report indicates there is new information, the instrument is instructed to send its data. Some systems also provide for key instruments to interrupt other transmissions to send urgent new data.

Duplexing also allows an operator to send control signals back to the site of a transmitting sensor. This can be done using a single transmission line in one of three ways:

1. *Full duplex* allows signals to pass in both directions at the same time.
2. *Half-duplex* allows signals to pass in both directions but only in one direction at a time.
3. *Simplex* allows signals to pass in only one direction.

Transmission Channels

Four types of transmission channels are regularly used by water utilities for transmitting telemetry signals. These channels include a privately owned cable, such as a wire between two buildings on the same property; a leased telephone line; a radio channel; and a microwave system. A system using space satellites is also available but is presently quite expensive, and new equipment is now available that can send signals over a cellular phone.

A leased telephone line is usually the least expensive transmission channel and is generally quite reliable and free of interference. Most modern telemetry transmitters are designed to operate over voice-grade lines. Radio channels may be in the VHF (very high frequency) or UHF (ultra-high frequency) bands. Both radio and microwave systems generally require line-of-sight paths with no obstructions such as buildings.

CONTROL SYSTEMS

The control of equipment can be accomplished in a variety of ways, from completely manual to completely automatic.

Direct Manual Control

Under complete manual control, each piece of equipment is adjusted by the water system operator by directly turning it on and off; for example, turning the handwheel on a valve. Manual control has the advantage of low initial cost and little complicated equipment to maintain. It may require more work for the operator, however, and proper operation of the equipment depends completely on the operator's expertise and judgment. If the equipment to be operated is at different locations, the operator must go to each location to perform the operation.

Remote Manual Control

Remote manual control still requires the operator to initiate each adjustment, but it is not necessary to go to the equipment location. Instead, the operator has a remote station, such as a switch or push button, which turns the equipment on and off. Examples of actuators for remote operation are solenoid valves, electric relays, and electric motor actuators. Proper operation of the equipment is still dependent on the judgment of the plant operator.

Semiautomatic Control

Semiautomatic control combines manual control by the plant operator with automatic control of specific pieces of equipment. For example, a circuit breaker will disconnect automatically in response to an overload but must then be reset manually.

Automatic Control

Automatic control systems turn equipment on and off or adjust their operation in response to signals from sensors and analytical instruments. The plant operator does not have to exercise any control. A simple example is the thermostat on a heating system. The two general modes of operation for automatic control are on–off differential and proportional control.

On–off differential control turns a piece of equipment either full on or off in response to a signal. For example, the same signal that activates a service pump can turn on a chlorinator. If there is any need to adjust the chlorine feed rate, it must be done manually by the plant operator.

Proportional control can adjust the operation of a piece of equipment in response to a signal in several ways.

Feedforward proportional control measures a variable and adjusts the equipment proportionally (Figure 18-6). An example is the adjustment of chlorinator feed rate from a flowmeter signal. The faster the water flows through the meter, the more chlorine that is fed. As long as the chlorine demand of the water remains constant, this method of operation is satisfactory.

Feedback proportional control measures the output of the process and reacts backward to adjust the operation of the piece of equipment. In the illustration in Figure 18-7, the chlorine residual analyzer is set by the operator to maintain a specific chlorine residual. It then adjusts the feed rate of the chlorinator to maintain the residual in spite of variations in both chlorine demand and changes in the water flow rate. This is also called *closed-loop control* because it is continuously self-correcting. The principal problem with this control system is that, if there are wide variations in water flow rate, the system will spend a lot of time seeking the correct value. If the flow rate increases, the analyzer will detect a low residual and will increase the feed rate. But it will probably overfeed for a short time, then underfeed again, and so on until it finds the correct feed rate.

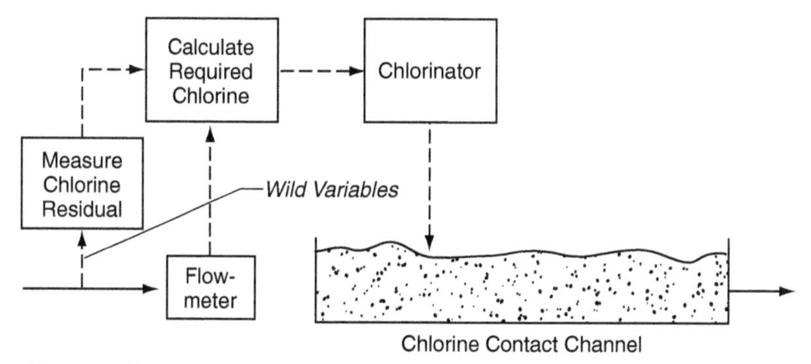

Figure 18-6 Feedforward control of chlorine contact channel

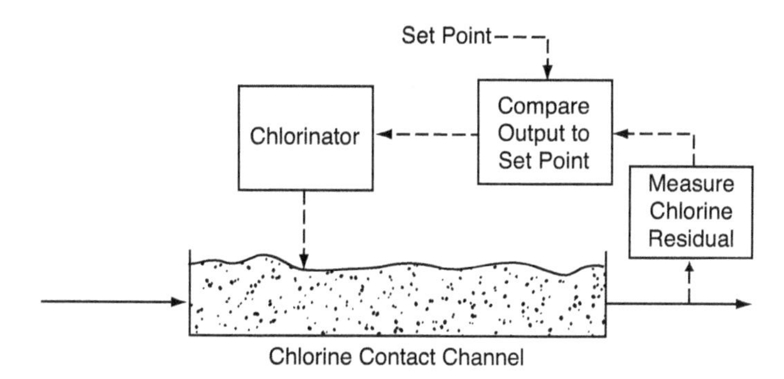

Figure 18-7 Feedback control of chlorine contact channel

The ultimate automatic control for this situation is a combination of both feedforward and feedback control. The chlorinator is set up to primarily adjust in response to changes in flow rate, but the analyzer then monitors the residual and makes minor adjustments in the feed rate as necessary to maintain the selected residual in the finished water.

SUPERVISORY CONTROL AND DATA ACQUISITION

Supervisory control and data acquisition (SCADA) is rapidly becoming the principal method of control in the water industry. A SCADA system consists of four basic components. The first two components have already been discussed—sensors (RTUs) to monitor the variables and telemetering to send the information to a central location.

The third component is a central control location that has equipment for monitoring the operation and sending back commands. The fourth component is equipment for reviewing the operation, giving commands to the remote equipment, and recording information for historical purposes. This generally consists of a video screen, a keyboard, and computer data storage equipment.

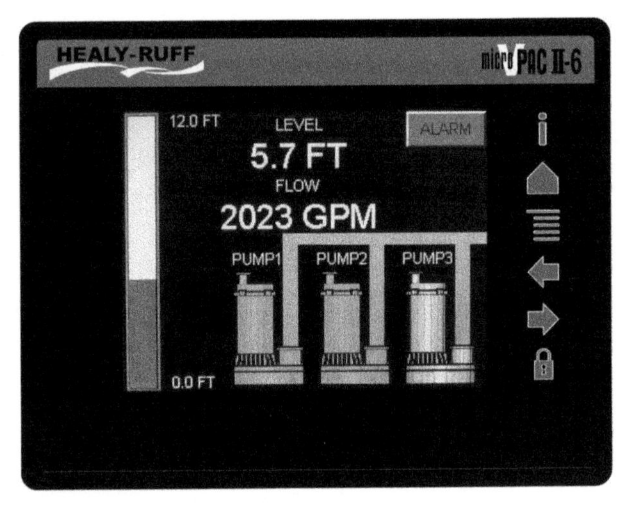

Courtesy of ICS Healy-Ruff.

Figure 18-8 Remote supervision of a reservoir

When a water system has everything controlled from a single computer, it is called *centralized computer control.* Early SCADA systems were of this type, having one big computer to control everything on the water system. Two problems with this type of operation are the great dependency on the computer and on the telemetry links to the remote locations. For example, if a remote pumping station is completely controlled from a central computer and there is loss of the telephone telemetry line, the station would shut down and be unusable. The only way to operate it is to have someone operate it manually from the station.

As smaller, more powerful, and less expensive computers have become available, SCADA systems called *distributed computer control* have been developed (Figure 18-8). It is now possible to have local computer control of subsystems and individual pieces of equipment. The trend is now extending to the development of "smart equipment," which adjusts itself and monitors its own operation.

State regulatory agencies typically have mixed views about operating water system equipment completely by computer control. On the plus side, the computer eliminates the possibility of human error—the computer cannot fall asleep on the job. On the other hand, breakdown of a computer system or an error in the software could result in loss of pressure or contamination of the system. State agencies will always want to see that an automated system is furnished with all possible monitoring and reporting capabilities, and programmed to fail-safe in the event of an emergency.

BIBLIOGRAPHY

AWWA. 2010. Principles and Practices of Water Supply Operations—*Water Transmission and Distribution*, 4th ed. Denver, Colo.: American Water Works Association.

AWWA. M2—*Instrumentation and Control*. Denver, Colo.: American Water Works Association.

Mays, L.W. 2000. *Water Distribution Systems Handbook*. Denver, Colo.: American Water Works Association; New York: McGraw-Hill.

Chapter 19

Distribution System Mapping and Information Management

The different types of maps and records described in this chapter have been developed over the years and are the standards for well-run water systems. Most water systems use computerized records, but some systems still use hardcopy records. The same information is required, but it is just stored and available in a different format.

DISTRIBUTION SYSTEM MAPS

Various types of maps are necessary to provide information on mains, hydrants, and services. Because most of the distribution system equipment is underground, taking time to keep careful records during construction and repair saves a lot of time in the long run by knowing where everything is and being able to quickly find it. Accurate distribution system records are also necessary to determine a valuation of the distribution system, and these maps are essential for engineers to design system improvements.

Comprehensive Maps

A map that provides an overall picture of the entire system is usually called a *comprehensive map*, or *wall map*, because a large copy of it is usually hung on a blank wall in the distribution center office for ready reference by everyone. Computerized maps often contain layers of detail that can be accessed by zooming in or out of the image. The following information should appear on the map:

- Street names
- Distribution water mains with the sizes noted
- Transmission mains shown in a code different than distribution mains
- Fire hydrants and valves with their designated numbers
- Reservoirs, tanks, and booster stations
- Water source connections and interconnection with other water systems

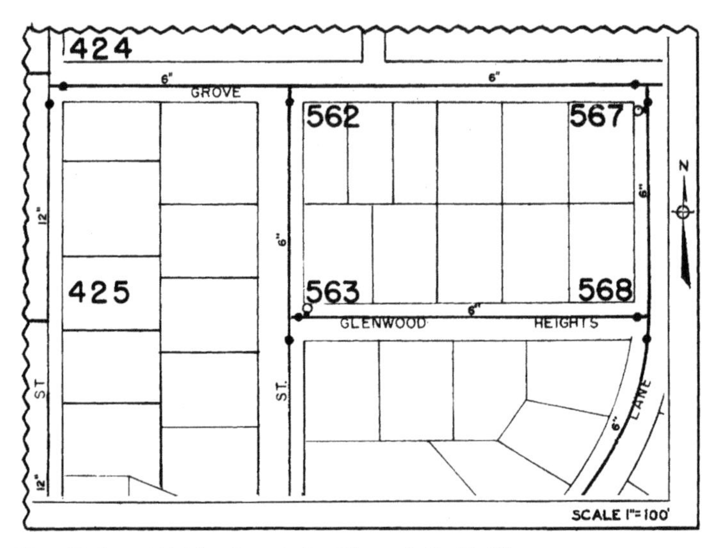

Courtesy of Water Works and Lighting Commission, Wisconsin Rapids, Wis.

Figure 19-1 Comprehensive map showing intersection numbers

- Pressure zone limits
- Notation of the street-numbering grid

Part of a typical comprehensive map is shown in Figure 19-1. The map should be as large as possible and should not be cluttered with unnecessary information that would be distracting. A good map scale for a small water system is 500 ft to 1 in. (6 m to 1 mm). Larger systems often use 1,000 ft to 1 in. (12 m to 1 mm).

Ordinary commercial maps are sometimes available to use as basic layouts, but they should be carefully checked for accuracy before they are used. After any corrections are made, they can be enlarged by a professional reproduction firm, but care must be taken that the scale is not seriously distorted in the process. Copies of the comprehensive map are often provided to the municipal engineering department and the fire department.

Sectional Maps

A sectional map (also commonly called a *plat*) is a series of maps covering sections of the water system, usually stored in a plat book. Plat books are either computer files or hardcopy collections of system maps. The maps are usually at a scale of either 50 or 100 ft to 1 in. (0.6 or 1.2 m to 1 mm) for small and medium-sized communities, and 200 ft to 1 in. (2.4 m to 1 mm) for larger systems. In addition to the information on the comprehensive map, the following details are usually included:

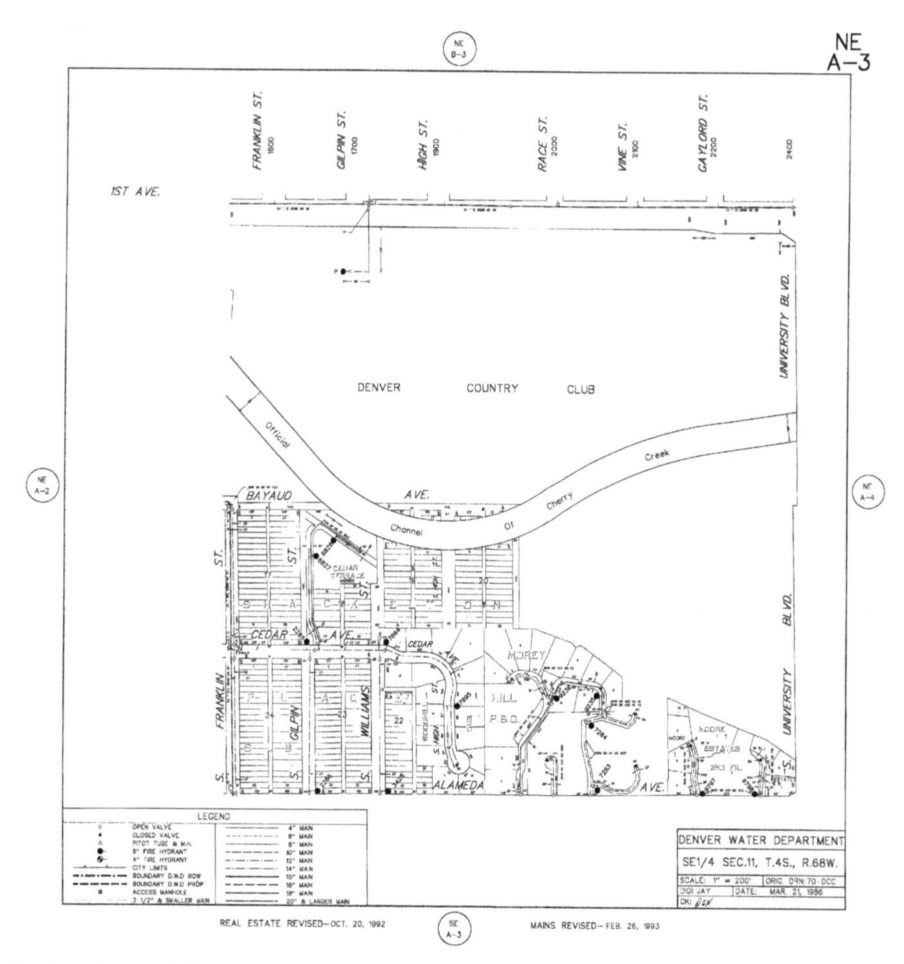

Courtesy of Denver Water.

Figure 19-2 Sectional map

- Type of main material (i.e., cast iron [CI], ductile iron [DI], polyvinyl chloride [PVC], or asbestos–cement [A–C])
- Main installation date
- Main distance from property line if other than standard
- Block numbers

If the scale of the map allows, details such as house numbers and curb boxes may be shown. However, if all this information would cause undue clutter on the map, it is best provided on other maps or records.

Sectional maps should not overlap each other. Each should have a definite cutoff line on each side so that one map butts to the next one (Figure 19-2). The maps should be indexed, and, for quick reference, the number of the adjoining map in each direction should be indicated. A common

1E	1D	1C	1B	1A
2E	2D	2C	2B	2A
3E	3D	3C	3B	3A

Figure 19-3 Small, comprehensive map divided into sections

method of numbering maps is shown in Figure 19-3. Water systems that may experience growth in any direction sometimes use a variation of this numbering, starting at a central location with four quadrants so that the map number would also have a designation such as SE or NW on it.

Sectional maps can often be developed from tax assessment maps, insurance maps, subdivision maps, or city engineering maps, but they should be carefully checked for accuracy before being used. The historical method of preparing sectional maps was to have the original copies drawn in ink on tracing cloth, and many older systems are still using this method. The reasoning behind this method was that copies could be made for everyday use to reduce wear on the originals. Computer file maps are preferred since they are not subject to wear. Backups of computer file maps, however, need to be kept in a safe and secure location, and these files need to contain the latest updates.

The originals should be kept in a safe place, such as a fireproof vault. Changes to the originals should preferably be made as soon as possible. New prints should be made and distributed whenever a significant number of changes have been made. One method of ensuring that the hardcopy information on sectional maps is never lost is to have microfilm copies made and stored at a separate, secure location.

Valve and Hydrant Maps

Many water systems also prepare valve and hydrant maps, either combined or as two separate maps. These maps are of sufficient detail to locate every valve and hydrant, and are primarily for use by field crews during maintenance or emergencies. Notes should also be included for special information, such as valves and hydrants that operate in reverse from most in the system.

Many systems use individual plat sheets of convenient size. Another alternative is the plat-and-list method, in which a relatively small map identifies the valves and hydrants by number. This is used in conjunction with a list that provides basic information, such as the type, make, size, street

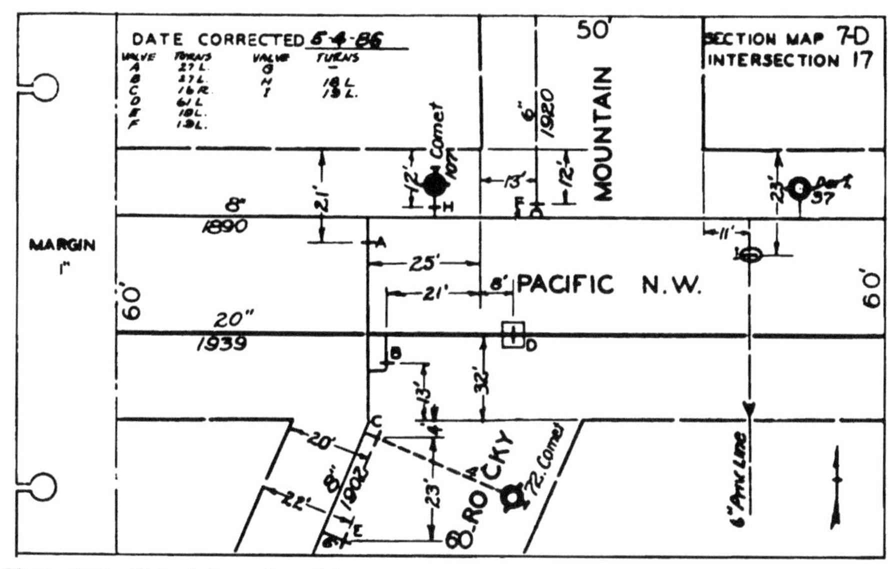

Figure 19-4 Valve intersection plat

location, and location measurements. Both of these map systems can be accommodated by computerized mapping programs.

Another variation is to use the basic location maps with detailed street intersection level computer maps (or cards where hardcopy systems are still used that are usually kept in a looseleaf book; Figure 19-4). One advantage of this system is that the location map rarely needs to be corrected because it shows only an overview of part of the system. The intersection plats have plenty of room for all required field information and are usually updated immediately in the field as changes are made. With this system, each intersection must be given an identifying number that is shown on the basic location map.

Plan and Profile Drawings

Plan and profile drawings are usually provided by an engineer for all new construction. They show the *proposed* location of the new main and appurtenances, and indicate how they connect with the existing system, and how it is anticipated they will avoid all obstructions, such as sewers, gas mains, cables, trees, and telephone poles. The engineers often use abbreviations on these drawings that are not explained. Some of these are

- POT—point on tangent,
- POC—point on curve,
- BC or PC—beginning of a curve,
- EC or PT—end of a curve,
- PI—point of intersection, and
- EL—elevation.

Distances are usually provided in terms of stationing, starting with 0+00 at the beginning connection. For example, 2 + 42 on the map will indicate 242 ft from the starting point.

Installations are rarely constructed exactly as anticipated on the drawings. Some unanticipated obstructions are usually encountered that require deviation from the plan, and these changes must be quickly recorded on a copy of the plan so they are not forgotten. Good practice is then to prepare a set of "as-built" drawings at the completion of the job to keep as a permanent record.

Information from the construction drawings should be transferred to the other water system drawings, and the plans should be identified so they may be easily found if needed in the future. The plans should be filed in a safe, dry, clean location.

Map Symbols

Every water utility should adopt a set of map symbols that are used on all maps and records. They must be simple, clear, and preferably the same as symbols used by other water systems. Commonly used symbols are shown in Figure 19-5.

EQUIPMENT RECORDS

Computerized records have replaced card files and are often kept for details that cannot be included on maps for each valve, hydrant, water service, and water meter.

Valve Records

A valve record usually has information on the make, size, type, and location of the valve on one side, and maintenance information on the other (Figure 19-6). Valves may be assigned numbers, which are referenced on the distribution maps, and are then filed numerically. It is wise to keep as much location information as possible so that the valve can be located quickly in an emergency. For example, although distances to trees are handy reference points in winter when there is snow on the ground, there should also be other measurements to more permanent markers, such as lot stakes, because the trees may sometime be cut down.

Hydrant Records

Like valve records, hydrant records should include information on both the hydrant and auxiliary valve on one side, and maintenance and repair information on the other (Figure 19-7). Although it is a job that almost everyone would like to put off, carefully kept hydrant maintenance records are particularly important. If a hydrant should ever fail to operate properly when there is a fire, and there is serious property loss from the fire, there is a

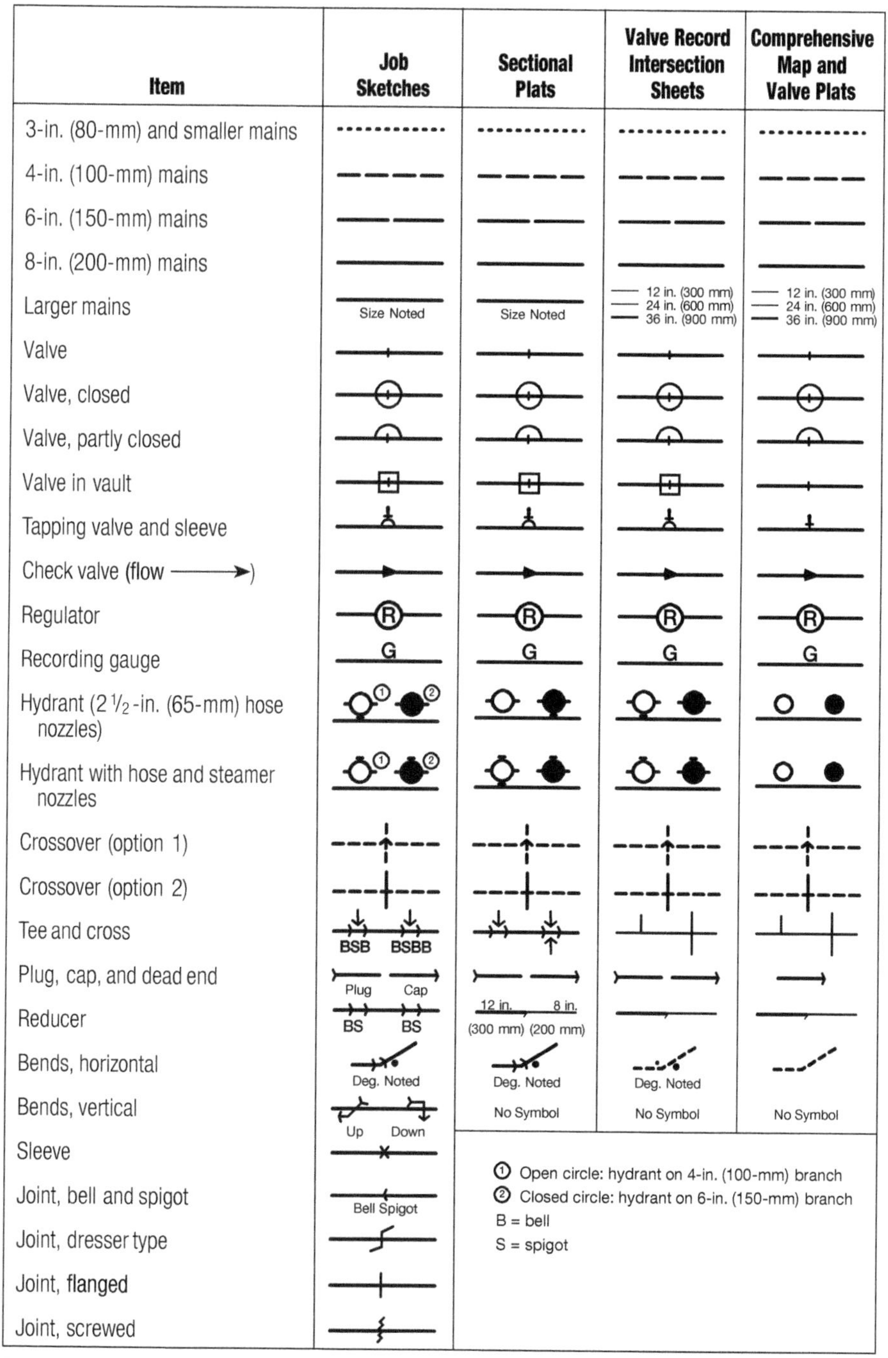

Item	Job Sketches	Sectional Plats	Valve Record Intersection Sheets	Comprehensive Map and Valve Plats
3-in. (80-mm) and smaller mains				
4-in. (100-mm) mains				
6-in. (150-mm) mains				
8-in. (200-mm) mains				
Larger mains	Size Noted	Size Noted	12 in. (300 mm) 24 in. (600 mm) 36 in. (900 mm)	12 in. (300 mm) 24 in. (600 mm) 36 in. (900 mm)
Valve				
Valve, closed				
Valve, partly closed				
Valve in vault				
Tapping valve and sleeve				
Check valve (flow ⟶)				
Regulator	R	R	R	R
Recording gauge	G	G	G	G
Hydrant (2 ½-in. (65-mm) hose nozzles)	① ②			
Hydrant with hose and steamer nozzles	① ②			
Crossover (option 1)				
Crossover (option 2)				
Tee and cross	BSB BSBB			
Plug, cap, and dead end	Plug Cap			
Reducer	BS BS	12 in. 8 in. (300 mm) (200 mm)		
Bends, horizontal	Deg. Noted	Deg. Noted	Deg. Noted	
Bends, vertical	Up Down	No Symbol	No Symbol	No Symbol
Sleeve				
Joint, bell and spigot	Bell Spigot			
Joint, dresser type				
Joint, flanged				
Joint, screwed				

① Open circle: hydrant on 4-in. (100-mm) branch
② Closed circle: hydrant on 6-in. (150-mm) branch
B = bell
S = spigot

Figure 19-5 Typical map symbols

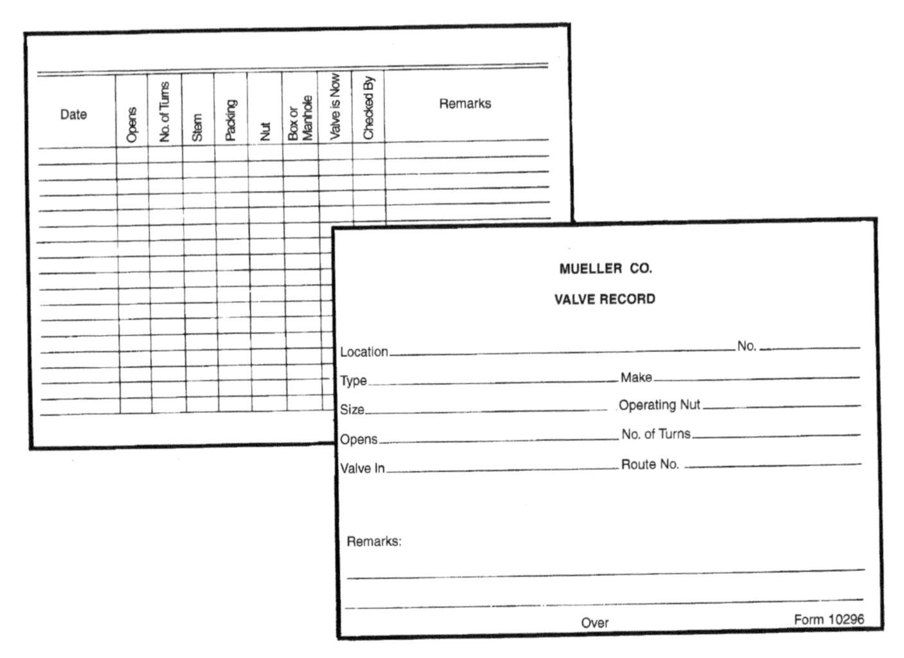

Date	Opens	No. of Turns	Stem	Packing	Nut	Box or Manhole	Valve is Now	Checked By	Remarks

MUELLER CO.

VALVE RECORD

Location ________ No. ________

Type ________ Make ________

Size ________ Operating Nut ________

Opens ________ No. of Turns ________

Valve In ________ Route No. ________

Remarks:

Over Form 10296

Courtesy of Mueller Company, Decatur, Ill.

Figure 19-6 Valve record information input document

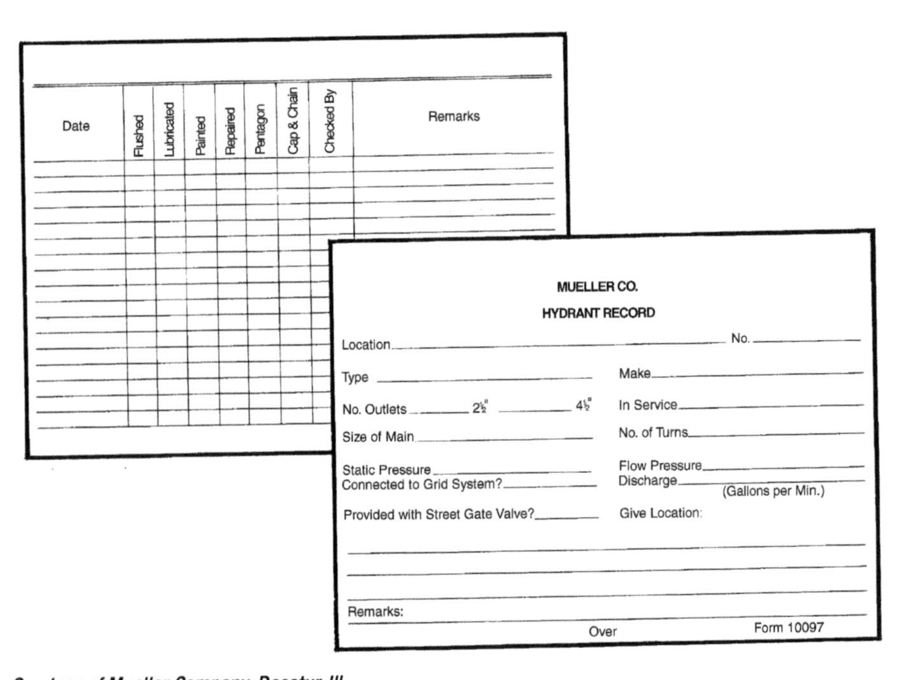

Date	Flushed	Lubricated	Painted	Repaired	Pentagon	Cap & Chain	Checked By	Remarks

MUELLER CO.

HYDRANT RECORD

Location ________ No. ________

Type ________ Make ________

No. Outlets ________ 2½" ________ 4½" In Service ________

Size of Main ________ No. of Turns ________

Static Pressure ________ Flow Pressure ________

Connected to Grid System? ________ Discharge ________ (Gallons per Min.)

Provided with Street Gate Valve? ________ Give Location:

Remarks:

Over Form 10097

Courtesy of Mueller Company, Decatur, Ill.

Figure 19-7 Hydrant record information input document

good chance that the property owner will sue the water utility for damages. The best defense the utility can have is records showing that there is a continuous, planned hydrant maintenance program.

Service Records

Many utilities also maintain records with details of each water service. Information includes installation date, name of the installing plumber or contractor, and details of all materials. The information often includes a sketch of the installation with dimensions, burial depth, point of entering the building, meter location, and anything else that may be of interest in future years. Details of the installation of plastic service lines should be particularly thorough because they are difficult to locate once they are covered. The best time to prepare the service information is during the plumbing inspection, before the contractor is allowed to backfill the trench.

Meter Records

Meter records are also often maintained by specialized system management programs. Information included identifies the meter make, size, model, and purchase date. Test data and installation location are also included.

Technical Information

Every water distribution system should develop a workable file system for technical information. An example would be to establish a file section for fire hydrants. Within that section there should be file folders with literature on equipment from each manufacturer, and additional files for purchases, warranties, correspondence, and factory and local representative phone numbers.

New catalogs should be dated when they are received, and as they are filed, old ones should be discarded unless they have some special historical value. Most files gradually grow to unreasonable size because nobody takes the time to delete old material.

GEOGRAPHIC INFORMATION MANAGEMENT SYSTEMS

The array of data needed to efficiently operate and manage a water distribution system is immense. Almost all data used for this purpose is related to a geographic location. Computerized programs are used to assemble this data and organize it for practical use. These systems are known as Automatic Mapping/Facility Management/Geographic Information Systems (AM/FM/GIS).

An AM/FM/GIS has locations and areas shown on a computerized map. This map has many layers with each containing related data such as surface elevations and features, streets, water distribution components, and even electric power facilities. There can be hundreds of factors associated with each location.

The objective of the AM/FM/GIS is to provide a fully integrated database to support operation, management, and maintenance functions. Some of the important applications are water resource management, such as water quality information from various entry points into the distribution system; customer services, like the location and frequency of complaints; operations to assist with directing water flow for fire fighting; water demand forecasting by including weather information; and water system modeling where new information can be used to refine the hydraulic model.

Maintenance management information can be included in the AM/FM/GIS or these records can be kept separately either manually or in a maintenance management system. The maintenance management process is greatly aided using an automated system that is integrated with other information. This provides for better planning and scheduling of work and to quickly access historical records.

Supervisory control and data acquisition (SCADA) system data can be very useful for the AM/FM/GIS. Operational data from the distribution system can be integrated for use in many ways. As indicated earlier, this data can help refine the system hydraulic model. It can also provide valuable data to analyze system performance and target quality improvement investigations.

Several other information systems are important to a distribution system operation. These include source of supply systems, treatment plant process control systems, laboratory information management systems, leakage control data, emergency response data, and customer information. Ideally, all of these systems are integrated so that the most important information is shared and available for analysis and system operation. Many utilities have chosen to integrate only selected data from various systems to make the integration easier and still get the benefit of the most important information.

BIBLIOGRAPHY

AWWA. 2010. Principles and Practices of Water Supply Operations—*Water Transmission and Distribution*, 4th ed. Denver, Colo.: American Water Works Association.

AWWA. M32—*Computer Modeling of Water Distribution Systems*. Denver, Colo.: American Water Works Association.

Mays, L.W. 2000. *Water Distribution Systems Handbook*. Denver, Colo.: American Water Works Association; New York: McGraw-Hill.

Shamsi, U.M. 2005. *GIS Applications for Water, Wastewater, and Stormwater Systems*. Boca Raton, Fla.: CRC Press.

Chapter 20

Public Relations

It is important for a water utility to maintain public confidence. Satisfied customers will pay their bills, are less likely to complain if they are temporarily inconvenienced during system maintenance and construction, and are more likely to be supportive of rate increases and bond issues. Operations personnel are often engaged with customers when they are investigating a complaint. Not only is this an opportunity to solve a problem and gain customer confidence, important information can be obtained on the system's condition.

ROLE OF DISTRIBUTION PERSONNEL

Distribution personnel are key to how the public perceives the water utility. A professional and courteous attitude is needed at all times when in contact with the public. Utility personnel are "on stage" whenever they are discussing water system matters, whether they are on duty or off. Effective public relations requires three primary elements:

1. *Good communications.* Listen to the customer. Provide information by answering questions, explaining policies, and providing tips about water supply and how to save water and money.
2. *Caring.* Show you care about the customer and your job. Present a professional appearance. Try to solve the customer's problem.
3. *Courtesy.* Be polite and respectful when engaging customers.

WORKERS WHO HAVE CONTACT WITH THE PUBLIC

Distribution system personnel generally have the closest contact with the public, and the impression they make can go a long way in maintaining a good public image for the water utility.

Meter Readers

Meter readers are in the unenviable position of being naturally associated with the water bill. If the customer feels the bill is too high for any reason, the meter reader is the most available target for complaint. In addition, the meter reader must deal with bad weather, unfriendly dogs, and, sometimes, unfriendly people.

Through all of this, the meter reader must remain informative and polite. Following are some guidelines for behavior of meter readers:

- Meter readers must maintain a neat appearance. Customers are understandably reluctant to admit into their homes people who are dirty, unshaven, or are wearing grubby clothes. Most utilities now provide uniforms, which not only ensure a neat appearance, but also make workers easily identifiable with the water utility.
- Customers are increasingly reluctant to let strangers into their homes. Meter readers should display name tags and carry credentials that can be produced if a customer asks for verification of employment.
- Meter readers do not have to answer all questions asked by customers. They should not be expected to know all the correct answers on all subjects, but they must listen as politely as possible. It is a good idea for the utility to have informative brochures available on current subjects that the meter readers can hand to the customers. A number of brochures of this type are available from the AWWA bookstore.
- If a leak on the water service is noticed, it should be reported to the customer immediately. If the customer is not home, a note should be left on the door or mailed reporting the problem.
- Meter readers should try to stay pleasant and polite and display enthusiasm for their work.
- Customers should be addressed properly, such as by "Miss," "Sir," "Mrs.," and so on.

Maintenance Workers

Although maintenance workers may not have as much direct contact with the public as do meter readers, they are highly visible. Their appearance and conduct can go a long way in forming a good public impression of the water utility. Some guidelines for behavior of maintenance workers are as follows:

- Although there may be times when it is difficult, they should stay as well groomed as possible.
- Local residents often ask maintenance workers what they are doing. If the work being done is obvious, there is a strong temptation to give an unwise answer such as "digging for gold." A few people may find this funny, but most will take offense. The best policy is to give a short, straightforward answer. If residents want more than short answers, it is generally best to refer them to a supervisor.
- Customers should be notified in advance of temporary shutoffs if at all possible. Even when shutting down for a serious main break, one or

more workers should be dispatched as soon as possible to warn residents and suggest that they fill some containers with water for drinking and washing.

- If the shutdown is planned in advance, letters can be printed on utility letterhead and distributed to each building, or a doorknob card such as the one illustrated in Figure 20-1 can be hung on a door to each building.
- Residents should also be warned that the water may be cloudy for a short time after it is turned back on. They should be advised to not wash clothes until the water has cleared.
- Work sites should be maintained as neatly as possible, and should be restored as close to their original conditions as possible when the work is completed.
- Residents generally do not like having workers lounging on their lawns while eating lunch or taking a coffee break. It is better to move to a more secluded location.
- Vehicles and other equipment should be kept clean.
- Vehicles should be parked out of the way of local residents if possible.
- Workers should not nap in a truck in a visible location during their break. The public is sure to get the wrong impression.

Vehicle Operators

Almost all water distribution system employees drive vehicles at least occasionally. Employees should be frequently reminded that their actions while driving can be very important in determining the public's opinion of the utility. Following are some important points for drivers:

- Careless driving is dangerous and causes accidents.
- Careless driving makes people angry; hence, they develop a poor opinion of the water utility as a whole.
- Conversely, courteous driving forms a good impression.
- Utility personnel must exercise good judgment on where they park vehicles. Parking a vehicle in front of a tavern is not a good idea, even though workers may actually be working at a house nearby. If vehicles are frequently seen in front of coffee or doughnut shops, the public typically draws the conclusion that this is where the workers are spending most of their time.

Some municipalities or water utilities have developed programs in cooperation with local police in which all vehicle drivers periodically attend a four-hour safe-driving course. Workers usually believe that they already know everything about safe driving and dislike attending, but experience has proven that the program is definitely worth the time invested.

EBMUD

SHUTDOWN NOTICE

IMPORTANT

FOR	CUSTOMER
BY	
TIME	DATE

We regret that it will be necessary to temporarily shut off water services to your premises between

________________ and ________________

on ________________________________

This shutdown is being made so that distribution facilities in your area can be worked on to improve your water service.

This temporary interruption in service will be as brief as possible. Your understanding and cooperation are appreciated. We suggest that you fill some containers with the water you may require during the shutdown.

THANK YOU.

EAST BAY MUNICIPAL UTILITY DISTRICT

For additional information call the Business Office checked below.

☐ **OAKLAND**
287-0825
2144 Poplar St.

☐ **RICHMOND**
287-0831
3999 Lakeside Dr.

☐ **SAN LEANDRO**
287-0838
589 E. Lewelling Blvd.

☐ **WALNUT CREEK**
287-0834
2551 N. Main St.

H-008 • 4/94 (94208)

Courtesy of East Bay Municipal Utility District.

Figure 20-1 Example of a customer service card

DEALING WITH THE MEDIA

On occasion, distribution system employees may be approached by reporters from the local media asking about the work they are doing. Although one must be courteous, the general rule in talking to reporters is, *Don't unless you absolutely have to!* There are many opportunities for being misquoted, and it is quite embarrassing for the utility to see the wrong information in print.

The best policy is to give a very brief explanation and offer no more information than is requested. Beyond that, the workers should say they are not qualified to go into more detail. Large municipalities and water utilities maintain a public relations department whose job is to deal with the media, so reporters should be referred there for additional details. If such a department does not exist, the reporter should be referred to the utility manager, city manager, or mayor.

CUSTOMER COMPLAINT RESPONSE

Responding to customer complaints is an important element in assessing system performance. Information from customers about water quality, reliability of service, and security can be helpful when these topics are of concern. A professional interaction with customers is often the most effective method to convey a positive view of the utility and gain customer satisfaction. Also, customer complaint reporting procedures may be required by some state or local agencies.

The objectives of a Customer Complaint Response program are to (1) address the customer's concerns; (2) gather information that can be used to identify system problems; and (3) catalog data that can be used to assess long-term system performance goals. System operators need specific training to achieve these objectives.

Addressing Customer Concerns

The first contact with the customer is usually by telephone (or e-mail in some utilities). It is important to listen to the customer and to be sympathetic with their concerns. At the same time, the contact information and complaint details need to be obtained and recorded. Many times the complaint can be addressed satisfactorily over the telephone. If not, an onsite investigation should be scheduled.

Prior to the onsite investigation, the investigator should review the distribution system maps for the area. This may provide clues about the situation. Also, operations logs should be checked to see if any maintenance or unusual operations have occurred recently.

The investigator should be dressed professionally. They should wear a utility shirt if this is available. They should also carry an official identification to show to the customer. A water quality testing kit should always be part of the investigation supplies. The kit should include, at a minimum, sample bottles for chemical and microbiological tests, a pH meter, a chlorine residual meter, and a temperature gauge. No matter the complaint, the water should be examined and tested. This provides a demonstration that this is a professional investigation.

Upon arrival, the investigator should listen to the customer's concerns and take notes. The water quality tests should be conducted and samples taken for examination at the laboratory. If the investigator knows what the problem is, he or she can tell the customer.

Before the investigator leaves, the situation should be reviewed and the customer told what will happen after the investigation is complete. The customer should be advised that they will receive a phone call with the water quality test results and should also be given a date for the call. This would be a good time to give them additional information about their water supply, so prepared materials should be readily available to hand to the customer. When the investigator calls the customer, they can advise what has been done and should then ask if the problem has cleared up. The water quality test results should assure the customer that the water is safe. Depending on the situation, a followup visit can be scheduled if deemed necessary.

Collecting Information

Information from customer complaints may indicate a serious distribution system problem. Therefore, the results of a complaint investigation need to be carefully recorded and conveyed to management for further evaluation.

Colored water, taste, or odor complaints may indicate pipeline corrosion, biological growths, or cross-connections. Documenting these (and other) problems can help identify areas that need flushing or other maintenance procedures.

State, public utility commissions, health departments, or other regulatory agencies may require certain documentation of customer complaints. Utility investigators must be aware of these requirements and ensure they are satisfied.

Customer Complaints as a Performance Measure

Most utilities strive to have few customer complaints. The frequency of complaints can be used as a measure of water quality performance. Utilities should establish a performance goal for technical complaints (complaints per 1,000 customer accounts, for example). Annual results should be compared to the performance goal, and operations adjusted if the goal is not reached. *Technical complaints* are complaints associated with water quality, taste, odor, appearance, pressure, main breaks, and disruptions of water service.

Distribution system operators may be assigned to investigate customer complaints. This provides an opportunity to solve a problem for the customer and, perhaps, discover a larger system problem. It is often a challenge to collect and share customer complaint information when several utility departments are involved. The complaint records should, therefore, be kept in a location accessible to all utility personnel who will use the information.

BIBLIOGRAPHY

ANSI/AWWA G200. *AWWA Standard for Distribution Systems Operation and Management.* Denver, Colo.: American Water Works Association.

AWWA. 2007. *Focus First on Service: The Voice and Face of Your Utility.* Denver, Colo.: American Water Works Association.

AWWA. 2010. Principles and Practices of Water Supply Operations—*Water Transmission and Distribution*, 4th ed. Denver, Colo.: American Water Works Association.

Bloetscher, F. 2009. *Water Basics for Decision Makers: Local Officials' Guide to Water and Wastewater Systems.* Denver, Colo.: American Water Works Association.

Deb, A.K., J. Yakir, F.M. Hasit, and F.M. Grablutz. 1995. *Distribution System Performance Evaluation.* Denver, Colo.: Awwa Research Foundation and American Water Works Association.

Kirmeyer, G.J., M. Friedman, J. Clement, A. Sandvig, P. Noran, K. Martel, D. Smith, M. LeChevallier, C. Volk, E. Antoun, D. Hiltebrand, J. Dyksen, and R. Cushing. 2000. *Guidance Manual for Maintaining Distribution System Water Quality.* Denver, Colo.: Awwa Research Foundation and American Water Works Association.

Lauer, W.C. 2004. *Water Quality Complaint Investigators Field Guide.* Denver, Colo.: American Water Works Association.

Smith, C., ed. 2005. *Water Distribution System Assessment Workbook.* Denver, Colo.: American Water Works Association.

Symons, J.M. 2010. *Plain Talk About Drinking Water: Questions and Answers About the Water You Drink*, 5th ed. Denver, Colo.: American Water Works Association.

Chapter 21

Distribution System Hydraulics

An understanding of hydraulics is necessary for the proper operation of a water distribution system. Some of the basic concepts used in the operation of water distribution systems are covered in this chapter. More complete coverage of the subject can be found in the referenced publications listed at the end of the chapter.

FLUIDS AT REST AND IN MOTION

Hydraulics is the study of fluids in motion or under pressure. In this handbook, the subject will be confined to the behavior of water in water distribution systems.

Static Pressure

Water flows in a water system when it is under a force that makes it move. The force on a unit area of water is termed *pressure*. The pressure in a water system is a measure of the height to which water theoretically will rise in an imaginary standpipe open to atmospheric pressure. The pressure can be *static*; that is, it exists although the water does not flow. Pressure can also be *dynamic*, existing as "moving energy."

All objects have weight because they are acted on by gravity. When a 1-lb brick is placed on a table with an area of 1 square inch (1 in.^{2}), it exerts a force of 1 pound per square inch (psi) on the table. Two stacked bricks on the 1-in.^{2} table would exert a force of 2 psi. But if the size of the table is doubled, the pressure is halved. And if the table size is tripled, the pressure in pounds per square inch is reduced by one third.

Likewise, a column of water 10 ft high exerts a total force of 4.33 psi. If you connect a pressure gauge at the bottom of a water tube with 10 ft of water in it, this is what it will read. If you also connect the gauge to the bottom of a larger-diameter column with 10 ft of water, the pressure at the bottom will still read the same (Figure 21-1). Water pressure is dependent only on the height of the column. On the other hand, the total weight exerted on the floor by the water in the large column will obviously be much more.

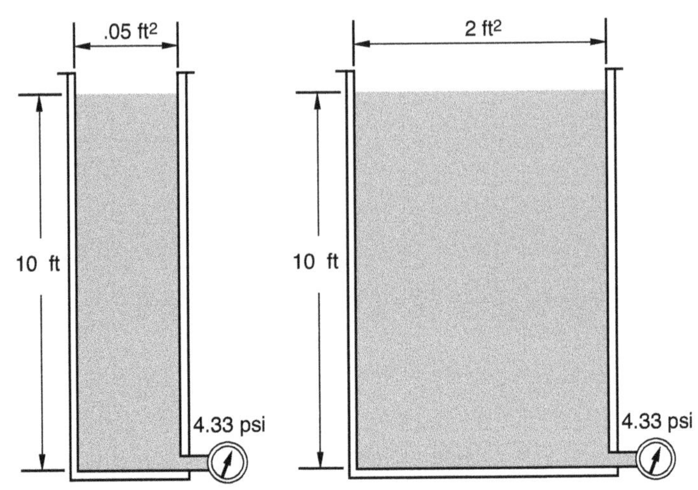

Figure 21-1 Hydraulic head depends only on column height

Dynamic Pressure

If the water in the column is permitted to empty horizontally from the bottom of a column, the water will begin to flow under the hydrostatic pressure applied by the height of the column. The flowing water will have little hydrostatic pressure, but it will have gained moving, dynamic pressure, or kinetic energy. The hydrostatic pressure is static potential energy converted into moving energy.

One can add energy to a water system and thereby increase hydrostatic and dynamic pressure. A pump does this when it pumps water into elevated storage. The hydrostatic pressure (height) to which the water can be pumped is equivalent to pressure (less losses) at the pump discharge.

Pressure is usually measured in either pressure in pounds per square inch or feet of head in US units, or as kilopascals (kPa) pressure or meters (m) head in metric units. A pressure of 1 lb/in.^{2} is equal to approximately 6.895 kPa.

Velocity

The speed at which water moves is called *velocity*, usually abbreviated *V*. The velocity of water is usually measured in feet per second (ft/sec) in US units and meters per second (m/sec) in metric terms. For comparison, a rapidly moving river might move at about 7 ft/sec (2.13 m/sec).

The quantity of water (*Q*) that flows through a pipe depends on the velocity (*V*) and the cross-sectional area (*A*) of the pipe. This is stated mathematically as the formula $Q = A \times V$. Or, in terms of velocity,

$$V = \frac{Q}{A}$$

For example, a flume is 2 ft wide and 2 ft deep, so the cross-sectional area of the flume is 4 ft^2. The flume is flowing full of water and the quantity is measured at 12 ft^3 in 1 second (12 ft^3/sec). The velocity of the water would therefore be

$$V = \frac{Q}{A}$$

$$= \frac{12 \text{ ft}^3/\text{sec}}{4 \text{ ft}^2} = 3 \text{ ft/sec}$$

Friction Loss

As water flows through a pipeline, there is friction between the water and the walls of the pipe. The friction loss causes a loss of head (pressure) as the water flows through the pipe. The amount of friction depends partly on the smoothness of the pipe walls. All new pipe is quite smooth, whereas old, badly corroded cast-iron pipe will have a very high friction factor. The degree of pipe roughness is commonly denoted by a *C* factor, which is a coefficient in the Hazen–Williams formula that has long been used for determining flow in pipe. High *C* values imply less friction.

The head loss due to friction also depends on the velocity of the water flowing through the pipe, the diameter of the pipe, and the distance the water travels through the pipe.

Figure 21-2 is a commonly used nomograph for approximating the flow in ductile-iron pipe. In the example shown by a dashed line, a 12-in. pipe is flowing at approximately 600 gpm and the pipe has a *C* factor of 140. A line is drawn from the 600-gpm point on the discharge line, through the point for 12-in. pipe, and to the pivot line. A line is then drawn from that point to 140 on the flow coefficient line. This line crosses the loss of head line at about 0.7, indicating this is the head loss per 1,000 ft of pipeline. If, for example, you are determining the loss of head in a pipeline 3,000 ft long with no valves or bends, the theoretical loss of head would be three times the indicated value (3 × 0.7 = 2.1 ft of head).

Pipe fittings also add a significant pressure loss in flow, and this is usually expressed as the equivalent length of straight pipe. To use the nomograph in Figure 21-3, a line is drawn from the pipe size to the point for each type of fitting, and the equivalent pipe length is read from the center scale. The total of all readings is then added to the actual length of the pipeline in determining the expected loss of head.

Referring to the dashed line in Figure 21-2, each medium sweep elbow in the previous 12-in. pipeline example would add friction loss equal to about 26 ft of pipe. So if the example pipeline has 20 elbows along the 3,000-ft length, it would add friction loss equal to an additional 20 × 26 = 520 ft of pipe. This would cause additional loss of head as follows:

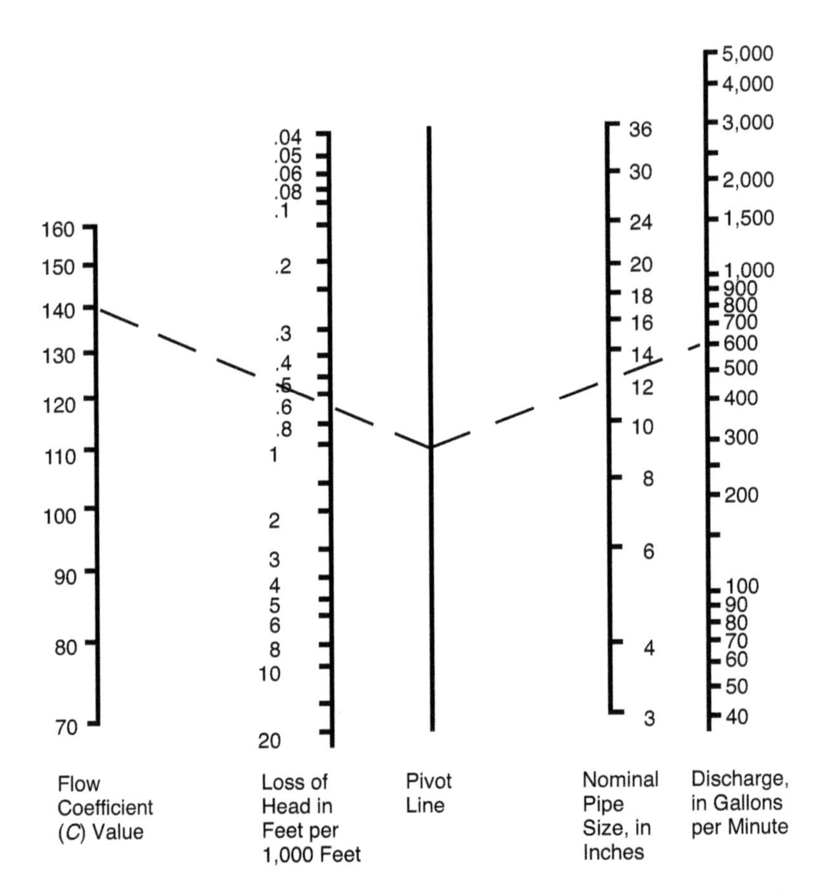

Figure 21-2 Flow of water in ductile-iron pipe

loss of head per 1,000 ft = 0.7 ft

$$\text{loss of head for 520 ft} = \frac{520}{1{,}000} \times 0.7 = 0.36 \text{ ft loss of head due to elbows}$$

This would be added to the loss of head determined for 3,000 ft of pipe, so the total loss would be 2.1 + 0.36 = 2.46 ft of head. If there are also tees, valves, and other fittings in the pipeline, the head loss that they cause can be computed and added to the total.

This example also illustrates that the loss in head can become quite significant over a long pipeline. If, for example, after adding up the losses caused by all the other fittings, the total loss of head in the pipeline is 5 ft, this loss in terms of pressure would be 5 ft × 0.433 lb/in.^{2}/ft = 2.17 psi.

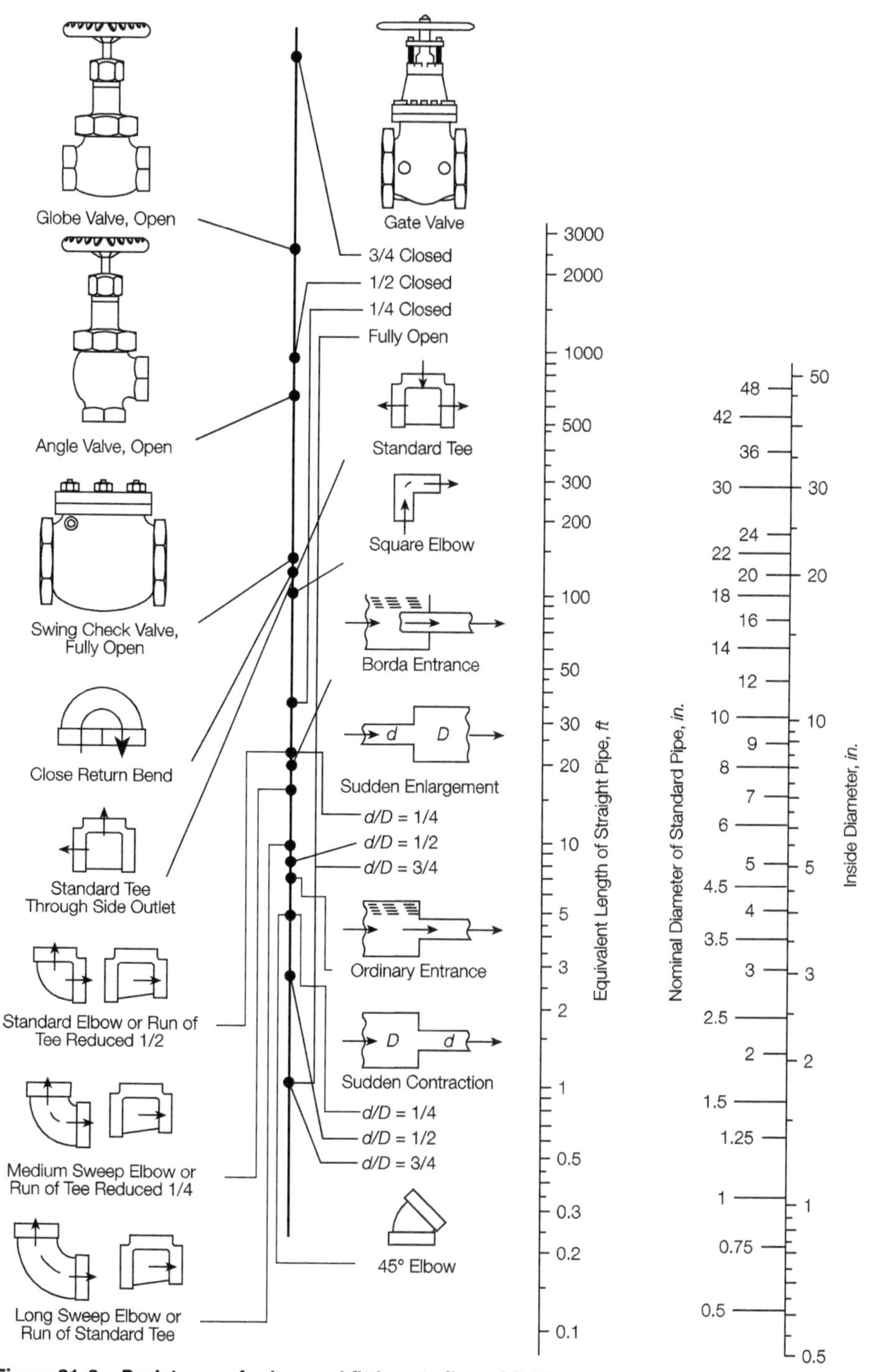

Figure 21-3 Resistance of valves and fittings to flow of fluids

Table 21-1 Designation of US pipe sizes to the metric system

Customary Inches	Proposed Millimeters	Customary Inches	Proposed Millimeters
¼	8	16	400
⅜	10	18	450
½	15	20	500
¾	20	21	525
1¼	25	24	600
1	32	27	675
1½	40	30	750
2	50	33	825
2½	65	36	900
3	80	42	1,050
3½	90	48	1,200
4	100	54	1,350
6	150	60	1,500
8	200	66	1,650
10	250	72	1,800
12	300	78	1,950
14	350	84	2,100
15	375		

In other words, if the pressure entering the pipe is 50 psi, the theoretical pressure at the far end would be reduced to 50 – 21.65 = 28.35 psi.

To convert the information on the nomographs for metric use, refer to Table 21-1 for the metric equivalents of US unit pipe sizes. Flow in gallons per minute (gpm) can be converted to liters per second (L/s) when multiplying by 0.06308. Head of water expressed in feet can be converted to meters of water when multiplying by 0.3048.

HYDRAULIC GRADIENT

The head of water at any point in a water system refers to the height to which water would rise in a freely vented standpipe. The head at each point would be the height of the water column. The imaginary line joining the elevations of these heads is called the *hydraulic grade line.* The slope or steepness of this line is called the *hydraulic gradient.*

A simple hydraulic gradient is illustrated in Figure 21-4. Assuming there is equal flow in all sections of the line, the gradient becomes steeper as the pipe becomes smaller because of the friction head loss. If there were no flow in the line, the water head at the end of the line would be at the same level as the water in the reservoir.

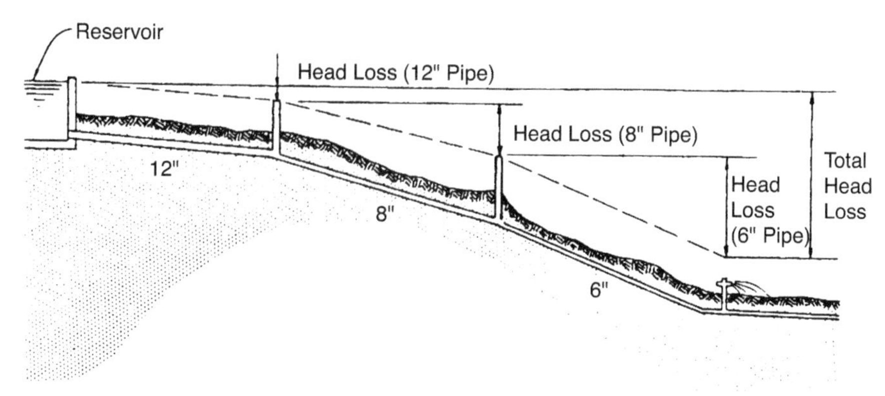

Figure 21-4 Pipe size affects hydraulic gradient

In the example in Figure 21-5, water flows from a reservoir toward a booster pump. The pump adds pressure and causes an upward slope of the hydraulic grade line, then the head falls as the water travels to the end of the system.

HYDRAULIC TRANSIENTS

Very brief but often drastic changes in flow within pipes can cause magnified pressure changes. These often short-lived (transient) events can result in broken pipelines, can damage pumping facilities, and may allow contaminants to enter the potable water system. Hydraulic transients can cause either high or low pressures and may last from milliseconds to several seconds in duration.

Water Hammer

The most common hydraulic transient encountered in water distribution systems is "water hammer." Water hammer occurs most often when a valve is closed too quickly. This causes a pressure wave to develop within the pipe. The magnitude of the maximum pressure is related to the velocity of the water, the pressure at the time, and the rate the valve was closed. Water hammer has caused rupture of pipe and pump casings; pipe collapse; vibration; excessive pipe displacement; pipeline fittings failure; and cavitation.

System operators should take care to avoid creating water hammer. Always operate valves slowly to eliminate water hammer. Educate anyone that may use fire hydrants to open and close with care. This will help reduce water hammer.

Surge Control

A specialized type of water hammer can be caused by sudden stoppage of system pumps. This often occurs during a power outage. When the pump stops, water from the system (perhaps from a filling storage reservoir) can

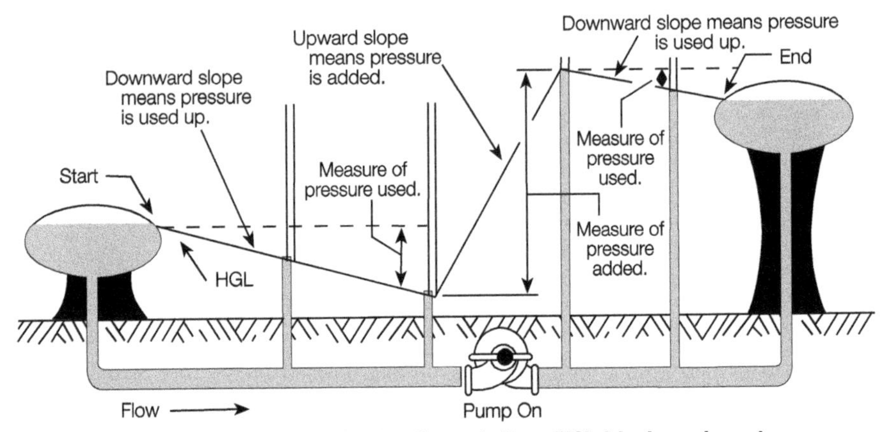

Figure 21-5 Five basic principles of hydraulic grade lines (HGLs) in dynamic systems

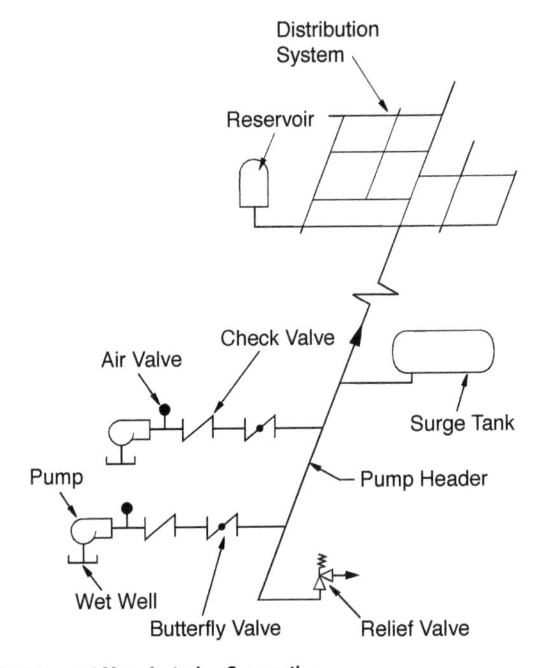

Courtesy of Wal-Matic Valve and Manufacturing Corporation.

Figure 21-6 Typical pumping/distribution system

surge back toward the pump. This "water hammer" can damage pumps, pipelines, fittings, and other equipment.

Surge control equipment should be installed on all distribution system pumps (see Figures 21-6 through 21-11). Typical surge control systems include pump control valves, surge tanks, air valves, relief valves, and check valves. These devices require regular maintenance according to manufacturer recommendations.

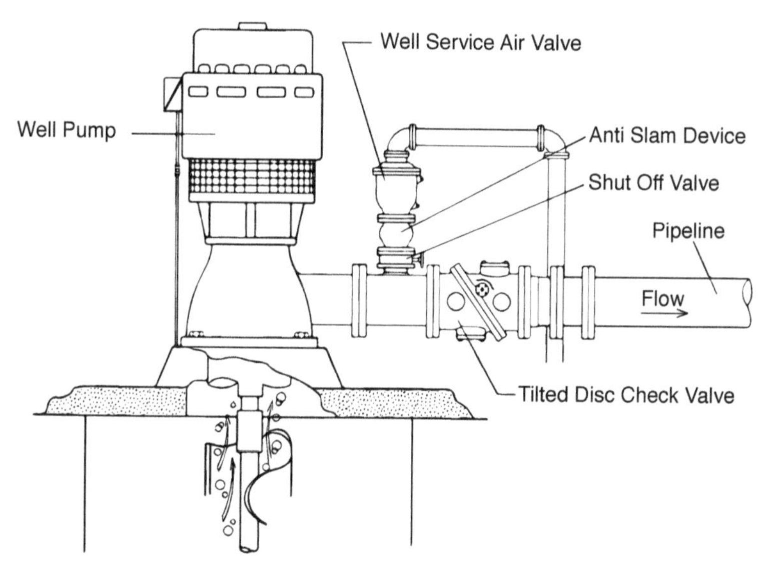

Courtesy of Wal-Matic Valve and Manufacturing Corporation.

Figure 21-7 Vertical turbine well pump with a well service air valve

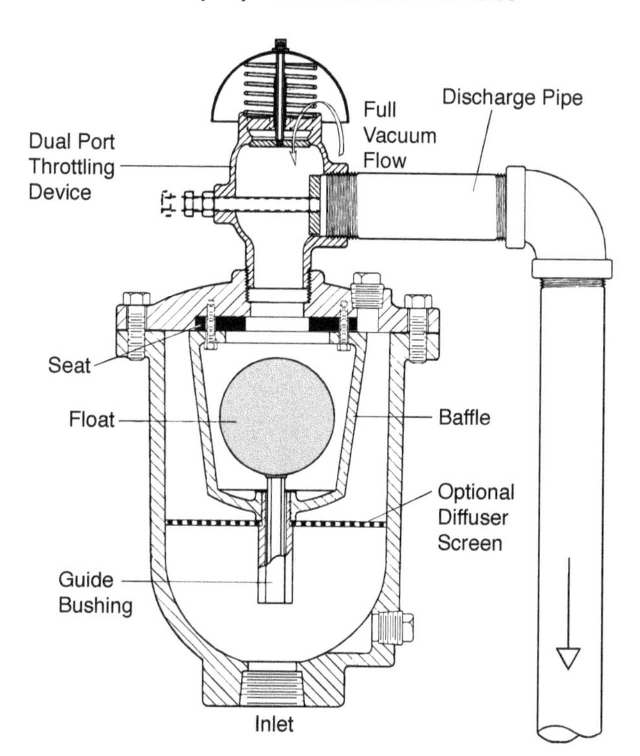

Courtesy of Wal-Matic Valve and Manufacturing Corporation.

Figure 21-8 Well service air valve

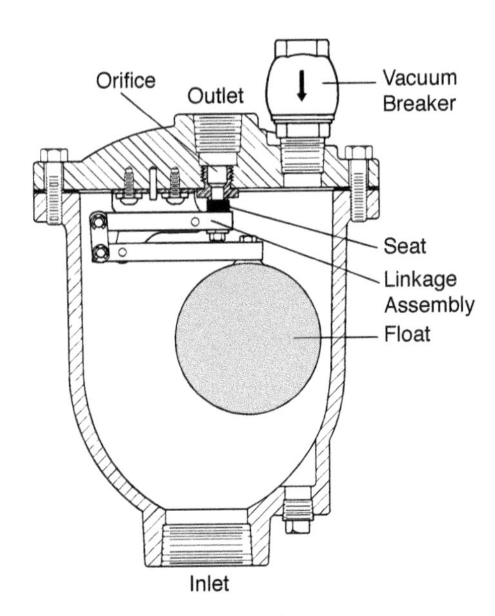

Courtesy of Wal-Matic Valve and Manufacturing Corporation.

Figure 21-9 Air-release valve

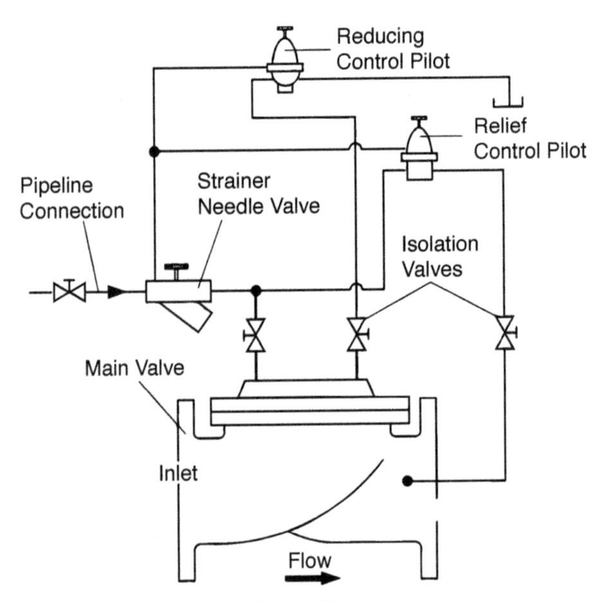

Courtesy of Wal-Matic Valve and Manufacturing Corporation.

Figure 21-10 Surge relief valve and anticipator valve

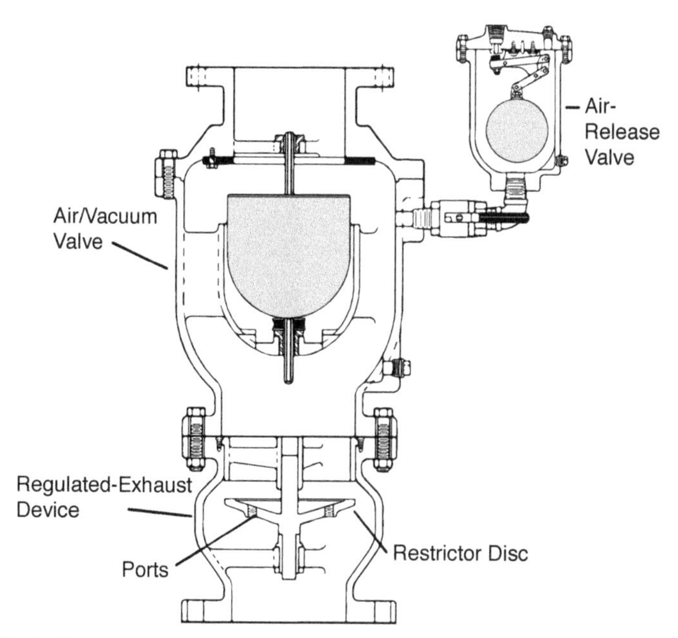

Courtesy of Wal-Matic Valve and Manufacturing Corporation.

Figure 21-11 Surge-suppression air valve

BIBLIOGRAPHY

Armstrong, L., ed. 2011. *Hydraulic Modeling and GIS.* Redlands, Calif.: ESRI Press.

AWWA. 2010. Principles and Practices of Water Supply Operations—*Basic Science Concepts and Applications*, 4th ed. Denver, Colo.: American Water Works Association.

AWWA. 2010. Principles and Practices of Water Supply Operations—*Water Transmission and Distribution*, 4th ed. Denver, Colo.: American Water Works Association.

Fleming, K.K., R.W. Gullick, J.P. Dugandzic, and M.W. LeChevallier. 2006. *Susceptibility of Distribution Systems to Negative Pressure Transients.* Denver, Colo.: Awwa Research Foundation.

Freidman, M., L. Radder, S. Harrison, D. Howie, M. Britton, G. Boyd, H. Wang, R. Gullick, M. LeChevallier, D. Wood, and J. Funk. 2004. *Verification and Control of pressure Transients and Intrusion in Distribution Systems.* Denver, Colo.: Awwa Research Foundation.

LeChevallier, M.W., R.W. Gullick, M.R. Karim, M. Friedman, and J.E. Funk. 2003. The potential for health risks from intrusion of contaminants into distribution systems from pressure transients. *J. Water Health* 1(1):3–14.

Nicklow, J.W., and P.F. Boulos. 2005. *Comprehensive Water and Wastewater Treatment Plant Hydraulics Handbook for Engineers and Operators. Denver, Colo.: American Water Works Association.*

Appendix A
Conversion of US Customary Units

Linear Measurement

fathoms	×	6	=	feet (ft)
feet (ft)	×	12	=	inches (in.)
inches (in.)	×	0.0833	=	feet (ft)
miles (mi)	×	5,280	=	feet (ft)
yards (yd)	×	3	=	feet (ft)
yards (yd)	×	36	=	inches (in.)

Circular Measurement

degrees (angle)	×	60	=	minutes (angle)
degrees (angle)	×	0.01745	=	radians

Area Measurement

acres	×	43,560	=	square feet (ft^2)
square feet (ft^2)	×	144	=	square inches ($in.^2$)
square inches ($in.^2$)	×	0.00695	=	square feet (ft^2)
square miles (mi^2)	×	640	=	acres
square miles (mi^2)	×	27,880,000	=	square feet (ft^2)
square miles (mi^2)	×	3,098,000	=	square yards (yd^2)
square yards (yd^2)	×	9	=	square feet (ft^2)

Volume Measurement

acre-feet (acre-ft)	×	43,560	=	cubic feet (ft^3)
acre-feet (acre-ft)	×	325,851	=	gallons (gal)
barrels water (bbl)	×	42	=	gallons (gal)
cubic feet (ft^3)	×	1,728	=	cubic inches ($in.^3$)
cubic feet (ft^3)	×	7.48052	=	gallons (gal)
cubic feet (ft^3)	×	29.92	=	quarts
cubic feet (ft^3)	×	0.000023	=	acre feet (acre-ft)
cubic inches ($in.^3$)	×	0.00433	=	gallons (gal)
cubic inches ($in.^3$)	×	0.00058	=	cubic feet (ft^3)
gallons (gal)	×	0.1337	=	cubic feet
gallons (gal)	×	231	=	cubic inches ($in.^3$)
gallons (gal)	×	0.0238	=	barrels (bbl)
gallons (gal)	×	4	=	quarts (qt.)
gallons, US	×	0.83267	=	gallons, Imperial

gallons (gal)	× 0.00000308	=	acre-feet (acre-ft)
gallons (gal)	× 0.0238	=	barrels (42 gal) (bbl)
gallons, Imperial	× 1.20095	=	gallons, US
pints (pt)	× 2	=	quarts (qt.)
quarts (qt)	× 4	=	gallons (gal)
quarts (qt)	× 57.75	=	cubic inches ($in.^3$)

Pressure Measurement

atmospheres	× 29.92	=	inches of mercury
atmospheres	× 33.90	=	feet of water
atmospheres	× 14.70	=	pounds per square inch ($lb/in.^2$)
feet of water	× 0.8826	=	inches of mercury
feet of water	× 0.02950	=	atmospheres
feet of water	× 0.4335	=	pounds per square inch ($lb/in.^2$)
feet of water	× 62.43	=	pounds per square foot (lb/ft^2)
feet of water	× 0.8876	=	inches of mercury
inches of mercury	× 1.133	=	feet of water
inches of mercury	× 0.03342	=	atmospheres
inches of mercury	× 0.4912	=	pounds per square inch ($lb/in.^2$)
inches of water	× 0.002458	=	atmospheres
inches of water	× 0.07355	=	inches of mercury
inches of water	× 0.03613	=	pounds per square inch ($lb/in.^2$)
pounds/square in. ($lb/in.^2$)	× 0.01602	=	feet of water
pounds/square foot (lb/ft^2)	× 6,954	=	pounds per square inch ($lb/in.^2$)
pounds/square in. ($lb/in.^2$)	× 2.307	=	feet of water
pounds/square inch ($lb/in.^2$)	× 2.036	=	inches of mercury
pounds/square inch ($lb/in.^2$)	× 27,70	=	inches of water
feet suction lift of water	× 0.882	=	inches of mercury

Weight Measurement

cubic feet of ice	× 57.2	=	pounds (lb)
cubic feet of water (50°F)	× 62.4	=	pounds of water
cubic inches of water	× 0.036	=	pounds of water
gallons water (50°F)	× 8.3453	=	pounds of water
milligrams/liter (mg/L)	× 0.0584	=	grains per gallon (US) (gpg)
milligrams/liter (mg/L)	× 0.07016	=	grains per gallon (Imp)
milligrams/liter (mg/L)	× 8.345	=	pounds per million gallons (lb/mil gal)
ounces (oz)	× 437.5	=	grains
parts per million		=	milligrams per liter (for normal water applications)
grains per gallon (gpg)	× 17.118	=	parts per million (ppm)
grains per gallon (gpg)	× 142.86	=	pounds per million gallons (lb/mil gal)
percent solution	× 10,000	=	milligrams per liter (mg/L)
pounds (lb)	× 16	=	ounces

pounds (lb)	×	7,000	=	grains
pounds (lb)	×	0.0004114	=	tons (short)
pounds/cubic inch (lb/in.^{3})	×	1,728	=	pounds per cubic foot (lb/ft^3)
pounds of water	×	0.0166032	=	cubic feet (ft^3)
pounds of water	×	2,768	=	cubic inches (in.^{3})
pounds of water	×	0.1198	=	gallons (gal)
tons (short)	×	2,000	=	pounds (lb)
tons (short)	×	0.89287	=	tons (long)
tons (long)	×	2,240	=	pounds (lb)
cubic feet air (@60°F and 29.92 in. mercury)	×	0.0763	=	pounds

Flow Measurement

barrels per hour (bbl/h)	×	0.70	=	gallons per minute (gpm)
acre-feet/minute	×	325.851	=	gallons per minute (gpm)
acre-feet/minute	×	726	=	cubic feet per second (ft^3/sec)
cubic feet/minute (ft^3/min)	×	0.1247	=	gallons per second (gal/sec)
cubic feet/minute (ft^3/min)	×	62.43	=	pounds of water per minute
cubic feet/second (ft^3/sec)	×	448.831	=	gallons per minute (gpm)
cubic feet/second (ft^3/sec)	×	0.646317	=	million gallons per day (mgd)
cubic feet/second (ft^3/sec)	×	1.984	=	acre-feet per day (acre-ft/d)
gallons/minute (gpm)	×	1,440	=	gallons per day (gpd)
gallons/minute (gpm)	×	0.00144	=	million gallons per day (mgd)
gallons/minute (gpm)	×	0.00223	=	cubic feet per second (ft^3/sec)
gallons/minute (gpm)	×	0.1337	=	cubic feet per minute (ft^3/min)
gallons/minute (gpm)	×	8.0208	=	cubic feet per hour (ft^3/h)
gallons/minute (gpm	×	0.00442	=	acre-feet per day (acre-ft/d)
gallons/minute (gpm)	×	1.43	=	barrels (42 gal) per day (bbl/d)
gallons water/minute	×	6.0086	=	tons of water per 24 hours
million gallons/day (mgd)	×	1.54723	=	cubic feet per second (ft^3/sec)
million gallons/day (mgd)	×	92.82	=	cubic feet per minute (ft^3/min)
million gallons/day (mgd)	×	694.4	=	gallons per minute (gpm)
million gallons/day (mgd)	×	3.07	=	acre-feet per day (acre-ft/d)
pounds of water/minute	×	26.700	=	cubic feet per second (ft^3/sec)
miner's inch			=	flow through an orifice of 1 in.^{2} under a head of 4 to 6 in.
miner's inches (9 gpm)	×	8.98	=	gallons per minute (gpm)
miner's inches (9 gpm)	×	1.2	=	cubic feet per minute (ft^3/min)
miner's inches (11.25 gpm)	×	11.22	=	gallons per minute (gpm)
miner's inches (11.25 gpm)	×	1.5	=	cubic feet per minute (ft^3/min)

Work Measurement

British thermal units: Formerly defined as the quantity of heat required to raise the temperature of 1 lb of water by 1°F; now defined as 1,055.06 joules.

British thermal units (Btu)	×	777.5	=	foot-pounds (ft-lb)
British thermal units (Btu)	×	39,270	=	horsepower-hours (HP-h)
British thermal units (Btu)	×	29,280	=	kilowatt-hours (kW-h)
foot-pounds (ft-lb)	×	1,286	=	British thermal units (Btu)
foot-pounds (ft-lb)	×	50,500,000	=	horsepower-hours (HP-h)
foot-pounds (ft-lb)	×	37,660,000	=	kilowatt-hours (kW-h)
horsepower-hours (HP-h)	×	2,547	=	British thermal units (Btu)
horsepower-hours (HP-h)	×	0.7457	=	kilowatt-hours (kW-h)
kilowatt-hours (kW-h)	×	3,415	=	British thermal units (Btu)
kilowatt-hours (kW-h)	×	1.241	=	horsepower-hours (HP-h)

Power Measurement

boiler horsepower	×	33,480	=	British thermal units per hour (Btu/h)
boiler horsepower	×	9.8	=	kilowatts (kW)
British thermal units/second (Btu/sec)	×	1.0551	=	kilowatts (kW)
British thermal units/minute (Btu/min)	×	12.96	=	foot-pounds per second (ft-lb/sec)
British thermal units/minute (Btu/min)	×	0.02356	=	horsepower (HP)
British thermal units/minute (Btu/min)	×	0.01757	=	kilowatts (kW)
British thermal units/hour (Btu/h)	×	0.293	=	watts (W)
British thermal units/hour (Btu/h)	×	12.96	=	foot-pounds per minute (ft-lb/min)
British thermal units/hour (Btu/h)	×	0.00039	=	horsepower (HP)
foot-pounds per second (ft-lb/sec)	×	771.7	=	British thermal units per minute (Btu/min)
foot-pounds per second (ft-lb/sec)	×	1,818	=	horsepower (HP)
foot-pounds per second (ft-lb/sec)	×	1,356	=	kilowatts (kW)
foot-pounds per minute (ft-lb/min)	×	303,000	=	horsepower (HP)
foot-pounds per minute (ft-lb/min)	×	226,000	=	kilowatts (kW)
horsepower (HP)	×	42.44	=	British thermal units per minute (Btu/min)
horsepower (HP)	×	33,000	=	foot-pounds per minute (ft-lb/min)
horsepower (HP)	×	550	=	foot-pounds per second (ft-lb/sec)
horsepower (HP)	×	1,980,000	=	foot-pounds per hour (ft-lb/h)
horsepower (HP)	×	0.7457	=	kilowatts (kW)
horsepower (HP)	×	745.7	=	watts (W)
kilowatts (kW)	×	0.9478	=	British thermal units per second (Btu/sec)
kilowatts (kW)	×	56.92	=	British thermal units per minute (Btu/min)
kilowatts (kW)	×	3,413	=	British thermal units per hour (Btu/h)
kilowatts (kW)	×	44,250	=	foot-pounds per minute (ft-lb/min)
kilowatts (kW)	×	737.6	=	foot-pounds per second (ft-lb/sec)
kilowatts (kW)	×	1.341	=	horsepower (HP)
tons of refrig. (US)	×	288,000	=	British thermal units per 24 hours
watts (W)	×	0.05692	=	British thermal units per minute (Btu/min)

watts (W) × 0.7376 = foot-pounds (force) per second (ft-lb/sec)

watts (W) × 44.26 = foot-pounds per minute (ft-lb/min)

watts (W) × 1,341 = horsepower (HP)

Velocity Measurement

feet/minute (ft/min) × 0.01667 = feet per second (ft/sec)

feet/minute (ft/min) × 0.01136 = miles per hour (mph)

feet/second (ft/sec) × 0.6818 = miles per hour (mph)

miles/hour (mph) × 88 = feet per minute (ft/min)

miles/hour (mph) × 1.467 = feet per second (ft/sec)

Miscellaneous

grade: 1 percent (or 0.01) = 1 foot per 100 feet

Appendix B
Metric Conversions

Linear Measurement

inch (in.)	× 25.4	= millimeters (mm)
inch (in.)	× 2.54	= centimeters (cm)
foot (ft)	× 304.8	= millimeters (mm)
foot (ft)	× 30.48	= centimeters (cm)
foot (ft)	× 0.3048	= meters (m)
yard (yd)	× 0.9144	= meters (m)
mile (mi)	× 1,609.3	= meters (m)
mile (mi)	× 1.6093	= kilometers (km)
millimeter (mm)	× 0.03937	= inches (in.)
centimeter (cm)	× 0.3937	= inches (in.)
meter (m)	× 39.3701	= inches (in.)
meter (m)	× 3.2808	= feet (ft)
meter (m)	× 1.0936	= yards (yd)
kilometer (km)	× 0.6214	= miles (mi)

Area Measurement

square meter (m^2)	× 10,000	= square centimeters (cm^2)
hectare (ha)	× 10,000	= square meters (m^2)
square inch ($in.^2$)	× 6.4516	= square centimeters (cm^2)
square foot (ft^2)	× 0.092903	= square meters (m^2)
square yard (yd^2)	× 0.8361	= square meters (m^2)
acre	× 0.004047	= square kilometers (km^2)
acre	× 0.4047	= hectares (ha)
square mile (mi^2)	× 2.59	= square kilometers (km^2)
square centimeter (cm^2)	× 0.16	= square inches ($in.^2$)
square meters (m^2)	× 10.7639	= square feet (ft^2)
square meters (m^2)	× 1.1960	= square yards (yd^2)
hectare (ha)	× 2.471	= acres
square kilometer (km^2)	× 247.1054	= acres
square kilometer (km^2)	× 0.3861	= square miles (mi^2)

Volume Measurement

cubic inch ($in.^3$)	× 16.3871	= cubic centimeters (cm^3)
cubic foot (ft^3)	× 28,317	= cubic centimeters cm^3)
cubic foot (ft^3)	× 0.028317	= cubic meters (m^3)

cubic foot (ft^3)	× 28.317	= liters (L)
cubic yard (yd^3)	× 0.7646	= cubic meters (m^3)
acre foot (acre-ft)	× 1233.48	= cubic meters (m^3)
ounce (US fluid) (oz)	× 0.029573	= liters (L)
quart (liquid) (qt)	× 946.9	= milliliters (mL)
quart (liquid) (qt)	× 0.9463	= liters (L)
gallon (gal)	× 3.7854	= liters (L)
gallon (gal)	× 0.0037854	= cubic meters (m^3)
peck (pk.)	× 0.881	= decaliters (dL)
bushel (bu.)	× 0.3524	= hectoliters (hL)
cubic centimeters (cm^3)	× 0.061	= cubic inches ($in.^3$)
cubic meter (m^3)	× 35.3183	= cubic feet (ft^3)
cubic meter (m^3)	× 1.3079	= cubic yards (yd^3)
cubic meter (m^3)	× 264.2	= gallons (gal)
cubic meter (m^3)	× 0.000811	= acre-feet (acre-ft)
liter (L)	× 1.0567	= quart (liquid) (qt)
liter (L)	× 0.264	= gallons (gal)
liter (L)	× 0.0353	= cubic feet (ft^3)
decaliter (dL)	× 2.6417	= gallons (gal)
decaliter (dL)	× 1.135	= pecks (pk)
hectoliter (hL)	× 3.531	= cubic feet (ft^3)
hectoliter (hL)	× 2.84	= bushels (bu.)
hectoliter (hL)	× 0.131	= cubic yards (yd^3)
hectoliter (hL)	× 26.42	= gallons (gal)

(Note: US gallons are listed above.)

Pressure Measurement

pound/square inch (psi)	× 6.8948	= kilopascals (kPa)
pound/square inch (psi)	× 0.00689	= pascals (Pa)
pound/square inch (psi)	× 0.070307	= kilograms/square centimeter (kg/cm^2)
pound/square foot (lb/ft^2)	× 47.8803	= pascals (Pa)
pound/square foot (lb/ft^2)	× 0.000488	= kilograms/square centimeter (kg/cm^2)
pound/square foot (lb/ft^2)	× 4.8824	= kilograms/square meter (kg/m^2)
inches of mercury	× 3,376.8	= pascals (Pa)
inches of water	× 248.84	= pascals (Pa)
bar	× 100,000	= newtons per square meter
pascals (Pa)	× 1	= newtons per square meter
pascal (Pa)	× 0.000145	= pounds/square inch (psi)
kilopascals (kPa)	× 0.145	= pounds/square inch (psi)
pascal (Pa)	× 0.000296	= inches of mercury (at 60°F)
kilogram/square centimeter (k/cm^2)	× 14.22	= pounds/square inch (psi)
kilogram/square centimeter (k/cm^2)	× 28.959	= inches of mercury (at 60°F)
kilogram/square meter (k/m^2)	× 0.2048	= pounds per square foot (lb/ft^2)
centimeters of mercury	× 0.4461	= feet of water

Weight Measurement

ounce (oz)	×	28.3495	= grams (g)
pound (lb)	×	0.045359	= grams (g)
pound (lb)	×	0.4536	= kilograms (kg)
ton (short)	×	0.9072	= megagrams (metric ton)
pounds/cubic foot (lb/ft^3)	×	16.02	= grams per liter (g/L)
pounds/million gallons (lb/mil gal)	×	0.1198	= grams per cubic meter (g/m^3)
gram (g)	×	15.4324	= grains
gram (g)	×	0.0353	= ounces (oz)
gram (g)	×	0.0022	= pounds (lb)
kilograms (kg)	×	2.2046	= pounds (lb)
kilograms (kg)	×	0.0011	= tons (short)
megagram (metric ton)	×	1.1023	= tons (short)
grams/liter (g/L)	×	0.0624	= pounds per cubic foot (lb/ft^3)
grams/cubic meter (g/m^3)	×	8.3454	= pounds/million gallons (lb/mil gal)

Flow Rates

gallons/second (gps)	×	3.785	= liters per second (L/sec)
gallons/minute (gpm)	×	0.00006308	= cubic meters per second (m^3/sec)
gallons/minute (gpm)	×	0.06308	= liters per second (L/sec)
gallons/hour (gph)	×	0.003785	= cubic meters per hour (m^3/h)
gallons/day (gpd)	×	0.000003785	= million liters per day (ML/d)
gallons/day (gpd)	×	0.003785	= cubic meters per day (m^3/d)
cubic feet/second (ft^3/sec)	×	0.028317	= cubic meters per second (m^3/sec)
cubic feet/second (ft^3/sec)	×	1,699	= liters per minute (L/min)
cubic feet/minute (ft^3/min)	×	472	= cubic centimeters /second (cm^3/sec)
cubic feet/minute (ft^3/min)	×	0.472	= liters per second (L/sec)
cubic feet/minute (ft^3/min)	×	1.6990	= cubic meters per hour (m^3/h)
million gallons/day (mgd)	×	43.8126	= liters per second (L/sec)
million gallons/day (mgd)	×	0.003785	= cubic meters per day (m^3/d)
million gallons/day (mgd)	×	0.043813	= cubic meters per second (m^3/sec)
gallons/square foot (gal/ft^2)	×	40.74	= liters per square meter (L/m^2)
gallons/acre/day (gal/acre/d)	×	0.0094	= cubic meters/hectare/day (m^3/ha/d)
gallons/square foot/day (gal/ft^2/d)	×	0.0407	= cubic meters/square meter/day (m^3/m^2/d)
gallons/square foot/day (gal/ft^2/d)	×	0.0283	= liters/square meter/day (L/m^2/d)
gallons/square foot/minute (gal/ft^2/min)	×	2.444	= cubic meters/square meter/hour (m^3/m^2/h) = m/h
gallons/square foot/minute (gal/ft^2/min)	×	0.679	= liters/square meter/second (L/m^2/sec)
gallons/square foot/minute (gal/ft^2/min)	×	40.7458	= liters/square meter/minute (L/m^2/min)
gallons/capita/day (gpcd)	×	3.785	= liters/day/capita (L/d per capita)
liters/second (L/sec)	×	22,824.5	= gallons per day (gal/d)
liters/second (L/sec)	×	0.0228	= million gallons per day (mgd)
liters/second (L/sec)	×	15.8508	= gallons per minute (gpm)
liters/second (L/sec)	×	2.119	= cubic feet per minute (ft^3/min)

liters/minute (L/min)	× 0.0005886	=	cubic feet per second (ft^3/sec)
cubic centimeters/second	× 0.0021	=	cubic feet per minute (ft^3/min)
cubic meters/second (m^3/sec)	× 35.3147	=	cubic feet per second (ft^3/sec)
cubic meters/second (m^3/sec)	× 22.8245	=	million gallons per day (mgd)
cubic meters/second (m^3/sec)	× 15,850.3	=	gallons per minute (gpm)
cubic meters/hour (m^3/sec)	× 0.5886	=	cubic feet per minute (ft^3/min)
cubic meters/hour (m^3/sec)	× 4.403	=	gallons per minute (gpm)
cubic meters/day (m^3/d)	× 264.1720	=	gallons per day (gpd)
cubic meters/day (m^3/d)	× 0.00026417	=	million gallons per day (mgd)
cubic meters/hectare/day	× 106.9064	=	gallons per acre per day (gal/acre/d)
cubic meters/sq meter/day	× 24.5424	=	gallons/square foot/day ($gal/ft^2/d$)
liters/square meter/minute	× 0.0245	=	gallons/square foot/minute ($gal/ft^2/min$
liters/square meter/minute	× 35.3420	=	gallons/square foot/day ($gal/ft^2/d$)

Work, Heat, and Energy

British thermal units (Btu)	× 1.0551	=	kilojoules (kJ)
British thermal units (Btu)	× 0.2520	=	kilogram-calories (kg-cal)
foot-pound (force) (ft-lb)	× 1.3558	=	joules (J)
horsepower-hour (hp-h)	× 2.6845	=	megajoules (MJ)
watt-second (W-sec)	× 1.000	=	joules (J)
watt-hour (W-h)	× 3.600	=	kilojoules (kJ)
kilowatt-hour (kW-h)	× 3,600	=	kilojoules (kJ)
kilowatt-hour (kW-h)	× 3,600,000	=	joules (J)
British thermal units per pound (Btu/lb)	× 0.5555	=	kilogram-calories per kilogram (kg-cal/kg)
British thermal units per cubic foot (Btu/ft^3)	× 8.8987	=	kilogram-calories/cubic meter ($kg\text{-}cal/m^3$)
kilojoule (kJ)	× 0.9478	=	British thermal units (Btu)
kilojoule (kJ)	× 0.00027778	=	kilowatt-hours (kW-h)
kilojoule (kJ)	× 0.2778	=	watt-hours (W-h)
joule (J)	× 0.7376	=	foot-pounds (ft-lb)
joule (J)	× 1.0000	=	watt-seconds (W-sec)
joule (J)	× 0.2399	=	calories
megajoule (MJ)	× 0.3725	=	horsepower-hour (hp-h)
kilogram-calories (kg-cal)	× 3.9685	=	British thermal units (Btu)
kilogram-calories per kilogram (kg-cal/kg)	× 1.8000	=	British thermal units per pound (Btu/lb)
kilogram-calories per liter (kg-cal/L)	× 112.37	=	British thermal units per cubic foot\(Btu/ft^3)
kilogram-calories/cubic meter ($kg\text{-}cal/m^3$)	× 0.1124	=	British thermal units per cubic foot (Btu/ft^3)

Velocity, Acceleration, and Force

feet per minute (ft/min)	× 18.2880	=	meters per hour (m/h)
feet per hour (ft/h)	× 0.3048	=	meters per hour (m/h)
miles per hour (mi/h)	× 44.7	=	centimeters per second (cm/sec)
miles per hour (mi/h)	× 26.82	=	meters per minute (m/min)
miles per hour (mi/h)	× 1.609	=	kilometers per hour (km/h)
feet /second/second (ft/sec^2)	× 0.3048	=	meters/second/second (m/sec^2)

inches/second/second (in./sec^2)	× 0.0254	= meters/second/second (m/sec^2)
pound force (lbf)	× 4.44482	= newtons (N)
centimeters/second (cm/sec)	× 0.0224	= miles per hour (mph)
meters/second (m/sec)	× 3.2808	= feet per second (ft/sec)
meters/minute (m/min)	× 0.0373	= miles per hour (mph)
meters per hour (m/h)	× 0.0547	= feet per minute (ft/min)
meters per hour (m/h)	× 3.2808	= feet per hour (ft/h)
kilometers/second (km/sec)	× 2.2369	= miles per hour (mph)
kilometers/hour (km/h)	× 0.0103	= miles per hour (mph)
meters/second/second (m/sec^2)	× 3.2808	= feet/ second/second (ft/sec^2)
meters/second/second (m/sec^2)	× 39.3701	= inches/ second/second (in./sec^2)
newtons (N)	× 0.2248	= pounds force (lbf)

Appendix C
Distribution Math Calculations

Distribution system operators need to be proficient in basic math calculations to satisfy the requirements for certification. The calculations listed in this appendix illustrate the most common calculations found on certification exams and that are encountered in daily job tasks. The references at the end of the appendix contain a more complete discussion of the math and the details involved in the problem solutions.

PRACTICAL UNIT CONVERSIONS FOR WATER

Cubic Feet (ft^3) to Gallons (gal)

1 ft^3 of water is equivalent to 7.48 gal at normal temperatures.

Example: If a residential water meter registers in units of 100 ft^3 and 25 units of water have passed through the meter, how many gallons is this? 1 ft^3 is 7.48 gal, so 25 × 100 × 7.48 gal/ft^3 = 18,700 gal.

Gallons (gal) of Water to Pounds (lb)

1 gal of water weighs approximately 8.34 lb.

Example: How many pounds of water are contained in a 50-gal day tank? 50 × 8.34 = 417 lb.

Gallons (gal) to Liters (L)

1 gal is equivalent to 3.785 L, or 3,785 mL.

Example: An operator transfers 36 gal of polymer to a day tank. How many liters is this? 1 gal is 3.785 L, so 36 × 3.785 = 136.26 L. This is 136,260 mL.

Grams (g) to Pounds (lb)

1 lb is equivalent to 453.6 g.

Example: An operator feeds 13.5 lb of chemical each hour. How many grams is this? 453.6 × 13.5 = 6,123.6 g.

Million Gallons per Day (mgd) to Cubic Feet per Second (ft^3/sec)

1 mgd is equivalent to 1.55 ft^3/sec.

Example: What is the average flow rate, in cubic feet per second, in a plant that produces 20 mgd? 1 mgd = 1.55 ft^3/sec, so 20 mgd × 1.55 = 31 ft^3/sec.

Million Gallons per Day (mgd) to Gallons per Minute (gpm)

1 mgd is equivalent to 694 gpm.

Example: If a water treatment plant is operating at 6.75 mgd, how many gallons per minute is this? 1 mgd = 694 gpm, so 6.75 × 694 = 4,684.5 gpm.

Cubic Feet per Second (ft^3/sec) to Gallons per Minute (gpm)

1 ft^3/sec = 448.8 gpm.

Example: How many gallons per minute are represented in a flow of 3 ft^3/sec? 3 ft^3/sec × 60 sec/min = 180 ft^3/min; 180 ft^3/min × 7.48 gal = 1,346.4 gpm.

Percent (%) Concentration to Milligrams per Liter (mg/L)

1% is equivalent to 10,000 mg/L.

Example: An operator makes a 0.75% polymer solution in a day tank. How many milligrams per liter is this? 1% = 10,000 mg/L, so 0.75% × 10,000 = 7,500 mg/L.

Feet (ft) of Head to Pounds per Square Inch (psi) Pressure

1 psi is equivalent to a column of water that stands vertically at 2.31 ft.

Example: How many feet of head are developed at the base of a water tank that is 135 ft high? 2.31 ft of head = 1 psi, so 135 ft/2.31 = 58.4 psi.

Pounds per Square Inch (psi) Pressure to Feet (ft) of Head

1 ft of head is equivalent to 0.433 psi.

Example: How many feet of head are developed in a pumping system that is registering 68 psi? 1 ft of head = 0.433 psi, so 68 psi divided by 0.433 = 157 ft.

Grains per Gallon (gpg) to Milligrams per Liter (mg/L)

1 gpg = 17.1 mg/L.

Example: What is a water hardness of 11 gpg in mg/L? 11gpg × 17.1 mg/L for each gpg = 188.1 mg/L hardness.

Time Conversions

1 day is equivalent to 24 hours, which is equivalent to 1,440 minutes. 1 minute is equivalent to 60 seconds. There are 86,400 seconds in 1 day.

PRACTICAL DISTRIBUTION SYSTEM EXAMPLE PROBLEMS

Area

What is the cross-sectional area in square feet of a 10-in.-diameter pipe? 10 in. = 10/12 ft = 0.833 ft. Area = πr^2 = 3.14 × 0.417^2 = 0.546 ft^2.

How many gallons of coating paint should you buy to cover the walls of a cylindrical tank that is 30 ft high and 10 ft in diameter? A gallon covers 200 ft^2. Circumference of a circle = πd, circumference of tank = 3.14 × 10 = 31.4 ft^2.

Area = circumference × height = 31.4 × 30 = 942 ft^2. 942/200/gal = 4.71 gallons. So, you would need to buy 5 gallons to do the job.

Volume

What is the volume, in gallons, of a rectangular reservoir that is 38 ft × 75 ft if the water depth is 14 ft? 38 ft × 75 ft × 14 ft × 7.48 gal/ft^3 = 298,452 gal.

How many cubic feet of water are contained in 1,000 ft of 8-in. water main? Volume = $\pi R^2 H$ = 3.14 × 0.33 ft × 0.33 ft × 1,000 = 341.95 ft^3.

What is the volume, in gallons, of a 60-ft-diameter tank when the water depth is 30 ft? Volume = $\pi R^2 H$ = 3.14 × (r^2) × depth × 7.48 gal/ft^3 = 3.14 × (30 ft)(30 ft) × 30 ft × 7.48 = 634,154 gal.

How many cubic yards of soil must be excavated from a trench 250 ft long, 3.5 ft wide, and 5.0 ft deep? V = lwd, convert feet to yards first. 250 ft = 83.33 yd, 3.5 ft = 1.17 yd, 5.0 ft = 1.67 yd. V = 83.33 × 1.17 × 1.67 = 162.82 yd^3.

Feed Rate

Booster chlorination is needed and liquid hypochlorite solution is used that is 15 percent as available chlorine. If the operator needs to treat a flow of 1.6 mgd with 6 mg/L chlorine, how many gallons per hour hypochlorite solution are needed? 1,600,000 gpd/24 h = 67,000 gph. (67,000 gph × 6 mg/L)/ 150,000 mg/L = 2.68 gph.

Detention Time

Detention time = V/Q, where V is the volume and Q = the flow.

If a 4.52-mgd pump is used to fill a rectangular basin that is 75 ft × 36 ft, how long will it take to bring the water level from empty to 8 ft? V = 75 ft × 36 ft × 8 ft = 21,600 ft^3. Q = 4.52 mgd × 1.55 = 7 ft^3/sec. Therefore, DT = 21,600 ft^3/7 ft^3/sec = 3085.7 sec, or 51.4 min.

What is the contact time if chlorine is fed into a 36-in.-diameter raw water force main 23,000 ft in length and the flow through the force main is 8 mgd? 8,000,000 gpd × 1 day/ 24 h × 1 ft^3/7.48 gal = 44,563 ft^3/h. DT = Vol/ flow = (3.14 × 1.5 ft × 1.5 ft × 23,000 ft)/44,563 ft^3/h = 3.64 hours

Well Drawdown

The level of the aquifer of a well before pumping is 130 ft. When the pump is operated, the aquifer level drops to 173 ft. What is the drawdown? 173 ft – 130 ft = 43 ft.

Well Yield and Specific Capacity of Wells

A total of 950 gal was pumped from a well in a period of 5 minutes. What is the well yield in gallons per minute? 950 gal/5 min = 190 gpm.

Using the well yield in the example above, calculate the specific capacity if the drawdown is 25 ft. 190 gpm/25 ft = 7.6 gpm/ft.

A well produces 178 gpm. The drawdown is 14 ft. What is the specific capacity in gpm/ft? Specific yield = production/drawdown = 178/14 = 12.7 gpm/ft.

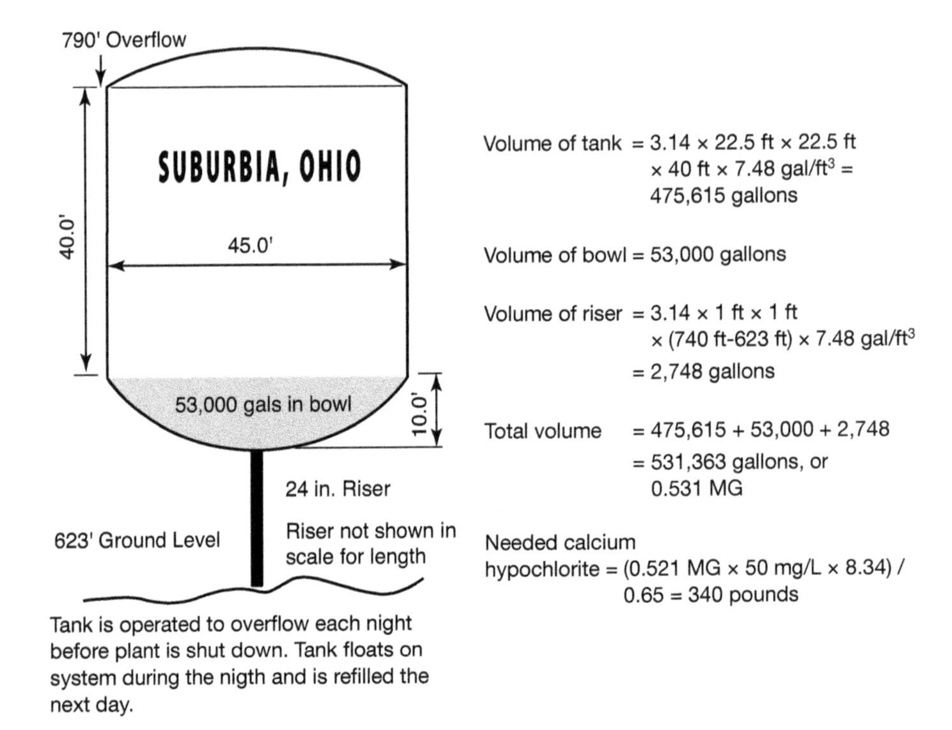

Figure C-1 Disinfection example

Disinfection of System Components

An operator wishes to disinfect a new well casing that has a 12-in. diameter. The desired chlorine dosage is 50 mg/L. The casing is 150 ft long, and the water level in the casing is 55 ft from the top of the well. How many pounds of calcium hypochlorite at 65 percent available chlorine are needed? The water column is 150 ft – 55 ft = 95 ft. Therefore, the volume is 0.785 × 1 ft × 1 ft × 95 ft × 7.48 = 558 gal. Dosage = 50 mg/L × (558/1,000,000) × 8.34 = 0.23 lb chlorine. Calcium hypochlorite is 65 percent available chlorine, so 0.23 lb chlorine/0.65 = 0.35 lb of calcium hypochlorite.

Determine the amount of 65 percent calcium hypochlorite needed to disinfect the water tower and riser (shown in Figure C-1) if a 50 mg/L dose is desired.

Head

The static water level of a well pump is 90 ft. The well drawdown is 29 ft. If the gauge reading at the pump discharge head is 3.9 psi, what is the total pumping head? (90 ft + 29 ft) + (3.9 psi × 2.31) = 128 ft.

Using the information depicted in Figure C-2, determine the pressure at the main. Mercury is 13.6 times as heavy as water. The pressure will be

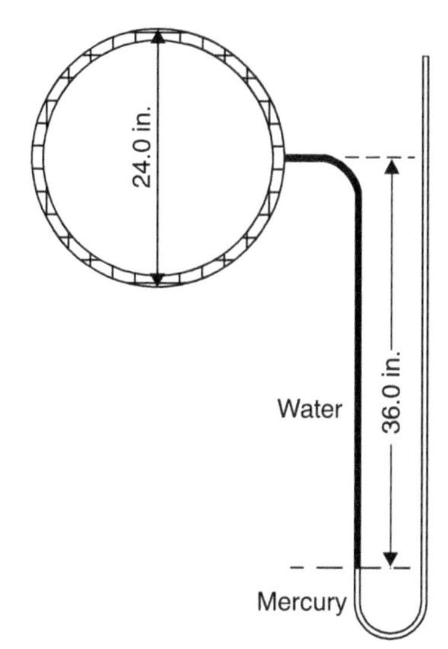

Figure C-2 Mercury manometer setup on a 24-in. main

the difference between the force created by the mercury and the force created by the water, so (3 × 13.6 ft) – (3 × 1 ft) = 37.8 ft of head. 37.8 ft × 0.433 = 16.4 psi.

Density and Specific Gravity

The specific gravity of mercury is 13.6. What does 1 gal of mercury weigh? 13.6 × 8.34 = 113.4 lb.

Flow Rate and Velocity

The flow rate calculation for water moving through a pipe is Q = A × V, where Q is the flow rate in cubic feet per second, A is the square foot area of the pipe or channel, and V is the velocity in feet per second.

Newly installed water mains are normally flushed at 2.5 ft/sec. How many cubic feet per second are needed to flush a 12-in.-diameter main? First, convert the 12-in. diameter to a radius in feet. 12 in./2 = 6 in. radius, and 6 in./12 = 0.5 ft. Q = 3.14 × 0.5 ft × 0.5 ft × 2.5 fps = 1.96 ft^3/sec.

The velocity in an 8-in. water main is 2 ft/sec. How many gallons per minute are flowing through it? The diameter of the main is 8 inches, so the radius in feet is (8 in. / 1 ft/12in.) / 2 = 0.34 ft^2 area. The velocity is 2 ft/sec × 60 sec/min = 120 ft/min. Q = A × V = 0.34 ft^2 × 120 ft/min × 7.45 gal/ft^3 = 305 gpm.

Weir Overflow and Surface Overflow Rates

Weir overflow rates are expressed as gallons per day per foot, or gallons per minute per foot, while surface overflow rates are expressed as gallons per day per square foot or gallons per minute per square foot. Weir overflow rates use the total linear distance of the weirs at the end of the basins

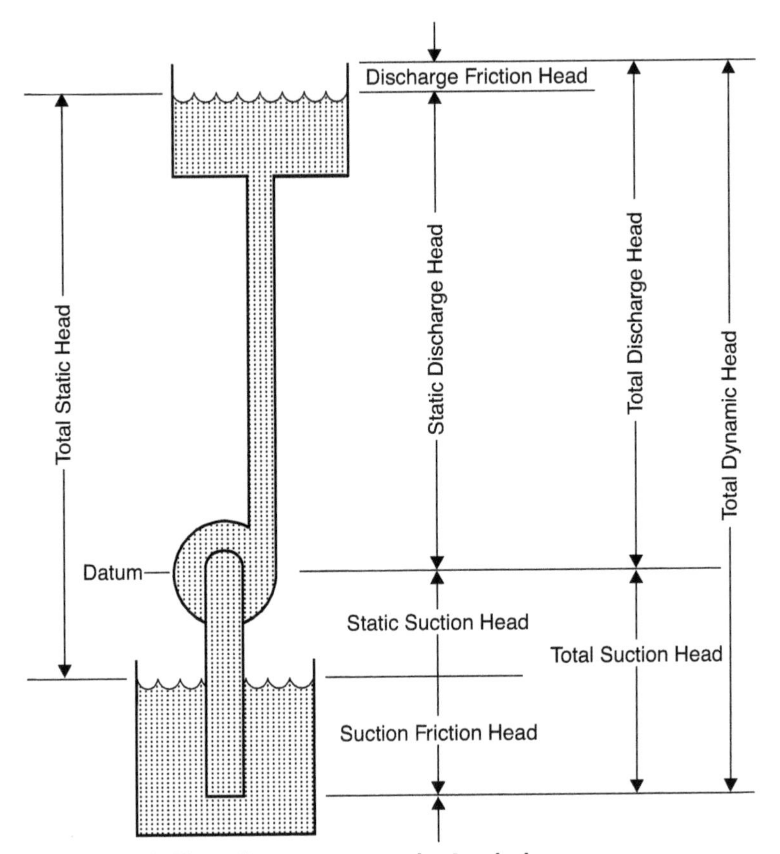

Figure C-3 Schematic illustrating common pumping terminology

for a calculation, and surface overflow rates use the square foot area of the surface of the basin. Both formulas work for rectangular and circular basins, and neither formula depends on basin depth.

Horsepower and Pumping

The power required to drive a pump is called water horsepower (WHP). Since pumps are inefficient, the power requirements for the pump are calculated as brake horsepower (BHP). The motors that drive pumps are also inefficient, so that power requirement calculation is called motor horsepower (MHP) (see Figure C-3).

What is the WHP requirement to pump 375 gpm against a head of 55 psi?

The formula for WHP = (gpm × head in feet)/3,960. Therefore (375 × (55 × 2.31))/3,960 = 12 hp.

If the pump in the above problem is 87% efficient, what is the BHP requirement? BHP = WHP/efficiency = 12/0.87 = 13.8 BHP.

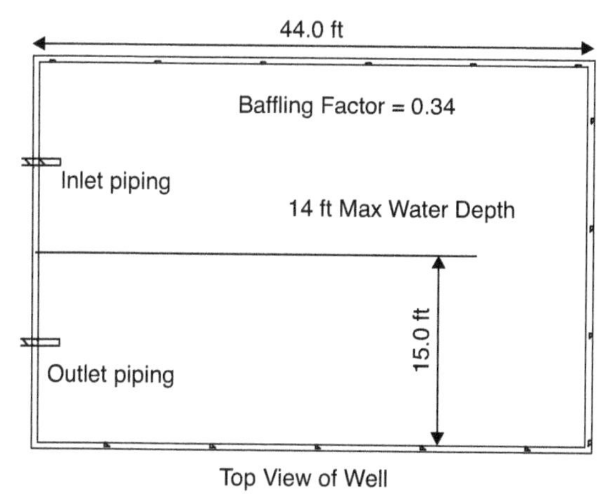

Figure C-4 Typical clearwell arrangement. Note baffle wall, which effectively increases the travel distance of the water and therefore the disinfectant contact time.

If the motor that drives the pump in the above problems is 92 percent efficient, what is the MHP requirement? MHP = BHP/efficiency = 13.8/0.92 = 15 MHP.

The above examples show that it always takes more energy to move water through systems because inefficiencies result from friction and head loss.

What is the WHP required to fill an elevated storage tank with a pump operating at 5,600 gpm if the static head is 85 psi and the suction head is 9 ft? 85 psi × 2.31 = 196.35 ft. 196.35 ft – 9 ft = 187.35 ft. Therefore, (5,600 gpm × 187.35) / 3,960 = 265 hp.

C × *T*

Using Figure C-4 and Table C-1, calculate the reportable $C \times T$ value in milligram minutes per liter. The highest raw water flow and lowest clearwell depth was at 1,800 hours. Therefore, $C \times T$ = (44 ft × 30 ft × 8.3 ft [water depth, WD])(7.48)(0.34 baffling factor)(1.6 mg/L)/300 gpm = 148.6 mg-min/L.

Calculate the CT for a clear well contact basin with a baffling factor of 0.5, a capacity of 2.4 million gallons, the plant flow of 5 mgd, and a chlorine residual at the basin outlet of 0.9 mg/L. Detention time = volume/flow, so T = 2.4 mg/5 mgd = 0.48 d × 1440 min/d = 691.2 min. Now we must multiply by 0.5 baffling factor so the allowed detention time is 345.6 min. The CT is calculated by multiplying this by the residual. CT = 0.9 mg/L × 345.6 min = 311 min-mg/L.

Floor Loading

The floor loading capacity in the chemical storage room is 438 lb/ft^2. How high can you stack 100-lb bags of soda ash if each bag is 30 in. × 18 in. × 6 in.? Each bag is 2.5 ft × 1.5 ft × 0.5 ft. Surface area of 1 bag

Table C-1 Plant operating data for *C* × *T* calculation

Time	Clearwell Depth	Cl_2 Residual	Raw Water Flow, gpm	Time	Clearwell Depth	Cl_2 Residual	Raw Water Flow, gpm
0000	11.0	1.5	220	1200	9.6	1.5	175
0100	11.0	1.5	220	1300	9.0	1.5	175
0200	10.9	1.5	220	1400	8.7	1.5	175
0300	10.9	1.5	220	1500	8.5	1.5	220
0400	10.6	1.4	220	1600	8.4	1.6	220
0500	10.5	1.5	220	1700	8.3	1.6	220
0600	10.0	1.5	280	1800	8.3	1.6	300
0700	9.8	1.7	270	1900	8.6	1.6	280
0800	9.7	1.7	270	2000	9.0	1.5	280
0900	9.7	1.7	270	2100	9.2	1.5	220
1000	9.9	1.6	220	2200	9.5	1.5	220
1100	10.0	1.6	230	2300	9.9	1.5	220

is 2.5 ft × 1.5 ft = 3.75 ft^2. Therefore, 3.75 ft^2 × 438 lb/ft^2 = 1,642.5 lb, and 1,642.5 lb × 1 bag/100 lb = 16.4 bags. 16.4 bags × 0.5 ft/bag = 8 feet maximum height.

Salaries and Budget

An employee receives an hourly wage of $15.35 plus overtime pay of 1.5 times the hourly wage for each hour worked over 40 hours per week. What is the weekly pay if the employee worked 45 hours (gross pay without any deductions)? Pay is 15.35 × 40 = $614.00 plus 15.35 × 1.5 × 5 = $115.13, so the total pay is 614.00 + 115.13 = $729.13.

Your distribution maintenance crew annual budget for salaries is $523,000. A 2 percent raise is given to all employees next year. What is the annual salary budget for your crew for next year? $523,000 × 0.02 = $10,460, so the annual total budget for salaries is $523,000 + 10,460 = $533,460.

BIBLIOGRAPHY

Giorgi, J. 2007. *Math for Distribution System Operators: Practice Problems to Prepare for the Distribution System Operator Certification Exams.* Denver, Colo.: American Water Works Association.

Giorgi, J. 2012. *Water Operator Certification Study Guide*, 6th ed. Denver, Colo.: American Water Works Association.

Pizzi, N.G., and W.C. Lauer 2013. *Water Treatment Operator Training Handbook*, 3rd ed. Denver, Colo.: American Water Works Association.

Index

Note: An *f.* following a page number indicates a figure. A *t.* indicates a table.

About the Author

William C. Lauer has more than 35 years of experience in drinking water quality and treatment process technology. He is an internationally recognized authority on distribution system operations, drinking water health effects, and treatment methods, authoring or editing 21 books and more than 70 articles on these subjects. He has served on many advisory panels such as the US Environmental Protection Agency (USEPA) Blue Ribbon Panel on Best Available Treatment Technologies, National Aeronautics and Space Administration Health Effects Issues for International Space Station, the National Academy of Science Health Effects of Recycled Water, and the Government of Singapore Reclaimed Water Expert Panel.

Distribution system and water treatment optimization for water quality improvement is a specialty. Mr. Lauer is an expert in water quality issues for water collection, distribution systems, and treatment. His projects have included those dealing with drinking water odors, corrosion, cross-connection control, disinfection, discolored water, customer relations, water system management, and particulate removal processes.

Mr. Lauer has recently completed serving more than 15 years as program manager for the Partnership for Safe Water. This program is supported by the American Water Works Association (AWWA), USEPA, Association of Metropolitan Water Agencies, National Association of Water Companies, Association of State Drinking Water Administrators, and Water Research Foundation. A new program to optimize the performance of drinking water distribution systems along with a self-assessment guidebook were both developed under his direction.

Mr. Lauer was the staff technical advisor for the development of AWWA utility management standards for Distribution System Operations and Water Treatment Plant Operation and Management. These programs seek to improve water and wastewater utility operations and the quality of services.

Mr. Lauer managed the water quality program for Denver Water, which serves more than one million customers. In this capacity, Mr. Lauer managed the cross-connection control program and supervised investigation of customer water quality complaints. Operation and maintenance of the department's distribution system to maintain water quality was a focus of this position.